A Century of X-rays and Radioactivity in Medicine

With Emphasis on Photographic Records of the Early Years

A Century of X-rays and Radioactivity in Medicine

With Emphasis on Photographic Records of the Early Years

Richard F. Mould

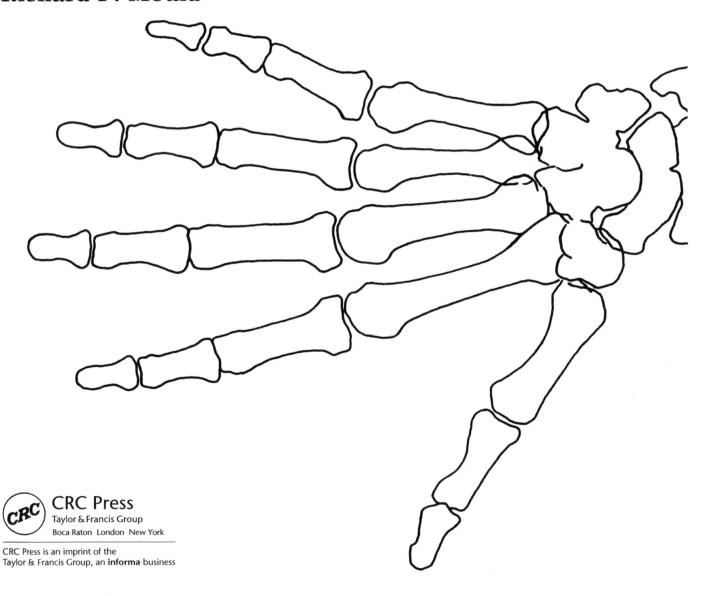

CRC Press
Taylor & Francis Group
Boca Raton London New York

CRC Press is an imprint of the
Taylor & Francis Group, an **informa** business

CRC Press
Taylor & Francis Group
6000 Broken Sound Parkway NW, Suite 300
Boca Raton, FL 33487-2742

First issued in paperback 2019

© Richard F. Mould 1993
CRC Press is an imprint of Taylor & Francis Group, an Informa business

ISBN-13: 978-0-7503-0224-1 (hbk)
ISBN-13: 978-0-367-40251-8 (pbk)

British Library Cataloguing-in-Publication Data

A catalogue record for this book is available from the British Library

Library of Congress Cataloging-in-Publication Data

Mould, Richard F. (Richard Francis)
 A century of X-rays and radioactivity in medicine : with emphasis
on photographic records of the early years / Richard F. Mould.
  p. cm.
 Includes bibliographical references and index.
 ISBN 0–7503–0224–0
 1. Radiology, Medical—History. 2. Nuclear medicine—History.
I. Title.
 [DNLM: 1. Technology, Radiologic—history. 2. Radiography—
history. 3. X-Rays. WN 11.1 M926c 1993]
 R895.5.M68   1993
 616.07′572′09—dc20
 DNLM/DLC
 for Library of Congress   93-19005
                                                                  CIP

Typeset in Great Britain by Integral Typesetting, Great Yarmouth

Visit the Taylor & Francis Web site at
http://www.taylorandfrancis.com

and the CRC Press Web site at
http://www.crcpress.com

To Imogen
with love from
Grandad

Contents

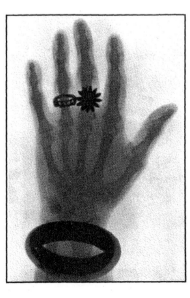

Contents

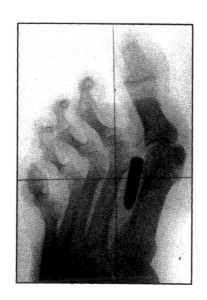

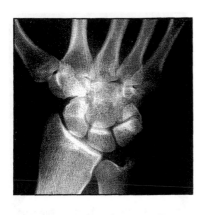

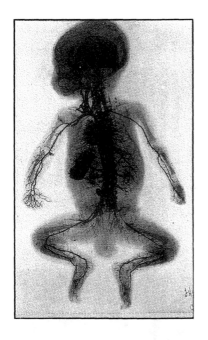

Contents

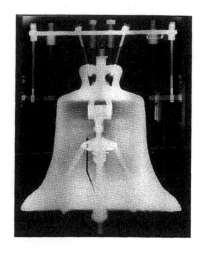

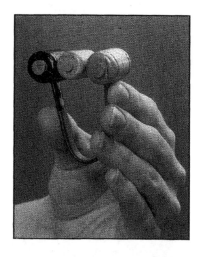

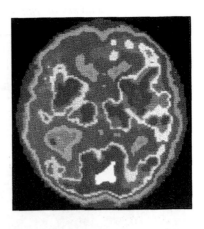

Drawn by] Professor Röntgen at work. [Walter E. Belgrave.

Figure 1. The world's first journalistic drawing of Röntgen in his laboratory. This was commissioned by the photographer and author on X-rays, Snowden Ward, and published in the April 1896 *Windsor Magazine*. It is a good likeness, but there is an obvious error on the left in the shape of the discharge tube which is of cylindrical design and not the famous pear shape.

Figure 2. The first illustration of an experimental arrangement in 1896 for the study of ionisation in air produced by X-rays, using a gold leaf electroscope as the measuring device. This was published by two Frenchmen, Benoist and Hurmuzescu.

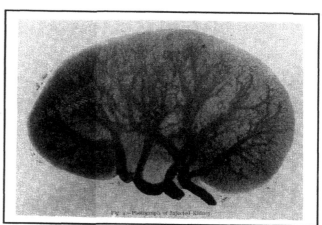

Fig. 4.—Photograph of injected kidney.

Figure 3. The first published arteriogram. Sydney Rowland was a medical student at St. Bartholomew's Hospital, London, at the time of the discovery of X-rays, and, unusually for one so young, was appointed 'Special Commissioner to the *British Medical Journal* for investigation of the applications of the new photography to medicine and surgery'. He published 26 reports between 8 February and 5 December 1896 and the BMJ included this 'photograph of an injected kidney' which was taken in Sheffield on 6 February 1896. Rowland went on to become editor of the world's first radiological journal in May 1896, The *Archives of Clinical Skiagraphy*, and the first issue was based largely on the BMJ reports; but for some unknown reason this arteriogram was never published in *The Archives* and is to be found only in the BMJ.

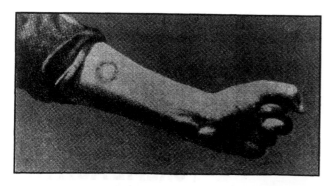

Figure 4. A photograph from a French newspaper of Pierre Curie's self-inflicted radium burn, 1901, which can be said to be the start of radiation biology and of radium therapy. (Courtesy: Muzeum Marii Sklodowskiej-Curie of the Polish Chemical Society, Warsaw.)

Figure 5. German 1995 postage stamp commemorating the centenary of X-rays.

Preface to the First Reprint

It is always a pleasure for an author when his book requires a reprint and I am most grateful to all those who have thought it worthy of purchase and to the many journal reviewers who have recommended it to healthcare personnel and to the general public.

The centenary of the discovery of X-rays is this year, 1995, and I have been fortunate to have been invited to speak at several congresses and university venues, but I would particularly like to take this opportunity to thank Professor Karsten Rotte and the *Physikalisch-medizinischen Gesellschaft of Würzburg* for their invitation in 1994 to deliver a lecture on *Marie Curie & Radium* to the very same society before which Röntgen gave his only public lecture on X-rays in January 1896; to Professor Andrzej Kulakowski for his invitation to speak in 1994 in Warsaw on *Marie Curie & Radium Therapy* at the commemoration at the *Maria-Sklodowska-Curie Memorial Cancer Center & Institute of Oncology* of the 60th anniversary of Marie Curie's death; and to Professor Axel Haase of the *Physical Institute of the University of Würzburg* who has invited me in 1995 to deliver an illustrated after-dinner speech at the Residenz Palace in Würzburg, for the magnetic resonance microscopy conference, and to speak on *Röntgen and X-rays in Medicine*. I would also like to thank Madame Eve Curie-Labouisse, the daughter of Marie Curie and also her biographer, for the invitation to visit her at her home in New York in 1994 for an afternoon of historical discussions. Such invitations in historical surroundings make an author's life very enjoyable!

I am also most grateful to those who have sent me additional historical material to that already contained in the following pages, and to the Libraries which continue to help me with my ongoing radiological history research. A single book is never large enough to contain all the illustrations one would wish, and since 1993 when it was first published I have found the following four 'firsts' from 1896 and I now take the opportunity to include them with this Preface together with the commemorative stamp from Germany for the X-ray centenary, Figure 5. I also list four invited review papers[1,2,3,4] published 1995–1996 and which complement the chapters in this book.

References

1 Mould R F, *The early years of radiotherapy with emphasis on X-ray and radium apparatus*, Brit. J. Radiology, **68**, 567–582, June 1995.
2 Mould R F, *Röntgen and the discovery of X-rays*, Brit. J. Radiology, **68**, In press, November 1995.
3 Mould R F, *The early history of X-ray diagnosis with emphasis on the contributions of physics 1895–1915*, Physics in Medicine and Biology, In press, November 1995.
4 Mould R F, *Archives of Clinical Skiagraphy and The Roentgen Ray 1896–1901*, **69**, In press, May 1996.

Dick Mould
London, 1995

Figure 6. On 20 April 1995 the coffins of Marie and Pierre Curie were reinterred in the Panthéon, Paris, from their initial resting place in Sceaux, see Figure 2.11. Marie Curie is the first woman ever to be honoured by burial in the Panthéon and the ceremony was attended by the Presidents of France and Poland, François Mitterand and Lech Walesa, together with the Mayor of Paris, Jacques Chirac, and other dignitaries. The Joliot-Curie family accompanied the coffins which were carried by soldiers of the Republic Guard. It was also appropriate that the ceremony closed with the playing of a nocturne by Poland's most famous composer, Chopin, and that the Panthéon is very close indeed to the Rue L'Homond where radium was discovered and to the Institut du Radium in the Rue d'Ulm. The row of young people standing behind the two coffins are holding symbols relating to the work of the Curies. 'Po' for the discovery of polonium is clearly seen, as is β, symbolising beta rays.

Preface

This book has been written to fill a gap in radiological history, namely the publication of a comprehensively representative and entertaining photographic album which may be dipped into at random or browsed from cover to cover, yet also be regarded as an authoritative source of historical information. It might perhaps be described as a 'coffee table book' but not if that means remaining pristine on the inside and dusty on the cover—the more well-thumbed the edges, the better pleased this author will be!

The predominantly photographic presentation, with full and informative captions supported by short introductory essays, has been further encouraged by the favourable response to my smaller 1980 book *A History of X-rays and Radium*, now long out of print but still regularly requested. With over 640 illustrations, grouped into 200 topics, it is not surprising that the present collection has taken some 30 years to complete. It intentionally concentrates on the early years of discovery and invention, diagnosis and therapy, with archival accounts complementing the photographic record of the pioneering scientists and physicians, and their achievements and misfortunes. In most chapters, the history is more quickly brought up to date so that the old methods may be compared, or contrasted, with newer technologies. It is interspersed with a variety of radiological anecdotes and historical snippets which have been used over the years to keep students awake and enliven lectures on professional topics.

The material has been obtained from libraries, museums, hospitals, universities and individuals in more than 20 countries. There is something within its pages for everybody: radiotherapist, nuclear medicine physician, diagnostic radiologist, physicist, engineer, radiographer, technologist, nurse, equipment manufacturer, and art gallery and museum visitor. The more general reader will be intrigued by such esoteric topics as an X-ray aeroplane, radium baths, the court martial of a *British Medical Journal* author, customs and smuggling, an X-ray 'zoo', and radiographs of Egyptian mummies, old Masters, a Mercedes car and the Liberty Bell. An unusual insight into the difficulties of early radiation dosimetry is provided by the diversity of quantities and units of measurement proposed for X-rays and radium in the 40 years following their discovery. Recorded interviews with Röntgen have been included together with photographs of Marie Curie not previously widely published. Early patient photographs before and after X-ray or radium treatment have been located, including one taken 70 years after X-ray treatment. Research has been based on original source material and it will be difficult to find an early X-ray or radium application which has not been included, whereas the text and illustrations are very comprehensively indexed to make it easy to locate or retrieve the wide variety of people, methods, apparatus and examples featured in this book.

By the 1990s much documentation from the pioneer era has been irretrievably lost and many of the early textbooks and journals can now be viewed in only very few libraries throughout the world, not all of which are now routinely accessible. This illustrated history seeks to redress this imbalance by blowing the dust off our fascinating radiological past and presenting to a wider audience much material that has been buried for at least 50 years. As we approach the centenaries of the discoveries of X-rays (in 1995), radioactivity (in 1996) and radium (in 1998), I therefore wish you enjoyable reading on X-ray cannons, radium bombs, chiroscopes and osteoscopes, skiagrams, patents, X-ray spectacles and cosmetic radium creams.

Dick Mould
London, 1993

Acknowledgements

A book of this nature, researched over a period of 30 years, could not have been completed without a great deal of help from many people, not to mention organisations with radiological archives. The latter have included the British Institute of Radiology (including the extensive Thurstan Holland Collection and the Isenthal Bequest), British Museum, Science Museum in South Kensington, Deutsches Röntgen-Museum in Remscheid-Lennep, Institut Curie in Paris, Muzeum Marii Sklodowskiej-Curie of the Polish Chemical Society in Warsaw, Germanisches Nationalmuseum in Nürnberg, American College of Radiology in Reston, Virginia, Physical Institute of the University of Würzburg, Fachhochschule Würzburg-Schweinfurt, Siemens Archives in Erlangen, Deutsches Museum in Munich, the Otis Archives of the Armed Forces Institute of Pathology, Walter Reed Medical Center in Washington, the Welcome Library for the History of Medicine in London, the National Radiological Protection Board in Harwell, and the libraries of Westminster Medical School and the Royal Marsden Hospital/ Institute of Cancer Research in London, and of the World Health Organisation in Geneva and the International Atomic Energy Agency in Vienna. In addition, many companies in the radiological field have been very kind in providing me with photographs and information on equipment and I am most grateful to Amersham International, CGR Medical, Cuthbert Andrews, EMI, General Electric, Nucletron, Philips, Siemens, TEM Instruments and Varian.

It would, though, be impossible to list all those who have assisted me in one way or another with illustrations, tracing references and searching hospital libraries and I apologise if I have been at fault with omissions among those acknowledged below and in the captions of various figures.

Mr Richard Andrews
Dr Bernard Asselain
Prof. Milos Bekerus
The late Prof. Sven Benner
Prof. Helmar Bergmann
Prof. W. Böhndorf
Mr R. Brandl
Dr Gregor Bruggmnoser
Frau Ursula Buchholz
Prof. E. J. Burge
Mr John E. Burns
Mr John B. Davies
Dr Larry Doss
Mme Françoise Fayard
The late Dr Gilbert Fletcher
Prof. Hermann Frommhold
Prof. Alain Gerbaulet
Mr Ulrich Hennig
Ms Fiona Henniker
Mr David Heywood
Mrs Myra Heywood
Dr Basil Hilaris
Mrs Gunnel Ingham
Dr W. Alan Jennings

Dr Reg Jewkes
Prof. Guo-Liang Jiang
Mr Stephen Johnson
Prof. Charles Joslin
Dr Nancy Knight
Mgr.Jnz Anna Koscielniak
Prof. Georgy Kovacs
Prof. Andrzej Kulakowski
The late Dr Manuel Lederman
Mr Jack Lewis
Mr Otha W. Linton
Prof. Liu Tai Fu
M. Zofia Lukawska
Mr Rex Mark
The late Professor W. V. Mayneord
Dr Jack Meredith
Dr Harm Meertens
Mme Marie-Claude Moutier
Prof. Reinhold Müller
Ms Rosemary Nicholson
Dr Dattatreyudu Nori
Dr Colin Orton
The late Dr W. Petzold
The late Prof. Edith Quimby

Mr L. J. Ramsey
Mr Michael Rhode
Prof. Jurgen Richter
Prof. Karsten Rotte
Dr Alex Rudic
Dr Robert Shields
Dr Heinrich Seegenschmiedt
Dr Margaret Snelling
M. Malgorzata Sobieszczak
Dr Burton Speiser
Ms Marilyn Stovall
The late Herr Ernst Streller
Herr H. Studtrucker
Dr Lauriston S. Taylor
Dr Nigel Trott
The late Prof. Dale Trout
Prof. Helmut Vahrson
Frau Helga Wagner
Prof. Rune Walstam
Mr John Wareing
Mr Roland Weigand
Prof. Hisao Yamashita
Prof. Yin Wei Bo
Dr Christian Zwicker

I would also like to express my appreciation to those at Institute of Physics Publishing in Bristol and Philadelphia, without whom this book could never have been published: Adrienne Fenton, Sean Pidgeon, Jenny Pickles and Tamara Isaccs-Smith, and especially to my production editor, Martin Beavis, who has cheerfully provided superb expertise in all the phases of production.

Finally, I would like to acknowledge my debt of gratitude to the early pioneers of the applications of X-rays and radium, both professionals and public alike, for their achievements under what cannot have been easy circumstances, and their foresight in recording their observations, results and theories. Their prodigious output has enabled me to rescue from late 20th century obscurity and from the passage of time, which has caused much to be irretrievably lost, a wide spectrum of unique illustrative material for inclusion in this record of virtually 100 years of X-rays and radioactivity applied to medicine.

Chapter 1

Discovery of X-rays

Wilhelm Conrad Röntgen

X-rays were discovered by Wilhelm Conrad Röntgen on November 8th 1895 in his laboratory at the Physical Institute of the Julius-Maximilians University of Würzburg in Bavaria. At the time Röntgen was investigating the phenomena caused by the passage of an electrical discharge from an induction coil through a partially evacuated glass tube. The tube was covered with black paper and the whole room was in complete darkness, yet he observed that, elsewhere in the room, a paper screen covered with the fluorescent material barium platinocyanide became illuminated. It did not take him long to discover that not only black paper, but other objects such as a wooden plank, a thick book and metal sheets, were also penetrated by these X-rays. More important though, he found that[1] 'Strangest of all, while flesh was very transparent, bones were fairly opaque, and interposing his hand between the source of the rays and his bit of luminescent cardboard, he saw the bones of his living hand projected in silhouette upon the screen. The great discovery was made.'

Although Röntgen published 55 scientific papers only three were on the topic of X-rays, none of which included any of his own radiographs. The first, 'Ueber eine neue Art von Strahlen', communicated on December 28th 1895 and set out in 17 numbered paragraphs, was published in the *Sitzungsberichte der Physikalisch-medizinischen Gesellschaft zu Würzburg* [1.13] and translations, some including radiographs made by other scientists, were soon available in English and in French:

January 23rd 1896 in *Nature* (London)
February 1896 in a special issue of *The Photogram* (London) entitled *The New Light and the New Photography* 'With many examples of Photography through "opaque" substances, wood, leather, ebonite, &c., and photography of the skeleton within the living flesh.'
January 24th 1896 in *The Electrician* (London)
February 8th 1896 in *L'Eclair Electricité* (Paris)
February 14th 1896 in *Science* (New York).

On January 1st 1896 Röntgen wrote to scientific colleagues in several countries enclosing some example radiographs, each marked with the stamp 'Physik Institut der Universität Würzburg' [1.14–1.16]. Two of these packages were sent to the United Kingdom, to Lord Kelvin (University of Glasgow) and to Sir Arthur Schuster (University of Manchester) of which only the Schuster set survives, donated by his daughter Dr.

Norah Schuster to the Wellcome Institute for the History of Medicine, London.

Röntgen's second communication to the Physikalisch-medizinischen Gesellschaft was on March 9th 1896 and was a continuation of the first with additional numbered paragraphs 18–21. His third and final communication on the subject of X-rays, entitled 'Further observations on the properties of X-rays' was submitted to the Royal Prussian Academy of Science, Berlin on March 10th 1897 and published in the *Annalen der Physik und Chemie*. An English translation appeared in the *Archives of the Roentgen Ray* in February 1899.

His first and only public lecture was delivered on January 23rd 1896 at the Physikalisch-medizinischen Gesellschaft in Würzburg and the demonstration that day was of a radiograph of the hand of the famous anatomist Albert von Kölliker [1.18], who then proposed the term 'Röntgen rays' and called for three cheers for Röntgen. The audience cheered again and again.

The most detailed biography of Röntgen was written in 1931 by Otto Glasser[1], who describes the reaction of the world literature: daily press, popular magazines and scientific journals including:

'Electrical photography through solid bodies' *Electrical Engineer* (New York), January 8th 1896
'Sensational worded story' *The Electrician* (London), January 10th 1896
'Illuminated tissue' *New York Medical Record*, January 11th 1896
'Searchlight of photography' *The Lancet*, January 11th 1896
'Photography of unseen substances' *Literary Digest*, January 25th 1896
'Remarkable discovery: photographing through opaque matter' *Daily Telegraph* (Sydney), January 31st 1896.

Not all the responses were favourable, however, and in 1896 the London *Pall Mall Gazette* stated 'We are sick of the röntgen rays ... you can see other people's bones with the naked eye, and also see through eight inches of solid wood. On the revolting indecency of this there is no need to dwell.' *Punch* magazine referred in a pessimistic poem to 'grim and graveyard humour' and X-rays were also linked with vivisection[2]. Such comment could not, however, halt the progress of such a useful medical tool.

Röntgen had previously worked in several universities before moving to Würzburg in 1888, including

Strassburg and Giessen, and early in 1895 he had refused the offer of a professorial chair by the University of Freiburg because the government of Baden was unable to fulfil his laboratory equipment requirements. He wanted 11,000 Deutschmarks for several pieces of physical apparatus and improvements for planned experiments. The negotiations with the Government of Baden were short but intensive. When he left Freiburg, he was reported[3] to have said at the railway station 'This small country is doing a lot for the three universities [which were within its borders], but I understand that they cannot spend such a lot of money to offer an appointment to a foreigner [Röntgen was not born in Baden]. The idea to come to Freiburg was a nice dream for both [himself and the university], but could not be realised like many other dreams.' Instead of Röntgen, the University of Freiburg appointed Franz Himstedt (1852–1933) whose equipment requirements totalled only 1,250 Deutschmarks[3]. The X-ray fame of Würzburg therefore has an economic aspect and Freiburg had to wait until the summer of 1896 for its first X-ray images, made by the physicist Ludwig Zehnder (see also [3.5]), one of Röntgen's former students.

Röntgen remained at Würzburg until 1900 when he left to work at the University of Munich. In 1901 he was awarded the first Nobel Prize in Physics [1.5] and donated the prize money to the University of Würzburg. On the tenth anniversary of his discovery in 1905 a group of prominent German physicists, including Kohlrausch, Planck and Wien, had a marble plaque placed at the Physical Institute, Würzburg, with the inscription 'In diesem Hause entdeckte W. C. Röntgen im Jahre 1895 die nach ihm benannten Strahlen' (In this house in 1895 W. C. Röntgen discovered the rays which were named after him)[4]. For many years, though, this large plaque was lost until it was found being used in the Institute as a laboratory table top!

Despite the recognition accorded to Röntgen, there was, even into the 1970s, some debate concerning the contribution towards the discovery of X-rays by other researchers of electricity in high vacua, although Röntgen's pre-eminence was not challenged. Comroe[5] in his 1977 book *Retrospectroscope* which gives insights into medical discoveries (and near-miss discoverers), also rejects the view that the discovery of X-rays happened fortuitously, and is in agreement with Röntgen's biographer Glasser[1], who posed the question 'Was Röntgen's discovery accidental?' and then gave the answer that it was the final step in a brilliant and logical correlation of a multitude of facts which had been disclosed by many scientists, including Hertz, Lenard, Hittorf and Crookes.

Some insight into the well ordered mind of Röntgen can be found in two quotations he used in his speech in January 1894 when Rector of the University of Würzburg[6].

'Nature often reveals astounding marvels in even the most unremarkable things, but they can be recognised only by those who, with sagacity and a mind created for research, ask counsel from experience, the teacher in all things' (Athanasius Kircher, born 1602).

'When a law of nature, hitherto hidden, suddenly emerges from the surrounding fog, when the key, long sought after, to a mechanical combination is found, when the missing link takes its place in a chain of thought, there is the elation of spiritual victory for the discoverer, which by itself alone richly rewards him and lifts him for a brief moment onto a higher plane' (Werner von Siemens, 1816–1892).

Although Röntgen did not suffer from exposure to X-rays, probably because his experiments were over relatively few years and his X-ray tubes were shielded within metal boxes, he was well aware of the harmful biological effects experienced by other early workers. His warning to British scientists was conveyed by the London instrument manufacturer A. W. Isenthal [13.7] following a meeting with Röntgen in 1898[13].

'In April 1898 I was asked by my colleagues on the Council of the Röntgen Society, to arrange, if possible, for an interview with Röntgen at Würzburg University. Obtaining Röntgen's consent I called on him at his laboratory in the Physical Department, where he explained to me the set-up of his apparatus when he was led to the discovery of a new form of radiation. Röntgen was a very tall man, with a scholarly stoop, his face somewhat pock-marked, stern but kindly, and very modest in his remarks upon his achievement. I felt, of course, greatly elated at being in the presence of this world-renowned scientist. I became even more so when he asked me to accompany him to his private residence, offering me tea and chatting to me about "this English Röntgen Society" which I represented. He could, however, not accept my semi-official invitation to come to England, owing to so many previous engagements. In the course of our conversation he enquired of me whether we, in England, knew of the biological effects of the X-rays, and from a large portfolio produced some telling photographs of skin affectations, and asked me to make the facts known over here.'

Isenthal concluded his report with: 'On my return to England, I reported to the Röntgen Society, however, without evoking much interest regarding Röntgen's warning . . . as for myself, however, I immediately took what precautions were then possible.'

It is noted, however, in Chapter 3, that at a meeting on March 1st 1898 the Röntgen Society selected a Committee of Inquiry 'on the alleged injurious effects of Roentgen rays', and therefore at least some members were considering the possibility of harmful effects of radiation. Unfortunately though, the 'telling photographs of skin affectations' have been lost to posterity.

The Physics Department of the University of Würzburg has now moved to the outskirts from its original site in the centre of the city, but has a few permanent display cases commemorating Röntgen. Exhibits include his Nobel Prize of 1901, the original radiograph of the hand of Frau Röntgen [1.14], the box of weights used as one of his first X-ray test objects and his double-barrelled gun with accompanying radiograph [17.1]. The old buildings of the Physical Institute [1.9] are now the Fachhochschule Würzburg-Schweinfurt, but Röntgen's laboratory [1.11] is retained as part of a special Röntgen Museum. Some of his apparatus is displayed together with medals awarded to him, reprints of his scientific papers, a plaster death cast of his hands [1.7] and a very rare copy from Vienna of a book containing reproduction radiographs, of exceptionally high quality, of small animals taken in 1896 by Eder and Valenta[7] [9.7, 9.11].

The most extensive Röntgen museum is, however, the

Deutsches Röntgen-Museum in Remscheid-Lennep, in the city of Röntgen's birth, which includes the only existing radiograph of Röntgen's own hand, as distinct from that of his wife which he had circulated widely. It not only contains material relating directly to Röntgen but also a wide range of interesting X-ray apparatus and many unusual radiographs. The Museum is both an educational centre for schools and an unrivalled resource for historical research.

To give Röntgen himself the final word, it was recorded[8] that when interviewed in 1896 by Sir James Mackenzie Davidson who asked 'What did you think?', referring to the original experiment with the tube covered by light-tight black paper and the fluorescence of the distant barium platinocyanide screen, Röntgen replied 'I did not think, I investigated.'

Röntgen: 1845–1923

[1.1] Röntgen's early life in Germany and Holland. Left: the house in the Westphalian town of Lennep (present-day Remscheid) in which Röntgen was born in 1845. Centre: the only child of a German textile merchant and a Dutch mother. (Courtesy: Deutsches Röntgen-Museum, Remscheid-Lennep.) Right: Röntgen as a student in Holland.

The house in Lennep is now part of the Deutsches Röntgen-Museum and houses Röntgen's library of books and scientific papers. He did, however, only live in Lennep for the first three years of his life, at which point in time his father sold this home and moved to Apeldoorn in Holland where Röntgen received his initial education. This was followed in 1862 by his entrance to the Utrecht Technical School, and then a period of 10 months at the University of Utrecht (1865), although not as a regular student, until he passed in November 1865 the entrance examination to the Polytechnic School at Zurich, when he moved to Switzerland to become a student of mechanical engineering, receiving his diploma in 1868. One year later, studying under the Professor of Physics, August Kundt, he obtained his PhD degree after submitting a thesis on *Studies on gases*. In 1871 he followed Kundt to Würzburg and then (1872) to Strassburg. In 1875 he became a Professor of Physics in Hohenheim, in 1876 moving back to Strassburg and then in 1879 to Giessen and in 1888 to Würzburg to succeed Friedrich Kohlrausch. His final move was in 1900 when he took over the Physical Institute of the University of Munich at the special request of the Bavarian Government. It was in Munich that Röntgen died of cancer of the intestines in 1923.

[1.2] Photographs of Professor and Frau Röntgen. This likeness of Röntgen (left), taken in 1896, appears in a selection of his scientific papers published by the Deutsches Röntgen-Museum[9] but a similar photograph appeared as the frontispiece of one of the earliest books[10] on X-rays. The photographs above are both halves of stereographic pairs. (Courtesy: The Science Museum, London.)

[1.3] The old harbour of Würzburg photographed by Röntgen between 1890 and 1900. His interests outside science included the outdoor life and photography, and his biographer Glasser[1] records how Röntgen and his wife often spent their vacations at Pontresian in the beautiful Engadin mountains, taking with them Josephine Bertha Ludwig, the daughter of his wife's only brother, whom they had taken into their home in 1887 when she was six and adopted during her twenty-first year. On one such visit she found Röntgen especially intriguing because 'he often walked around with a great box and black piece of cloth and took photographs.'

[1.5] The Nobel Prize document awarded to Röntgen in 1901. (Courtesy: Physical Institute, University of Würzburg.)

[1.4] The only portrait in oils of Röntgen; it is by Wilhelm Reitz and dated 1895. It is unlikely, however, that the portrait was actually painted in 1895 since X-rays were only discovered in November of that year and it would have been out of character for him to have had his portrait painted. In addition, there is no record of any such event in any of his biographical material.

[1.6] Bust of Röntgen (far right) sculpted by Reinhold Felderhoff in 1896. (Courtesy: Deutsches Röntgen-Museum, Remscheid-Lennep). The statue of Röntgen, holding an X-ray tube in his right hand, was also sculpted by Felderhoff, in 1898, at the command of the German Emperor, and was sited on the Potsdam Bridge in Berlin. During World War II, in 1942, it was removed and smelted down for the metallurgical industry. This photograph formed part of the frontispiece of the 1907 textbook by Kassabian[14].

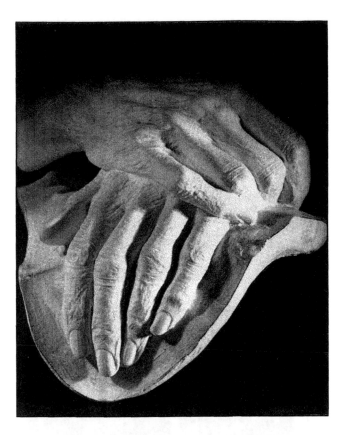

[1.7] Left: plaster cast of Röntgen's hands made immediately after his death in 1923. Unlike the hands of other X-ray pioneers [3.35], it is seen that Röntgen's are undamaged. (Courtesy: Deutsches Röntgen-Museum, Remscheid-Lennep.) Right: Röntgen's family grave in the Old Giessen Cemetery which is now a public park. Röntgen was Professor of Physics at the University of Giessen from 1879 to 1888 when he left for Würzburg. In later years, he often said[1] that except for his years in Würzburg the most pleasant period of his life was his time in Giessen. Both his parents died whilst he lived there, his mother in 1880 and his father in 1884 and both their names can be seen on the gravestone, together with that of his wife Bertha who died in 1919. (Courtesy: Prof. H. Vahrson.)

[1.8] Röntgen is commemorated on many postage stamps; the example shown was issued by Danzig Free State (now Gdansk in Poland) in 1939 bearing the slogan 'Fight Cancer, Cancer is Curable'. The two banknotes depicting Röntgen were emergency money issued in Lennep and Wellheim after World War I. Röntgen's birthplace is seen on the left of one of the notes.

[1.9] The main building of the Deutsches Röntgen-Museum was formerly the Lennep Rathaus and now contains an extensive radiological collection. The museum's collection of Röntgen's books and papers is held elsewhere in the town, in Röntgen's birthplace [1.1].

Röntgen's apparatus: 1895–1896

[1.10] The frontage of the Physical Institute, Würzburg, in 1896[10]. The building is now the Fachhochschule Würzburg-Schweinfurt.

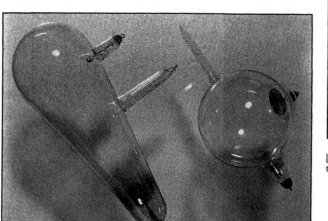

[1.12] Röntgen's laboratory in the University of Würzburg. (Courtesy: Deutsches Röntgen-Museum, Remscheid-Lennep.)

[1.11] Two X-ray tubes of 1895. The pear-shaped tube was known as a Hittorf tube and one of a similar design was being used by Röntgen when he made his discovery. See also [5.9] for a selection of tubes used by Röntgen. (Courtesy: Deutsches Museum, Munich.)

Röntgen's first communication and X-ray pictures

[1.13] Front covers of Röntgen's first communication, submitted for publication on December 28th 1895, under the title 'On a new kind of ray'. The paper cover of the first print run of his communication included the headline caption 'Professor Röntgen's wichtige Entdeckung!' (Important Discovery by Professor Röntgen!), a caption with which he disagreed and thus caused the cover to be reprinted without the offending headline. The original copy (left) is probably the only one remaining, and is now kept under lock and key in a safe in Würzburg. It cost 50 pfennigs whereas the reprint (centre) was priced at 60 pfennigs.

The Russian translation was printed in St. Petersburg in 1896 and includes the words 'Appended one phototypic plate made from a photograph recorded by Professor I. I. Borgmann and A. L. Gershun.' Many different language versions are exhibited in the Deutsches Röntgen-Museum, including a Japanese translation dated 1888.

[1.14–1.16] Three examples of Röntgen's radiographs mentioned in 'Ueber eine neue Art von Strahlen' (1895). Figures [1.15] and [1.16] were sent to Sir Arthur Schuster in Manchester on January 1st 1896. (Courtesy: Wellcome Trustees.) Figure [1.14] is inscribed by Röntgen to Professor Zehnder of Freiburg im Breisgau. (Courtesy: Deutsches Röntgen-Museum, Remscheid-Lennep.)

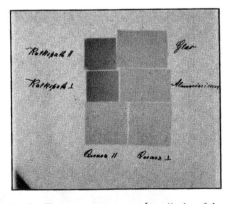

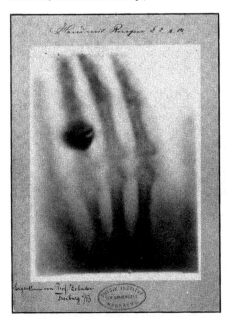

[1.14] Hand of Frau Röntgen. (Courtesy: Deutsches Röntgen-Museum, Remscheid-Lennep.)

[1.15] 'The justification for calling by the name "rays" the agent which proceeds from the wall of the discharge apparatus I derive in part from the entirely regular formation of shadows, which are seen when more or less transparent bodies are brought between the apparatus and the fluorescent screen (or the photographic plate). I have observed, and in part photographed, many shadow-pictures of this kind, the production of which has a particular charm.' The compass was one such example. (Courtesy: Wellcome Trustees.)

[1.16] This experiment was described as follows. 'These results lead to the conclusion that the transparency of different substances, assumed to be of equal thickness, is essentially conditioned upon their density; no other property makes itself felt like this, certainly to so high a degree. They show, however, that the density is not the only cause acting. I have examined, with reference to their transparency, plates of glass, aluminium calcite and quartz, of nearly the same thickness, and while these substances are almost equal in density, yet it was quite evident that the calcite was sensibly less transparent than the other substances which appeared almost exactly alike. No particularly strong fluorescence of calcite, especially by comparison with glass, has been noticed.' (Courtesy: Wellcome Trustees.)

THE UNIVERSITY,
GLASGOW. January 17. 1896.

Dear Prof. Röntgen,
When I wrote to you thanking you for your kindness in sending me your paper and the photographs which accompanied it I had only seen the photographs and had not had time to read the paper. I need not tell you that when I read the paper I was very much astonished and delighted. I can say no more just now than to congratulate you warmly on the great discovery you have made and to renew my thanks to you for your kindness in so early sending me your paper and the photographs.
Believe me,
Yours very truly,
Kelvin

[1.17] Letter from Lord Kelvin thanking Röntgen for his 'photographs', three of which are shown in [1.14–1.16].

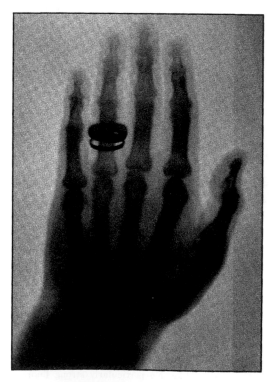

[1.18] Hand of the anatomist Albert von Kölliker radiographed by Röntgen on January 23rd 1896 after his only public lecture on the discovery of X-rays. (Courtesy: Deutsches Röntgen-Museum, Remscheid-Lennep.)

Reaction from the public: 1896

CHRISTIANSEN & BURNS,
MANUFACTURERS OF AND DEALERS IN
Bicycle Sleds,
Factory No. 782, 784 and 786 Seneca St.
Office No. 718 North Broadway.
Leavenworth, Kas., the 9th Jan. 1896.

[1.19] What have Bicycle Sleds to do with X-rays? This is from one of several letters to Röntgen written in January 1896 to ask for patents of the new discovery. Despite such approaches, Röntgen never patented his discovery because he believed it should be freely available for the benefit of society and that scientific discoveries should not be exploited for great personal gain. The summary below is derived from an exhibition at the Deutsche Röntgen-Museum which was prepared for the 1989 RSNA Congress in Chicago[12].

December 28th 1895 Röntgen handed a short paper entitled 'A new kind of rays—preliminary communication' to the Secretary of the Würzburg Physical–Medical Society for publication in the *Sitzungsberichte* (Proceedings).

January 1st 1896 Röntgen mailed reprints and photographs to various colleagues, including Professor Exner in Vienna. Exner informed others, one of whom was Dr Lechner who passed the information to his father, a journalist of *Die Presse*, a small paper in Vienna.

January 5th First public notice in *Die Presse*: 'A sensational discovery'. A journalist from the *Daily Chronicle* in London read this report and got somewhat confused, spelling Röntgen as Routgen and believing that the discovery was made in Vienna and not Würzburg.

January 6th This erroneous *Daily Chronicle* report was sent worldwide by cable from London.

January 8th First report in the American papers and first reprint request to Professor 'Routgen' from a Dr Robert L. Watkins of New York: '*I read in the paper that you can render opaque bodies transparent and thus photograph their inside. Will you kindly send me any description you may on the subject or inform where more information can be obtained.*'

January 9th First article in the local Würzburg newspaper and a letter from a John Baynes on letter paper from the Grand Union Hotel, New York. After commenting on the press reports Baynes then went on to declare his particular interest. '*I am interested in certain arts, in which such a light would be of use; and if the matter is in such a position as to be readily utilized, I should be glad of such particulars as you may feel disposed to communicate. What I should like to know is, whether the light is portable, and what its cost is; what apparatus is necessary and what conditions are needful to its successful operation? It is hardly necessary to say that I do not seek for the discovery of any secret or that I have any expectations of dealing with you in any other way than a proper basis commercial or otherwise.*'

January 9th Röntgen received a telegram from John P. Arnold, editor of the *World*, a New York Sunday paper, who declared himself a member of the German–American Association of Writers and Journalists.

January 9th The 'Bicycle Sled letter' was written by E. Christiansen MD in German. He asked to have the new light patented in the USA, offered Röntgen 25% of the total income for the duration of the patent (17 years) promising this to be 'a sizeable yearly income' and ended his letter with: '*Your invention is too valuable to science not to be exploited, it is on the other hand too dangerous to be allowed in everyone's hands.*'

January 10th Another request for an interest in any patent that 'Routgen' might register was sent by J. A. Throckmorton of Sidney, Ohio, whose letterhead declared him to hold the patents for the 'latest & best invention known in dentisty': patent gold clasps. '*Your marvelous triumph of a new light, attracts my attention. America needs it. Will you have it Patented here? If so I will be glad to take an interest in it. I can give you references, as to my commercial standing, and integrity, by referring you to the German American Bank, of Sidney, Ohio, USA.*'

January 14th E. & H. T. Anthony, manufacturers of photographic apparatus, of Broadway, New York asked 'Routgen' for a written report. '*We enclose herewith a clipping from one of the American papers, which clipping has been going the rounds and has attracted no little attention over here. We write to ask you whether you will be good enough to forward us some account of the work therein alluded to, with a picture illustrating it for publication in our journal "Anthony's Photographic Bulletin". If there is anything in the published reports which is really a notable advance or likely to prove of practical value, and if same is patentable, we should be glad to communicate with you with the idea of negotiating with you for the purchase for the rights for this country.*'

January 23rd First public presentation by Röntgen at the Physikalisch-Medizinischen Gesellschaft in Würzburg.

THE NEW ROENTGEN PHOTOGRAPHY.
" Look pleasant, please."

[1.20] Cartoon from the February 1896 magazine *Life*. Unfortunately the artist got it wrong: X-rays are not reflected back from the subject towards the X-ray tube and photographic plate!

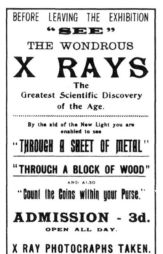

[1.21] Public lectures and demonstrations of X-rays soon became a popular entertainment. This leaflet (left) was distributed at an 1896 exhibition at the Crystal Palace, London. Complete sets of 'Röntgen apparatus' were available from several manufacturers of electro-medical equipment, as offered in these advertisements from a French pamphlet of 1896.

[1.22] As an example of an 1896 lecture synopsis the following is reproduced from H. Snowden Ward's 1896 textbook on *Practical Radiography*. '*A popular lecture with full demonstration, prepared in the Spring of 1896 and delivered with great success before large audiences in London and many provincial towns [13.6]. The Lecture is extremely interesting, though accurate and up-to-date. The demonstration includes the work of Crookes and others that led up to radiography; with examples of results obtained by Thermography, Electrography, etc, which have been erroneously attributed to the X Rays. The exhaustion of various vacuum tubes, the use of fluorescent and phosphorescent screens and the radiographing of various objects is shown and explained. About 50 lantern slides of the results of the foremost workers are included, and the work is on a scale that can be well shown in a hall containing 1,500 people.*'

[1.23] Probably the earliest photograph of an X-ray lecture or demonstration in the United Kingdom. It was specially commissioned by the *Windsor Magazine* for its April 1896 issue to illustrate an article by the professional photographer Snowden Ward entitled 'Marvels of the new light'. It was taken before February 10th 1896 and shows the engineer Campbell Swinton surrounded by his apparatus used in lecturing before the Royal Photographic Society. Positioning of the retort stand in which the pear-shaped tube is clamped required a series of books and is typical of the makeshift experimental arrangements which were initially used in the early months of 1896. The radiographs which accompanied this article were termed 'electrographs' and Snowden Ward concluded his paper with a comment concerning Salvioni's work on a fluoroscope (published in Perugia, February 6th 1896, see [10.1]). 'The most wonderful advance recorded at the time of writing is the invention of Professor Salvioni, of an instrument by which Röntgen rays may be *seen*. In the absence of all detail it is impossible to comment upon such an announcement beyond saying that if the invention is not exactly what we expect, the direction is one in which we may hope for practical results.'

Chapter 2

Discovery of Radioactivity and Radium

Henri Becquerel and Marie Curie

This is really a story of two discoveries: of radioactivity by Becquerel and of radium by Marie and Pierre Curie. The discovery of radioactivity was made in Paris on March 1st 1896 by Antoine Henri Becquerel, a Professor of Physics, when he developed photographic plates upon which he had placed crusts of the double sulphate of uranium and potassium a few days previously. It was a discovery which occurred at that particular point in time because of the weather!

Following Röntgen's discovery of X-rays, several scientists initially correlated fluorescence directly with the production of X-rays, and Henri Poincaré suggested that it might be worthwhile investigating whether or not rays similar to X-rays might be produced by ordinary fluorescent or phosphorescent substances. The widely accepted account of Becquerel's discovery is that, following Poincaré's suggestion, he placed fluorescent mineral crusts on photographic plates wrapped in light-tight black paper, exposed them to sunlight, and did indeed observe an image on the plates. One one occasion in February 1896 poor weather prevented exposure to sunlight and Becquerel put both the prepared plates and the minerals away in a drawer. On March 1st he removed them and, for some unknown reason, immediately developed the photographic plates before any exposure to sunlight. To his surprise, he saw silhouetted images of the crust shapes on the developed plates [2.2]. He concluded that neither sunlight, fluorescence nor phosphorescence was necessary to produce this effect. He also found that the radiation could penetrate thin strips of aluminium and copper as well as sheets of black paper. A somewhat different, and rather more credible, version of this story is given by Becquerel's son, Jean [2.2].

Becquerel presented his discovery the following evening at the weekly Monday meeting of the French Academy of Sciences, and publication followed within ten days in a paper in the *Comptes Rendus*[1] entitled 'On visible radiations emitted by phosphorescent bodies'.

Several suggestions were proposed to explain this new phenomenon, and that by Marie Curie[2] stated 'it could be imagined that all space is constantly crossed by rays similar to Röntgen rays, but much more penetrating and which could be absorbed only by some heavy atomic weight elements such as uranium and thorium.' However, some experiments performed by Pierre Curie and Albert Laborde in 1903 established the impossibility that radioactive substances were storehouses for certain forms of extra-terrestrial energy, such as from the sun.

The speed of publication achieved by Becquerel was fortunate, in that at the same time Sylvanus Thompson in London was working with uranium nitrate. He found that when the uranium salt was exposed to sunlight while resting on a shielded photographic plate, film blackening occurred directly beneath the location of the uranium nitrate. The President of the Royal Society, replying to Thompson on February 29th 1896, suggested that he should publish his findings without delay, but Thompson had been pre-empted by Becquerel. It was by such a short margin of time that the new Système International radiation unit of activity, defined in the 1980s, is now a becquerel and not a thompson!

Becquerel continued to study radioactivity during 1896 and 1897, and within three months of his discovery had demonstrated that uranium itself was the source of the radiation. He later collaborated with the Curies[3] and jointly with them was awarded the 1903 Nobel Prize for Physics. He died in 1908 at the age of 55 years. However, it was Becquerel's discovery of radioactivity that led to the Curies' announcement[4] in Paris on December 26th 1898 of the discovery of radium. Pierre Curie and his brother Jacques were already the discoverers of the phenomenon of piezoelectricity and designers of the instrument known as the Curie electrometer, which was based on the piezoelectric effect of quartz crystal.

Following the use of ionisation methods to measure the intensity of X-rays, Marie Curie realised that the radiations discovered by Becquerel could also be measured using techniques based on the ionisation effect. The Curie electrometer equipped with an ionisation chamber proved to be a viable alternative to photographic plates and Marie Curie showed that the intensity of the radiation was proportional to the amount of uranium. From the many substances studied, she found that only thorium compounds were similar to uranium and that the uranium mineral pitchblende showed a higher radiation intensity than could be explained merely by the presence of thorium and uranium compounds. The Curies then concentrated their efforts on pitchblende and were rewarded in July 1898 by the discovery of the radioactive element polonium which was associated with the bismuth extract of the ore.

Although a knowledge of the existence of the new element, radium, had been gained, it was an immense problem to refine it in any quantity. An idea of the difficulties involved in separating radium from pitchblende can be imagined when reading the reports that Marie Curie initially obtained only one-tenth of a grain of radium from two tons of pitchblende[5] and that it was considered to be an achievement by Giesel when his yield was four grains of radium bromide per ton of ore[6]. One grain of radium is equal to 65 mg and a 1960s radium treatment for cancer of the cervix often involved the use of 100 mg of radium. Thus in the first decade of this century, some half a ton of pitchblende ore would need to be processed to provide the radioactive content for a single patient treatment in terms of 1960s clinical practice. The radium yield improved but when it was 15 grains (which equals 1 g) per ton, the processing required was still enormous, involving five tons of various chemicals and 50 tons of water[7].

The pitchblende ore from which the Curies first isolated radium was mined at Joachimsthal (now Jáchymov) in Bohemia, where the mines had been famous for silver deposits from the 16th century. Until 1921 this mine remained one of the major sources of radium, and during the Stalin era, when mining of various minerals continued, labour was provided by some of those sentenced in the Communist courts; today the area is an environmental disaster. After World War I the Union Minière du Haut Katanga[8] was able to exploit mineral deposits rich in uranium which had been discovered in 1913 in the then Belgian Congo, now Zaire. The minerals were extracted from the Chinkolobwe Mine, Katanga, were processed [2.19] at Oolen, near Antwerp, and eventually made Belgium the premier radium producing country. Uranium ores were also discovered in many other parts of the world besides Bohemia and Katanga. For example, in the United States[9,10] the mineral carnotite was found in Utah and Colorado and in 1922, before the Katangan mines were in full production, the USA was producing four-fifths of the world supply of radium, although only 15 grains of radium were obtained from 500 tons of carnotite[11]. France obtained some of its radium from deposits in Portugal and Madagascar[8]. Experiments with English radium from Redruth in Cornwall were reported in 1904 and radium was also found in Bath and Buxton around the hot springs[6]. Other minor radium deposits were found in Hungary, Turkey, Canada, Ceylon, India[7], Saxony and Tibet[12].

The history of the Curies has been well documented, notably by their daughter Eve Curie[13] and by Robert Reid[14] and more recently by the Université Libre de Bruxelles[15]. Marie Curie's office and laboratory, within what is now the Institut-Curie in the rue d'Ulm in Paris, is now a permanent museum, although it had to be decontaminated of radium before being opened. Her desk is as she left it with pens, spectacles and notepaper; an enormous library of newspaper and magazine cuttings is also in her office. Her library of books contains one by Jacques Danne[16] who was responsible for preparing many of Marie Curie's radium sources and who described himself on the flyleaf as 'Préparateur particulier de M. Curie à l'école de physique et de chimie industrielles de Paris', her 1910 *Traité de radioactivité*[17] and a little known paperback book[18] she wrote on *La radiologie et la guerre* in 1921, styling herself Mme Pierre Curie. It is one of her few published commentaries on the applications of X-rays and the eight chapters are as follows:

Les rayons X
Comment peut-on produire des rayons X
Installation dans les hopitaux et voitures radiologiques
Travail radiologique dans les hopitaux
Personnel radiologique
Rendement et résultats
Organisation d'après-guerre
Radiothérapie et radiumthérapie

They are drawn from her experiences in World War I, when within ten days of the start of hostilities in 1914 she received from the Minister of War a formal request to equip operators for radiographic work and was given the official title of Director of the Red Cross Radiology Service[14]. At the outbreak of war the French army only had a single radiological car but Marie Curie was eventually responsible for putting 20 on the road and installing 200 radiological posts, which by the end of the war had examined more than a million men[13]. The first car was ready in October 1914 and was equipped with a 110 volt, 15 amp dynamo, a Drault X-ray unit, photographic equipment, curtains, some primitive screens and a few pairs of protective gloves[14]. The cars were nicknamed 'little Curies' in the army zones[13].

Marie Curie's contributions to science, both together with her husband Pierre, who died tragically in a street accident in 1906 at the age of 47 years, and on her own, were of course far more wide ranging than only the medical applications of radium which are the subject of this book. However, her interests in medicine are well documented and her legacy continues at the Institut-Curie in the rue d'Ulm. The Institute has been extended by the construction of a new Hospital opened in 1991 and named after Claudius Regaud[19] whose association with Marie Curie began in 1912. After several years of lobbying, an agreement had been reached in 1912 between the Pasteur Institute and the Sorbonne that a new laboratory should be constructed devoted to the study of radioactivity. This was to be the Institut du Radium (now the museum) and was divided into two parts. One was under the direction of Marie Curie herself and dealt with physics and chemistry research whereas the other was directed by Regaud and was devoted to medical and biological research[20]. Marie Curie was also instrumental in the building of the Radium Institute in her native Warsaw [2.21–2.23].

In 1921 Marie Curie travelled to the United States to receive from President Harding the gift of one gram of radium, worth $100,000 raised from public subscriptions by American women, to be taken back to Europe for medical purposes [2.19]. The mahogany and lead container used for its transportation is inscribed 'Presented by the President of the United States on behalf of the women of America to Madame Marie Sklodowska Curie in recognition of her transcendent service to science and to humanity in the discovery of radium', and remains in the museum at the Institut-Curie.

Not only was Marie Curie awarded two Nobel Prizes,

one in physics (jointly with Pierre Curie and Henri Becquerel in 1903) and one in chemistry (1911) but a third member of the Curie family, her daughter Irène Joliot-Curie (jointly with her husband Frédéric), was awarded a Nobel Prize for chemistry in 1935. The couple discovered artificial radioactivity in January 1934 when they bombarded boron with alpha particles [2.10] and produced radioactive nitrogen-13 with the emission of a neutron: $B^{10} + He^4 = N^{13} + n^1$. The N^{13} disintegrated with the emission of positrons to give a stable carbon-13 nuclide. In the case of aluminium and magnesium, they found[21] that radioactive isotopes of phosphorus and silicon, respectively, were produced. Marie Curie's daughter Eve[13] records her mother's comments to one of her former assistants, Albert Laborde, at a Physical Society meeting at which Irène and her husband explained their work: 'Bonjour! They talked well, didn't they? We're back again in the fine days of the old laboratory.'

Sadly, Marie Curie died soon afterwards, on July 4th 1934 at the age of 67 years, of aplastic anaemia which was directly related to her long exposure to radium, without, according to her daughter[13], taking adequate safety precautions when conducting her experiments, although she insisted on them being taken by her colleagues and students. She was twice a Nobel Prize winner, the most respected woman scientist of her time and arguably of the entire 20th century.

Henri Becquerel: 1852–1908

[2.1] Antoine Henri Becquerel in his laboratory. (Courtesy: Museum Marii Sklodowskiej-Curie, Warsaw.)

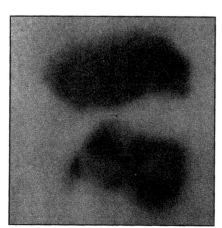

[2.2] First photographic exposure which proved the existence of spontaneous radiation from uranium to Henri Becquerel. This illustration was published in 1918 by his son Jean Becquerel[22] who was then working in the Physical Laboratory of the Natural History Museum of Paris. He also described in detail his father's discovery of radioactivity. Becquerel's experiment, using a photographic plate placed within a light-tight 2 mm thick aluminium cassette, was to study the fluorescence properties of the double sulphate of uranium and potassium by placing the sulphate crystals on top of the cassette for one day in strong sunlight, thus obtaining a constant fluorescence. When he processed the photographic plate he found only a feint amount of blackening, repeating the experiment several times but always obtaining the same result. Poor weather stopped the experiment repetitions for a few days and the cassette and crystals were stored in a drawer. When it was later exposed to the one day's sunlight a much greater blackening was observed leading Becquerel to conclude that the blackening was independent of any external circumstances but was proportional to time and that the finding was similar to that produced by the recently discovered X-rays.

[2.3] Photographic print and sketch prepared by Becquerel and dated March 2nd 1896, recording the experiment by which he made the discovery of radioactivity. This was given by Becquerel to Sir William Crookes and is now displayed in the British Institute of Radiology, London.

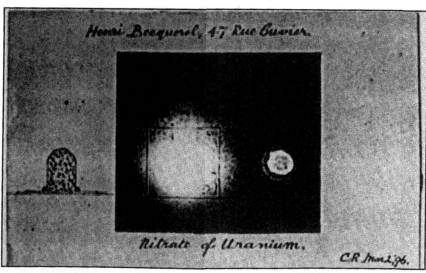

Marie Curie: 1867–1934

[2.4] Marie Curie's early life. The upper photographs show Marie at the age of 16 and a page from her personal notebook, from 1885, including her inscription of a poem by Heine and a sketch of her pet dog 'Lancet'. The family group shows Marie (centre) with her brother and sisters Jozef and Helena (right) and Zofia and Bronislawa (left) in 1871. Freta Street in old Warsaw was photographed at the end of the 19th century. The house in which Marie was born is the second from the left with the circular window in the top gable (restored in the 1950s following its destruction in August 1944). It was a school at the time of her birth, her father being a school master, and it now houses the Muzeum Marii Sklodowskiej-Curie [2.24] which is administered by the Polish Chemical Society. (Notebook illustration courtesy: Fondation Curie-Institut du Radium, Paris. Others courtesy: Museum Marii Sklodowskiej-Curie, Warsaw.)

[2.5] Pierre and Marie Curie, 1904. (Courtesy: Fondation Curie-Institut du Radium, Paris.)

[2.6] A cartoon entitled 'Radium' which appeared in the December 22nd 1904 issue of *Vanity Fair*. (Courtesy: Wellcome Trustees.)

[2.7] Double Nobel laureate: the 1903 Noble Prize for Physics was awarded jointly to Henri Becquerel, for his discovery of radioactivity, and (above) to Pierre and Marie Curie for their work on that phenomena. (Courtesy: Fondation Curie-Institut du Radium, Paris.) Marie also received the 1911 Nobel Prize for Chemistry for her discovery of radium and polonium. The decorative border of the Nobel document is reproduced on the postage stamp (right), honouring Marie Curie's two awards, issued in 1967 to mark the centenary of her birth.

[2.8] Top: the First Solvay Conference, 1911, in Brussels, which was attended by Marie Curie, Albert Einstein (second from right, standing), Lord Rutherford (fourth from right, standing), Langevin, Poincaré, Lorentz, Planck and many other eminent scientists. (Courtesy: Fondation Curie-Institut du Radium.) Bottom: the Fifth Solvay Conference, 1927, showing Marie Curie three places from the left on the front row. (Courtesy: Muzeum Marii Sklodowskiej-Curie, Warsaw.)

A. PICCARD   E. HENRIOT   P. EHRENFEST   Ed. HERZEN   Th. DE DONDER   E. SCHRÖDINGER   E. VERSCHAFFELT   W. PAULI   W. HEISENBERG   R.H. FOWLER   L. BRILLOUIN
P. DEBYE   M. KNUDSEN   W.L. BRAGG   H.A. KRAMERS   P.A.M. DIRAC   A.H. COMPTON   L. de BROGLIE   M. BORN   N. BOHR
I. LANGMUIR   M. PLANCK   Mme CURIE   H.A. LORENTZ   A. EINSTEIN   P. LANGEVIN   Ch.E. GUYE   C.T.R WILSON   O.W. RICHARDSON
Absents : Sir W.H. BRAGG, H. DESLANDRES et E. VAN AUBEL

[2.9] Marie Curie at the wheel of her radiological 'car' in Hoogstade, Belgium, 1915. Of the 20 cars she put into service she kept one for her personal use, a flat-nosed Renault with a body like that of a lorry[13], painted in regulation grey with a red cross and a French flag on its plates. She used this car at the war front from the very beginning, travelling at full speed of '20 miles per hour average' towards such battlefields as Amiens, Ypres and Verdun[13]. On arrival she would unpack the equipment at a chosen 'radiological hall', run the cable from the car to the X-ray unit and start work with the surgeon in a dark room 'where the apparatus was surrounded by a mysterious halo'[13].

[2.10] Not only did Marie Curie's daughter Irène and her husband Frédéric Joliot discover artificial radioactivity in 1934 but Joliot, in the Nuclear Chemistry Laboratory of the College of France, Paris, took in January 1939 the first photograph of the phenomenon of nuclear fission, which had been discovered by Hahn and Strassmann in 1938. (Courtesy: Fondation Curie-Institut du Radium, Paris.)

[2.12] The statue of Marie Curie, by Ludwika Nitsch, unveiled in Warsaw on September 6th 1935, the first anniversary of her death. It is also featured on the 1967 stamp commemorating the centenary of her birth.

[2.11] The Curies' grave in Sceaux. Respecting her will, Marie was buried in the grave together with her husband Pierre and his parents. At the time of her death she had been awarded 10 scientific prizes, 16 medals and 108 honorary titles. (Courtesy: Muzeum Marii Sklodowskiej-Curie, Warsaw.)

[2.13]   National heroine: Marie Sklodowska-Curie is commemorated on this Polish banknote of 1989 and, more than any other scientist, on postage stamps from many countries. The four stamps reproduced in this chapter were issued by her native Poland. Here, the radium stamp of 1992 incorporates the chemical symbol Ra, the atomic number 226, atomic weight 226.025 and valence electron configuration s^2, while the 1982 issue depicts Marie Curie in later life.

The laboratory in Paris: 1898

[2.14]   The shed at the School of Physics on the rue Lhomond where radium was discovered in 1898. (Courtesy: Fondation Curie-Institut du Radium, Paris.)

[2.15]   Marie Curie in her laboratory, April 1921 (standing) and in 1923 (seated, using the Curie electrometer). (Courtesy: Fondation Curie-Institut du Radium, Paris.)

Radium and Marie Curie: 1904–1923

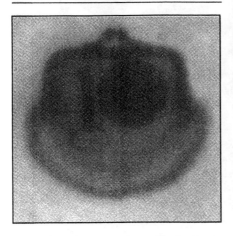

[2.16] Contents of a purse taken by Marie Curie[23] before 1904 using a radium source, for comparison with X-ray imaging. This was the only half-tone illustration in the German-translated *Experiments about radioactive material* which formed the first in a series of *Natural science and mathematical monographs*.

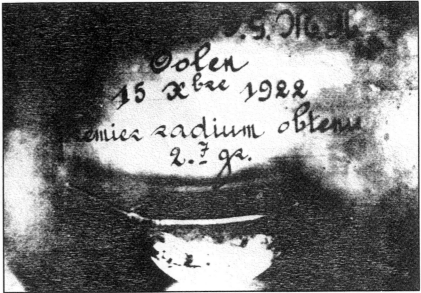

[2.17] Capsule containing 2.7 g of radium bromide 'photographed by its own radiation' in 1922 at Oolen in Belgium, the head office of the Union Minière du Haut Katanga. This photograph was used in the book published by the Fondation Curie for 'Le Vingt-Cinquième Anniversaire de la Découverte du Radium (1898–1923)' which was published to encourage generous donations for the continuation of the work of the Fondation Curie, with copies personally signed by Drs Appell, Roux, Marie Curie and Claudius Regaud.

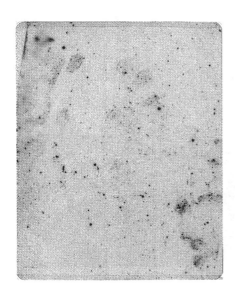

[2.18] Autoradiograph of a radium source certificate of calibration (superimposed on the document), signed by Marie Curie, that was brought to the Netherlands Cancer Institute in 1914. The source strength was 10.37 mg radium bromide, was manufactured by La Maison Armet de Lisle and the calibration was made at the Laboratoire de Physique Générale, Faculté des Sciences de Paris. The autoradiograph shows very clearly the contamination of the document and some fingerprints can be recognised at the top right. They could possibly be Marie Curie's whilst she was completing the certificate. A similarly contaminated document is displayed in the Curie-Institut museum in Paris. (Courtesy: Dr Haarm Meertens, Netherlands Cancer Institute.)

[2.19] President Harding, Marie Curie, Mrs Harding and Mrs William Brown Meloney photographed at the White House in 1921 just after the President, in the name of the women of America, had presented Mme Curie with the gram of radium purchased by voluntary subscription.

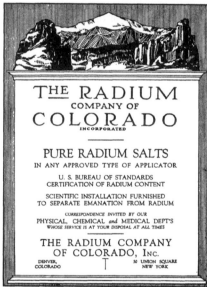

THE RADIUM
COMPANY OF
COLORADO
INCORPORATED

PURE RADIUM SALTS
IN ANY APPROVED TYPE OF APPLICATOR

U. S. BUREAU OF STANDARDS
CERTIFICATION OF RADIUM CONTENT

SCIENTIFIC INSTALLATION FURNISHED
TO SEPARATE EMANATION FROM RADIUM

CORRESPONDENCE INVITED BY OUR
PHYSICAL, CHEMICAL and MEDICAL DEPT'S
WHOSE SERVICE IS AT YOUR DISPOSAL AT ALL TIMES

THE RADIUM COMPANY
OF COLORADO, Inc.

DENVER,   50 UNION SQUARE
COLORADO   NEW YORK

[2.20] Most of the early textbooks on radium included an illustration of pitchblende ore: the photograph (top) is reproduced from Turner's 1911 book[5].

The Austrian Government soon realised the commercial value of a monopoly on the supply of radium and prohibited further export of the Bohemian pitchblende. French and British prospectors immediately began a worldwide search for radium ores and in 1899 two French mineralogists found immense deposits of a uranium-bearing ore in Colorado and Utah, which they called 'carnotite' in honour of the French chemist, Carnot. Its radium content was, however, much lower than that of pitchblende and commercial exploitation did not begin until 1913, but by 1922 American companies, such as the Radium Company of Colorado, had become the world's foremost producers of radium.

The uranium ore in the Belgian Congo was discovered just before the 1914–1918 war by the Union Minière du Haut Katanga whose main mining interests were in copper. However, because of the war they kept their discovery secret and only started mining on an extensive scale in 1922. This Katangan ore gave a radium yield similar to that of pitchblende. As a direct result of this, in 1923 the price of radium fell from US$120,000 to US$70,000 per gram and some of the American mines closed because they were unable to compete with the yield from this richer grade of ore. The photograph (above) shows the separation of radium from pitchblende at the Union Minière plant in Oolen, Belgium, during a visit by Marie Curie (seen on the left of the group) in March 1923. The smaller photograph (left) shows the reaction vats and pumps in the Oolen factory in 1929[8].

Radium Institute, Warsaw: 1925–1936

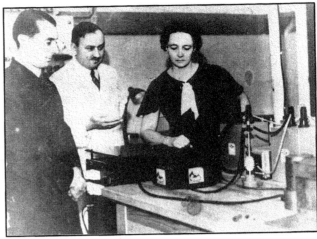

[2.23] Frédéric Joliot-Curie and his wife Irène with Professor Cezary Pawlowski in his Physics Department in the Radium Institute, Warsaw, 1936. Pawlowski was a student and co-worker of Maria Sklodowska-Curie. (Courtesy: Prof. Andrzej Kulakowski and The Maria Sklodowska-Curie Memorial Cancer Center and Institute of Oncology, Warsaw.)

[2.21] It was Marie Curie's dream to build a scientific institute in Warsaw similar to the Radium Institute of Paris and this was finally achieved after she had laid the corner stone in 1925 and taken part in the opening ceremony in 1932. That this was possible, was due to her own and to her sister Bronya's initiative, together with governmental and social help (a Diploma, as illustrated above, was given to those who contributed to the building funds) as well as the sum of US$53,460 given by American women. The work of this Institute continues to this day, still in Wawelska Street, but with a change of name in keeping with modern times, the Maria Sklodowska-Curie Memorial Cancer Centre and Institute of Oncology.

Marie Curie Museum, Warsaw

[2.22] Maria Sklodowska-Curie and Claude Regaud welcomed by Franciszek Lukaszczyk, Director of the Radium Institute, Warsaw, on the occasion of the opening ceremony of the Institute on May 29th 1932. (Courtesy: Prof. Andrzej Kulakowski and The Maria Sklodowska-Curie Memorial Cancer Center and Institute of Oncology, Warsaw.)

[2.24] 1990s view of the birthplace in Freta Street, Warsaw, of Maria Sklodowska-Curie (see also [2.4]) which now houses the Maria Sklodowska-Curie Museum of the Polish Chemical Society containing excellent reference material of all aspects of her life. The commemorative plaque states that she was the discoverer of both polonium and radium.

Chapter 3

Early Days of X-rays and Radium

Diagnosis, Therapy and Experiment

Most of the very first radiographs (called shadow-pictures by Röntgen and later termed skiagrams, skiagraphs, radiograms, röntograms and several other variants until the term radiographs finally became universally accepted) were of coins, keys and similar objects X-rayed through some opaque material; small mammals and hands (although seldom feet). The earliest classification termed hands and animals (such as frogs) 'living' radiographs whereas the remainder were not prefaced by an adjective.

In England the first radiograph was taken on January 8th 1896 by an engineer, A. A. Campbell Swinton [3.2] and in the USA the first radiographic image was published, in the *New York Medical Record*, on February 15th 1896. This was in an editorial 'Photographing the invisible' but the radiograph was German, not American, taken in Hamburg on January 17th 1896. This Hamburg radiograph also appeared in other publications, such as the pamphlet by one August Dittmar, published in Glasgow[1] with the caption 'Living hand, taken with X-rays after Professor Röntgen in the State Physical Laboratory at Hamburg. The plate was enclosed in a wooden case so that the rays penetrated the hand and the wood cover.'

The Glaswegian publisher, F. Bauermeister, describing himself as a 'Foreign Bookseller', offered several radiographs (mostly taken in Hamburg) for a sale price of one shilling each or four shillings and six pence post free for a set of six. The 'living hand' was one; the others included the 'Fracture of an arm healed after dislocation', a 'Right foot deformed by tread of horse', and a 'Young pig (one day old): taken in the Physical Institute at Munich University by Prof. Dr. L. Graetz.'

Early radiographs were also for sale on the other side of the Atlantic and in one of the earliest of the 1896 textbooks[2], by William Morton of New York, a series of 47 were available ranging in cost from $2 for 'Infant nine weeks old, life size' to several 50 cent and 60 cent radiographs including 'Trout showing bones', 'Foreign body in scrotum', 'Human teeth in situ' and 'Child's hand'. A fully documented X-ray workload from June to December 1896 was published by Thurstan Holland[3] in Liverpool, and consisted of 261 X-ray plates. These included a matchbox containing various items such as keys, deformities of the hand and other bony structures, the oesophagus of a boy who had swallowed a coin [4.5] (whose symptoms had caused a diagnosis of TB!) and an attempt to discover whether a set of false teeth had been swallowed [4.5].

Thurstan Holland also took radiographs of a diamond star and a paste brooch [13.14] and located a mummified bird in a red brick brought to him by a visitor to Egypt who wondered if it might contain something [9.5]. X-rays and gems were sometimes linked in popular magazines and the July 1896 issue of *The Strand Magazine* told the following tale to illustrate 'The Adventures of a Man of Science'.

A man who was suspected of having stolen and swallowed a diamond is lured into the laboratory. 'I desired him to strip and then, after some difficulty, arranged him in such a position that the rays should pass through his body. I turned off the light in the room, my electrical battery worked well, the rays played admirably in the vacuum tube. I removed the cap from the camera, . . .'. The result was 'an excellent plate showing the diamond just below the region of the ileo-caecal valve.' Unfortunately the author did not realise that paste stones, but not real diamonds, appear opaque on an X-ray plate!

Suggestions for the therapeutic uses of X-rays were also forthcoming and the earliest appears to have been by E. H. Grubbé[4], variously described as an X-ray manufacturer and a pharmacist-homeopath[5]. J. E. Gillman in Chicago sent a breast cancer patient to him for treatment on January 29th 1896 and on the following day he also treated a patient with lupus vulgaris. Two other early treatments were of patients with a cancer of the nasopharynx, February 1896 in Hamburg by Voigt[6], and a cancer of the stomach, July 1896 in Lyon by Despeignes[7].

However, it is generally accepted that Freund[8] in Vienna was the first to use X-rays logically and scientifically within the limits of the age. His first patient, treated in December 1896, was a five-year-old girl with hirsuites (Naevus pigmentosus pilosus). Schiff[9], who was the doctor responsible for this patient, describes the case in which an artificial alopecia was produced over the whole of the back which had been exposed to X-rays for two hours per day for 16 days. The hair started to fall out after 12 days and after total epilation a violent dermatitis set in which healed only very slowly. Schiff remarked that 'this accident was full of instruction' and in future he reduced the exposure to 10 minutes duration. This case history [3.12] was the beginning of radiotherapy.

The epilatory effect of X-rays established them as a direct agent for producing biological change. (In April 1896 John Daniel at Vanderbilt University observed and reported the first case of epilation on the scalp of his University Dean—who had volunteered for a radiograph of his brain![10]) Indeed, evidence for the harmful effects of the radiation was known within a few months of their discovery.

For example, Thomas Edison[11] reported in March 1896 that his eyes were sore after experimenting with X-rays [10.3]. On April 18th 1896 an English physician L. G. Stevens[12] reported in the *British Medical Journal* that 'those who work with X-rays suffer from changes of the skin which are similar in effect from the sun burn' and a subsequent note in the same journal suggested that X-rays were a normal constituent of sunlight, presumably because of the analogy between sunburn and X-ray erythema. T. C. Gilchrist[13] of Johns Hopkins Hospital reviewed 23 cases of X-ray injury which had been reported in the literature before January 1897, and in May 1897 N. S. Scott[14] of Cleveland reviewed 69 reports of X-ray injury. Röntgen himself warned in 1898 of the adverse biological effects (page 2).

Injuries were attributed to electrical effects, ultraviolet, and platinum particles from X-ray tubes. Red silk and thin rubber sheet were advocated as preventatives, as well as more substantial items.

In April 1898, the Röntgen Society appointed a committee (the first ever British radiological committee) to collect data on the harmful effect of X-rays and this 'Committee on X-ray Injuries' listed in the May 1898 issue of the *Archives of the Roentgen Ray* the following questions to be answered by the Committee, of which 1–6 were termed 'medical questions' and 7–12 were termed 'electrical questions':

1. Nature of injurious effects.
2. Description of case radiographed.
3. Part exposed to rays.
4. Condition of subject:
   a. Well nourished or emaciated.
   b. Temperature nervous or phlegmatic.
   c. Diathesis of patient.
   d. Local condition of part exposed.
5. Did the patient complain of any feeling of warmth, tingling, or other sensation during or after exposure?
6. Duration of effects, temporary or permanent.
7. Apparatus employed, influence machine, or induction coil, spark length, voltage and amperage used.
8. Form of tube: length from terminal to terminal.
9. Distance from patient's body.
10. Number of exposures: interval, if any, between exposures; duration of each exposure.
11. Situation of tube with regard to body or limb of patient, i.e. position of anode and cathode.
12. What covering or garment, if any, was used?
    a. Material of which it was composed.
    b. Rough or smooth.
    c. Colour if dyed.

Review papers discussing the hazards of overexposure to radiation continued to appear in the radiological literature in the first decade of this century[15–18], but in spite of these warnings of the dangers of superficial injuries through indiscriminate exposure to X-rays, the radiation protection facilities for many workers were rudimentary or non-existent for many years.

This was no doubt due in part to the latent period between exposures and the appearance of permanent ulcers or malignancies. It is also noted that some of the patients were doubtful about the effect of the radiation[9]: 'an occasional slight erythema (a reddening of the skin) is the sole visible result of treatment, so that patients are apt to become very sceptical of its success.'

Generalised radiation protection measures were eventually adopted, a British X-ray and Radium Protection Committee was formed in 1921 and an International Radiation Protection Committee in 1928. The first consideration by a national protection organisation of the principle of maximum tolerance dose of X-rays was by the United States Advisory Committee on X-ray and Radium Protection in 1934. This dose was derived essentially from a 1921 recommendation that 'a sort of grand average of the protective measures could be gleaned from the working conditions of a number of experienced radiologists, who had escaped injury and still enjoyed normal health.'

However, general acceptance of protection procedures was too late to prevent radiation injuries to superficial tissues, blood and internal organs of some of the early workers in the radiation field[19–21]. Many of them are remembered by a monument erected by the German Röntgen Society in Hamburg in 1936 to the X-ray and radium martyrs. This memorial originally contained 169 names from 15 different nations[22] but by 1959[23] the total had risen to 360. Röntgen was not among them, possibly because his experiments were mainly radiographic as opposed to fluoroscopic, and his X-ray tubes were housed within a metal box. The memorial does, though, include Marie Curie.

It is also noted that Pierre Curie and Henri Becquerel suffered radiation burns on their hands and at the finger tips with which they had carried radium tubes[8].

Indeed, some of the X-ray and radium procedures carried out in the early part of this century must have been extremely dangerous by today's standards. Pure radium bromide was burnt in a bunsen flame to determine the colour of the flame, which was carmine[24]. The effect of radioactive substances on cholera germs and on the bacilli of tuberculosis, diphtheria and typhus were studied, in which '5 cc of a very poisonous broth was injected under the skin of both ears of a rabbit and one ear exposed to uranium rays: no inflammation occurred in the treated ear'[8].

Experiments were also conducted with radium on animals with rabies[25] and baths of radium-bearing water were recommended for use at home 'as the radioactivity is permanent'[24].

There were several studies of the effect of radium on the eye, and in 1899 it was first reported[26] that radium stimulated the retina of the dark-adapted eye. These results were later confirmed by several authors[27,28]. One of the experiments described makes interesting reading[27].

'If two perfectly similar little bags of light-proof paper are prepared, the one filled with radium and the other with a corresponding quantity of sand, and, if these are placed alternately on the eyes, it is impossible, after about 30 repetitions of the experiment, for anyone to tell on which eye the radium and on which eye the sand is placed'.

It was also hoped at one time, by Pierre Curie and others, that radium might help to restore sight to the blind[29].

Self-exposure experiments, usually on the forearm, were a common procedure for many years and one American author[30] as late as 1927, who with many others used the erythema dose to determine X-ray patient exposure, recommended that 'because of medicolegal complications it is better for the operator to use his own skin'.

Radium therapy, as distinct from X-ray therapy, began with the accidental radium burn received by Henri Becquerel in April 1901, ten days after he had carried a tube of radium, loaned to him by the Curies, in his shirt pocket for six hours[8]. The effect was recognised by Besnier as being identical with X-ray dermatitis and prompted the Curies to loan some radium to Danlos at St. Louis Hospital, Paris.

This was then used for the treatment of lupus and other dermatological conditions, and Becquerel remarked in his Nobel Prize Lecture of 1903 that the use of radium was being explored for the treatment of cancer. The work of Danlos led in 1906 to Wickham and Degrais establishing a centre for radium therapy called the Biological Laboratory in Paris and publishing[31] the first textbook of radium therapy, lavishly illustrated in colour with patients' benign or malignant conditions before and after radium therapy, together with accompanying case histories of the radiation dose and fractionation. It was these authors who established the principle of the cross-fire technique [3.11, 3.16] as an alternative to a single radiation field technique and who first tabulated the geometrical and radioactive parameters of a series of radium applicators: apparatus shape, dimensions, surface for application, activity of radium salt, weight of radium salt, total activity, percentage of α, β and γ emissions.

Radiobiological experiments and theories began very early and in 1906 Bergonié and Tribondeau[32] stated what was to become known as their Law 'that sensitivity varies directly with the reproductive capacity of the cell and inversely with its degree of differentiation'—although there are exceptions.

Mice and rats were used as experimental animals by many of those investigating biological effects of radiation and one early example of such work was by Lawrence[33] in Melbourne, who published his results in the *Australian Medical Journal* of July 20th 1910. He used 30 mg of radium in mica and rubber applicators and then in lead applicators (termed 'interceptors') and varied the time of exposure. His main difficulty was stated to be the rapid growth rate of the adenocarcinoma in the mice: 'the growth frequently reaches a size equal to that of the mouse itself in about six weeks from the date of innoculation'. He found that tumour growth was greatly delayed 'but due to the intense action of the rays the treatment was not continued'.

The time factor (dose-fractionation) schemes in both external beam radiotherapy and in brachytherapy are still as much of interest in the 1990s[34–37] as when the topic was first investigated in 1918 by Kröning and Friedrich[38] in Freiburg. They found that when a definite dose was given to the skin, in some cases over a period of 690 minutes and in others in 86 minutes, the resulting erythemas were very different. The reaction to the shorter, intense exposure was much more severe than the other. The significance of this fact for routine clinial practice was not fully appreciated for some years but was eventually accepted and, for many years now, protracted radiotherapy schemes have been adopted worldwide.

A major influence in this was Claudius Regaud[39–42] who was the first to clearly demonstrate that since different kinds of tissues do not have the same powers of recovery, it is to be expected that, with the prolonged fractionation method of radiation dose delivery, some will be more affected than others. He introduced a radium needle into a ram's testis and found that after a short high dose (15 dose units in 5.5 hours) the tissue immediately surrounding the needle was necrosed, although peripheral regions of the organ remained normal. But after a prolonged low dose (4.6 dose units in 38 days) the result was complete sterilisation without massive necrosis. The growing sperm cells were killed while the interstitial tissue was unharmed.

Radium was also investigated as a possible 'magic bullet' against cancer, as recorded in 1909 in the *Deutschen Medizinischen Wochenschrift*[43] which reported on an advertisement in the weekly journal of the Charlottenburg Röntgen Society. Ampoules of radium containing 'Radiogenol' were offered for injection into tumours using injection points over a 2 cm grid to cover a tumour. According to the advertising company, a high percentage of cases treated by Czerny of Heidelberg achieved complete healing and tumour debulking. Czerny stated that his experience was with both animals and with humans and that no serious side effects occur with his recommended dosage but that 'the patients will radiate after four months.' Necrosis was observed but Czerny also stated that no single cure was observed and that 'radium is not yet the drug against cancer and where possible surgery should be the first choice.'

To conclude this chapter on the early days of X-rays and radium I reproduce below the reminiscences[44] by an author of note whose books describe his life and times in the Liverpool of the early 20th century, when he visited a radium clinic before 1910. When he recalled this episode in his life he was well over 80 years old.

'I recall an interesting experience many years ago when I was a young soldier before the First World War in 1914. Although I can't remember the name of the doctor, he was a Birmingham specialist (since identified as John Hall-Edwards[45]) struggling, shall I say, with radium, which he knew was powerful and was something he'd got to handle carefully as he touched his patients endeavouring to treat them with it.

'It was very early days and he offered to try what he could do for any patients who would accept his help. There was a chap with me in our Company whose young sister suffered with lupus. All kinds of things had been tried without success and our Medical Officer, hearing of it, arranged for her to let the Birmingham doctor try, and I went with the chap, his mother and the girl to the doctor in New Hall Street, Birmingham.

'We went several times and he had what seemed like a small stone at the end of a bamboo cane. The cane had been split to hold it and it had been squeezed in the two ends. Primitive indeed, wasn't it? He touched the skin with it very carefully, but it wasn't effective, rather defective. It made her slightly out of her mind though I believe she got that back in later years.

'But as I've said, he was handling something which he knew was dangerous but didn't know to what extent. A few years later he lost both hands[60] as a result of contact with it [3.35].

'Well, there are better means of doing things now. No more split bamboos. But there's a long way to go, isn't there?'

X-ray reports: 1896

[3.1] In the United Kingdom, one of the first professional societies to become interested in 'the new photography' of X-rays was, not surprisingly, the Royal Photographic Society, as demonstrated by the following list of X-ray reports appearing in the 1896 volume of its *British Journal of Photography*.

January 17th Campbell Swinton reporting on his experiments, after taking the first radiographs in the United Kingdom.

January 24th Announcement that Newton & Company of Fleet Street are distributing slides of Campbell Swinton's X-ray pictures.

7th February Report that X-rays should be called Röntgen's rays.

7th February Are X-rays the same as Draper's rays which were reported in 1840?

21st February Mr C. Simon of New York stated that he had invented a method to photograph his own brain 'with platinum plates'. This was also reported in the *British Medical Journal* of February 22nd.

28th February Radiography of Egyptian mummies by Dedekind.

6th March Use of X-rays in workhouses.

20th March Medico-legal implications of bad bone-setting. Report that if X-rays had been available earlier, at least one patient would have taken his surgeon to court, but the surgeon is now dead.

20th March A professional X-ray laboratory available: W. E. Gray of 92, Queen's Road, Bayswater, London.

20th March Early lawsuits involving X-rays.

20th March Use of X-rays to detect objects such as coins in postal packets, by J. Hall-Edwards of Birmingham.

27th March Report of comment on X-rays by the *Pall Mall Gazette* about the poor taste of looking at someone's bones.

3rd April Mr Brooke claims to have made X-ray pictures in London in 1877, using a magnet.

5th June X-ray tube design.

12th June Queen Amelia of Portugal X-rayed her court ladies to demonstrate the evils of tightly laced corsets.

12th June Amusing story about X-rays and piracy and the use of an X-ray picture for a skull and crossbones flag.

19th June Description of an X-ray set as a 'wonder camera of the Würzburg Professor'.

19th June Photography of the soul by Dr. Baraduc and the use of X-rays to cure alcoholics and cigarette smokers.

26th June X-ray of the Sphinx.

10th July Roentgen work for profit: an answer to enquiries by the *British Journal of Photography*.

17th July Lawsuit in Liverpool; also reported in June by the *British Medical Journal*.

31st July Enthusiasm for X-rays.

31st July M. Gaudoin set up a practice for beauty treatment using X-rays for unwanted hair, e.g. moustaches. Initially he was beseiged by many ladies for this service, and later by the same ladies for a return of their money.

31st July Bactericidal effect of X-rays.

11th September X-ray pictures in a bootmaker's window near Ludgate Circus and St. Pauls Cathedral, to demonstrate the influence of shoes on foot growth.

15th September Enthusiasm for X-rays.

23rd October Lawsuit involving X-rays as evidence: a case of attempted murder.

30th October X-rays of parchment in court: for evidence.

13th November Epilating effect of X-rays.

13th November Denial of X-ray pictures for poor patients.

15th December Humorous poem on X-rays.

X-ray pictures: 1896

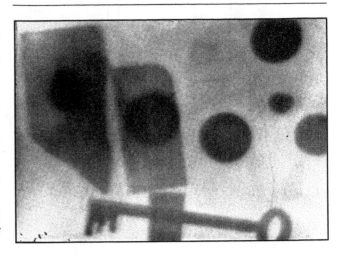

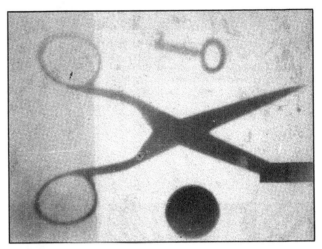

[3.2] Top: the first radiograph taken in England by A. A. Campbell Swinton on January 8th 1896. (Courtesy: The Science Museum, London.) Bottom: radiograph taken on February 15th 1896 by C. E. S. Phillips at the Royal Cancer Hospital, London, described as: '$1\frac{3}{4}$ hour exposure, 5″ induction coil, through $\frac{1}{8}$″ mahogany'.

[3.3] Living frog through a sheet of aluminium, taken in February 1896 using a 20 minute exposure. For comparison, see the frog in [9.1] and observe the improvement in image quality between 1896 and 1898. This was one of the radiographs published by *The Photogram* in a special February 1896 issue entitled *The New Light and the New Photography*. (Courtesy: Wellcome Trustees.)

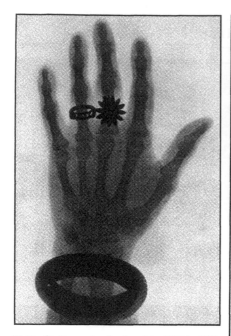

[3.4] Radiograph of a hand in 1896 by William Morton[2] of New York. The X-ray exposure was 2 seconds and copies of the radiograph were on sale for 60 cents.

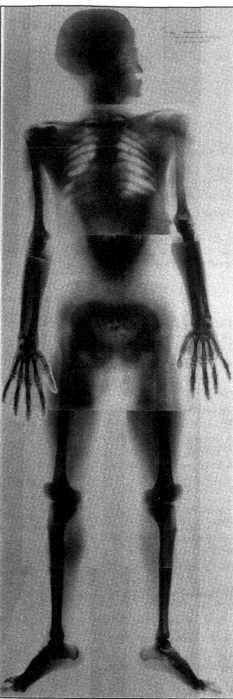

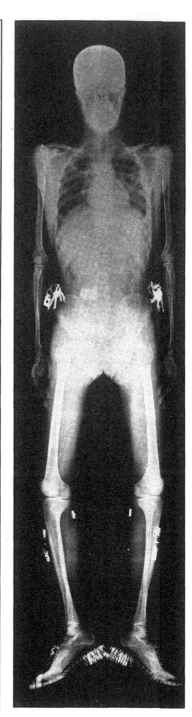

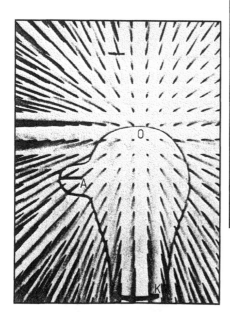

[3.6] In a report[58] to the Imperial Academy of Science in St. Petersburg in March 1896, Galitzin and Karnojitzky gave the results of their experiment using a board containing a regular pattern of nails and a photographic plate beneath the board. Some 13 different X-ray tubes were used and in their analysis of the resultant shadow patterns they demonstrated that the X-rays were produced from the target, which in the pear-shaped tube was the glass, marked 0, opposite the cathode. The anode is marked A and the cathode K.

[3.5] This whole body radiograph of a dead soldier (left) was taken, in nine sections, by Ludwig Zehnder at the University of Freiburg in 1896 and measures 1.84 metres in height. The exposure time was approximately 5 minutes per film. The faint writing, with an arrow pointing towards the forehead, reads 'small arms projectile located in the facing temple at a distance of 20 cm from the dry plate'. Zehnder had been a student of Röntgen in Giessen and Würzburg and was one of the colleagues to whom Röntgen mailed his earliest radiographs [1.14–1.16]. In 1919 Zehnder was appointed Professor of Physics at the University of Basel, retiring in 1936. The original life-size radiograph moved with him, from Freiburg to Basel, where it is still displayed almost a century after it was taken.

The other radiograph (right) is a single-exposure whole body image taken by a Dr Mulder in Bandung, Java about 1900, presumably of a living person, wearing knee-length boots with bunches of keys attached to an unseen belt, and may be compared with the earlier radiograph by William Morton shown in [4.4]. (Both radiographs courtesy: Deutsches Museum, Munich.)

Radium pictures: 1904

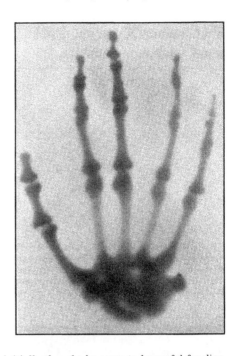

[3.7]   Radiumgraph of a mouse[48] published in 1904.

[3.8]   Radium was initially thought by some to be useful for diagnostic purposes and a few radiumgraphs are in the early literature. This is a 1904 radiumgraph of a hand[46] which was obtained after six hours exposure using 10 milligram of radium. The investigator's comments were 'While there is considerable permeation, the contrasts are poor and another disadvantage is that it takes hours to represent an image'. See also [23.2] for a comparison of X-ray and radium images.

Radium experiment in a New York garden: 1907

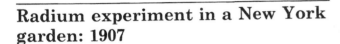

[3.9]   Robert Abbe of New York, writing in the February 1907 *Archives of the Roentgen Ray* tells how he answered a friend's question 'What difference is there between light from the stars and radium rays?' His reply that 'Radium rays move in absolutely straight lines, without deviation by atmosphere, water, lenses or prisms, and nothing is opaque to them. They would even penetrate that stone column near you [they were in Abbe's garden], which light does not', was received with incredulity. He then conducted the impromptu experiment illustrated here using 'a small glass tube containing a bit of radium about the size of a grain of rice, 60 milligram pure radium bromide, German'. Lead letters and a photographic plate were placed beneath the wooden board. The exposure time was three days. Abbe's letter to the *Archives* concluded with the comment 'This same demonstration has been made by Röntgen rays, but it is far more impressive to witness the active energy incessantly given off by the innocent looking little tube of yellow salt'.

Not all early experiments were as clear cut as that of Abbe. For instance in 1904 it was recorded[47] that when 1 gram of radium chloride is placed underneath a 1.5 gram lead block, this apparently causes the lead block to lose 0.000035 gram in weight!

Patients: X-rays, 1901 and radium, 1907

[3.10] Rodent ulcer treated with X-rays at the London Hospital in 1899. The photographs were published in the *Archives of the Roentgen Ray* in March 1901 and show the patient before and after treatment.

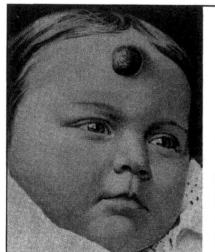

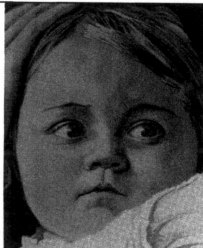

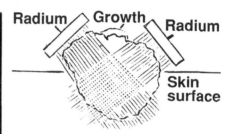

[3.11] First cancer patient treated by Wickham and Degrais[31] using their cross-fire technique (above). The tumour was an erectile angioma projecting 2 cm on the forehead of a seven month old child. The patient is shown before and after treatment.

1896 patient, Vienna: 70 years later

[3.12] I am indebted to Prof. H. Vahrson of Giessen for bringing the following to my attention and for providing this English translation from the *Strahlentherapie* paper by Fuchs and Hofbauer.[49]

Röntgen discovered X-rays on November 8th 1895 and the birth of X-ray therapy was one year later in November 1896. By a rare chance we[49] are able to report on the late results of this X-ray therapy applied 70 years ago since this irradiated patient has been in our care for the last 10 years. She was born in 1892 with a naevus pigmentosus piliferus which covered the whole surface of the back and was largely symmetrical. Her intelligence was normal. Her parents brought the child to the 'I Offentliche Kinderkranken-institut' where at that time Freund worked in the dermatological department. Because of the failure of every other therapy which had been attempted, Leopold Freund (1868–1943) made an attempt with X-rays. He later said of his thoughts at the time: 'In June 1896 I read in a Vienna newspaper the joke news that an American engineer who was intensively engaged in X-ray examinations lost his hair because of business. This notice interested me very much'[50].

The irradiations began on 24 November in the Graphischen Lehr- und Versuchsanstalt in Vienna where already at that time X-ray apparatus was at his disposal. Each irradiation treatment lasted two hours and the first series was given on the neck: the cranial part of the 36 cm long naevus. The first loosening of the hairs

was noticed after the tenth treatment. Some days later the irradiated skin was epilated but did not show any alarming alterations. Shortly thereafter a second series of treatments was given, which covered the lower part of the naevus: the skin and the lumbar spine.

To decide the questions as to whether the electric field of the high voltage generator also produced biological effects, this treatment field was protected by an aluminium plate which was positioned in the path of the X-rays: but at the same time, the number of treatments was increased to take into account the loss of radiation dose due to the plate. This explains the considerable overdose in the lumbar region. Some days after the last radiation treatment the hairs were lost but in the following weeks an extended ulcer developed in the region of the lumbar X-ray field. This persisted for about six years and was cured in 1902 but developed a scar. It healed after the application of sea water.

In 1930 another ulcer developed in the region of the old ulceration scar and this was managed with elutions of Zinnkraut (note: This is a plant which was used in 'popular' medicine) and other ineffective agents. Another ulcer developed in 1944, a total of 48 years after the initial X-ray treatment. This was cured after treatment in a waterbed.

Freund gave a last presentation mentioning this patient, in January 1937, at the Society of Physicians in Vienna but did not mention whether she was alive or dead. It was therefore a big surprise to us[49] when in 1956 an elderly woman entered our Institute and declared herself to be the child who had been irradiated 60 years earlier by Freund. She was 64 years old at this time and presented because of osteoporosis of the vertebral column in combination with back pain. Examination showed kyphosis of the thoracic spine with curvature in the lumbar region. The entire skin of the back was hairless and covered by small hyperketatotic changes. In the lumbar region a scar was found and the skin surrounding the scar showed atrophy. No damage was found to inner organs beneath the skin and we assumed therefore that the radiation was of a very soft quality.

Because of the historic interest in this patient she was asked to attend for follow-up at the age of 75 years and apart from back pain was found to be in good health. The X-ray ulcer is cured, as seen in the photographs. The osteoporosis is not thought to be radiation related but due to an ageing process. She told us that menarche occurred at the age of 13 years and that her only son is now 48 years old and in good health, as is her 13 year old grandson.

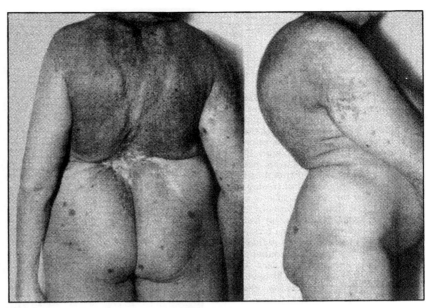

Real and fake
diamonds: 1896–1898

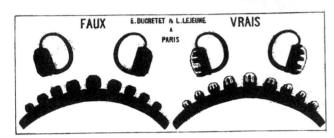

[3.13] Radiographs of jewellery were very popular before 1900 and these diamonds were radiographed in Paris sometime before 1898. This is because they were also printed in the first Italian book[51] on X-rays which was published in 1898 and had this radiograph overprinted with the Italian words FALSI and VERI to cover up the French version of FAUX and VRAIS.

[3.14] Radiograph by Thurstan Holland[3] in 1896. He described how, after an 1896 lantern slide lecture on X-rays, people were invited after payment of a fee to some charity, to view their hands on a fluorescent screen. He continued by describing the reaction of a 'very overdressed lady' who found that the stones in a large ring were densely opaque. Holland recorded that 'she made quite unprintable and vitriolic remarks' and observed that 'he had done someone a very bad turn'. This illustration is of a diamond star and a paste brooch.

Radium applicators,
tubes and needles

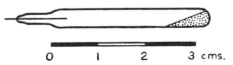

[3.15] The British Radium Standard at the National Physical Laboratory. In March 1912 an International Radium Standard of 21.99 milligram of pure radium chloride was adopted. It was prepared by Marie Curie and preserved at Sèvres near Paris. Secondary Standards were obtained by various nations and the British Standard was 21.10 milligram contained in a small glass tube, through which a platinum wire was inserted to dissipate accumulated electric charge[52].

It is seen that the glass tube is not completely filled with radium chloride and this was the design of some of the early therapeutic radium tubes. The disadvantage was that when they were implanted vertically into a tumour the radium salt fell to the bottom of the tube.

[3.16] Radium apparatus of Wickham and Degrais[31] who in 1906 established the Biological Laboratory of Paris. The flat applicators were for surface treatments and the cylindrical applicators, numbers 10 and 11, were used for treatment of the urethra, anus or uterus.

When they recorded the first radium treatment using the cross-fire technique [3.11] they described it as follows. 'Process of bombardment of the tissues by rays emitted from different points at the same time, and therefore crossing each other, the radiation being chiefly composed of projectiles travelling at immense speed.' (A similar picturesque description of gamma rays[53] was 'projectiles fired out by the radium as though it were a perpetual Maxim gun.')

Wickham and Degrais instituted this technique in 1905 for the treatment of deep-seated sciatic pains and in March 1907 used it for the treatment of a tumour [3.11].

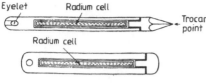

[3.17] Although there was a large range of linear radium sources, tubes and needles, in terms of geometrical and active length and linear activity (milligram radium/cm), the basic design from the early 1920s was essentially that shown in the schematic diagram. Eyelets were used to suture a needle to the patient's tissue to prevent the possibility of its involuntary removal before the end of treatment, or to attach string threads to a tube to enable removal, such as from the uterus or vagina after gynaecological brachytherapy had been completed. Some designs had removeable screw-on ends so that different radium cells could be placed in the outer container, depending on the required treatment. Union Minière du Haut Katanga, the Belgian company, supplied such sources in the 1920s, but by the 1950s the Radiochemical Centre, Amersham (later Amersham International), who were then the major supplier of radium sources, had no screw-on end designs and certainly did not supply any radium cells which, because of their very thin walls, were of a much greater hazard that the sealed sources illustrated here. These are the well known G-tubes which were 2 cm geometrical length, and available with different milligram radium contents

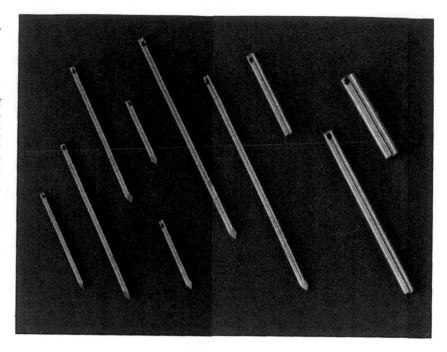

such as 10 mg, 15 mg, 20 mg and 25 mg. When radium was replaced by artificially produced radionuclides, such G-tubes then contained caesium-137 with an actively specified in terms of milligram radium equivalent [20.12, 20.56].

X-ray apparatus constructed 1896–1897

[3.18] X-ray apparatus constructed by Russell Reynolds in 1896–1897 for experimental work, but subsequently used by him in medical practice during 1905–1907. (Courtesy: The Science Museum, London.)

N-rays: Nancy, 1903–1904

[3.19] The most famous scientific fraud involving 'rays' was perpetrated by Rene Blondlot of the University of Nancy (hence the name N-rays) and corroborated by several French colleagues during 1903–04[59]. These new N-rays were produced not only by a Crookes tube, but also by ordinary sources of light and heat, by many metals, though never by wood, and, according to A. Charpentier, by muscle, nerves and the brain, and from the dried bodies of frogs which have been killed by curare. Much was attributed to N-rays, including superiority over X-rays. It was claimed that they enabled an observer to see more clearly in the dark and could be used to diagnose many diseases by changes in the bodily pattern of N-rays. By the summer of 1904 over 50 papers had been published, announcing new sources and further properties of the rays. They could be diffracted, photographed, and their spectrum determined using an aluminium spectrometer. Scientists in other countries, however, were unable to replicate successfully any of the N-ray experiments described by Blondlot and his fellow conspirators, and elected the American optical physicist, R. W. Wood, to visit Nancy and witness Blondlot's experiments in person.

Therein lay the debunking of N-rays, for when Wood was observing one demonstration by Blondlot he surreptitiously replaced the hardened-steel N-ray emitter with a (non-emitting) wooden ruler and, during the operation of the novel spectrometer, he removed the crucial aluminium prism and put it in his pocket. Unaware of this sleight of hand, Blondlot continued to 'observe' the same results as previously claimed. The well organised N-rays fraud was promptly exposed by Wood in a letter to *Nature*.

The motive for this extensive deception is difficult to imagine. It is suggested that Blondlot may have suffered from some form of delusion or 'self hypnosis' but that cannot account for the ready collaboration of so many other scientists, including Jean Becquerel, at the time of his father's Nobel recognition, who claimed that N-rays could be transmitted over a wire and published ten papers supporting the deception.

X-ray frauds: 1896–1916

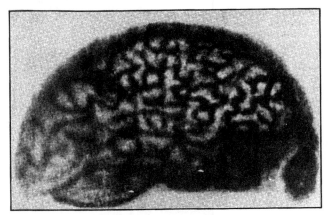

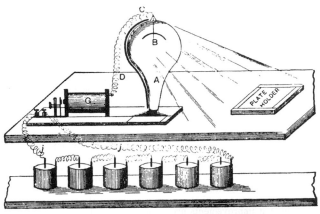

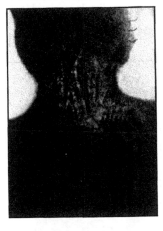

[3.20] Claims to have achieved brain photography were made in the 1890s and one claim actually tested was that of Hilbert L. Falk, President of the X-ray Shadowgraph Company of New Orleans, who published these X-ray pictures of the human brain and the muscular and vascular system of the neck. His rather unusual X-ray apparatus has only the positive terminal of the induction coil connected to an old 16-candle lamp and an aluminium reflector was placed on the lamp. Between the head of the subject and the lamp was a sheet of tin foil with a number of wires attached at various points, all joined to the negative terminal of the coil. An exposure lasted seven minutes during which Falk grasped the bulb and held it over the plate holder keeping it in motion continually. Two weeks later he was exposed in the same journal which carried the original report, the *Electrical Engineer* of New York, as 'perfect humbug' and 'an ignorant young fellow who has been an itinerant photographer and has dabbled a little in electricity'. He was outwitted when a test case anterior–posterior view of the brain turned out to be a lateral view. The 'brain' photograph is thought to have been of animal intestines.

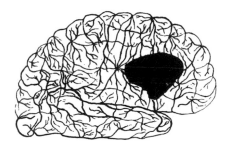

[3.21] A unique military anecdote was obtained from the pages of the *British Medical Journal* of September 30th 1916, from a report entitled 'Despatches from the Western Front'. It is perhaps the only case on record of an author being given a Court Martial for writing rubbish in a medical journal! In 1916 Sergeant James Shearer of the Royal Army Medical Corps described in the *BMJ* his results for delineation of internal organs by an electrical method. This impressive illustration (left) was a brain picture from a gunshot wound case. The *BMJ* noted that the new work appeared to succeed where X-ray photography fails, in producing pictures of structures hidden far below the surface of the body. An eye witness account of the technique stated that the exposure lasted 60 seconds, a little clicking was heard in a cupboard nearby, no darkened rooms, flashing lights or crackling spark gaps were required and the record of the organ was conveyed on to a wax sheet before being printed on photographic paper. Over three pages of *BMJ* print were devoted to Shearer's Delineator, and he also had the approval of the Director General of the Medical Services in France. His downfall came later, when he made a further claim for his apparatus that it was able to detect and identify enemy airplanes flying over the trenches at night. The Army Intelligence Department succeeded where the *BMJ* editors had failed, and exposed the work as fraudulent. Shearer's reward was a Court Martial.

[3.22] A West Virginian jeweller, A. G. Davis, was one of a number of investigators who in 1896 claimed that they could photograph the invisible without the use of X-rays. This photograph from the April 1st 1896 *Electrical Engineer* (New York) illustrated an item entitled 'Opaque objects made transparent by rays passing through chemical solutions.' Davis claimed he could seee the key through a solid block of steel 12″ × 18″ × 25″ between two wooden boxes housing two bottles containing chemicals with a secret formula.

X-rays on wheels: 1907–1919

[3.23]   Electricity generation for early X-ray apparatus was sometimes a problem, particularly when the equipment had to be mobile. This led to several innovative ideas, especially in military radiography, using the horse-gear dynamo of [8.9], pedal power as in [8.1], the airplane of [8.13], and the opportunistic mill power of [8.8]. As early as 1899, the April 15th issue of the *Sheffield Weekly Telegraph*, under the caption 'A new use of electric vehicles' printed the following item. 'A case was reported recently of a clever application of the electric storage battery of an automobile. A woman had received a complicated fracture of her arm, too complex for the physician to accurately locate. He then decided to make use of a Roentgen ray apparatus but found the patient too weak to be removed. He obtained the apparatus, but having no source of electricity convenient to operate it, called an electric cab by telephone. The current from the battery in the cab was conducted to the apparatus by special wires, which successfully operated it, and enabled the physician by the usual observations to locate the fracture in the arm and set it quickly. It is said that improvements are to be introduced in these vehicles whereby they can be made immediately serviceable to doctors in emergency cases.'

TROLLEY EQUIPMENT.

Made to the design of Dr. W. Steuart, for use in the Wards of St. Bartholomew's Hospital, for radiographing patients unfit to be moved.

Consists of a 16-inch Jointless-section (patent) Coil, on an oak Trolley, worked from the 200 volt electric light wiring, by means of a coal-gas Mercury Break, and a Series Resistance. Can be attached to any lamp plug, or socket, that will carry 5 amperes.

As shown, including the Valve Tube, and an X-Ray Tube to the value of £5.

Price £75.

LESLIE MILLER,
66, Hatton Garden, London, E.C.
Cables: "MILLEXRAY" LONDON.

The tricycle was used by a Berlin practitioner to transport X-ray equipment in 1907 but the use of motor vehicles became far more popular as they could also incorporate the generator, the cars illustrated here being used in 1913 (above) and 1919 (below). Between those dates X-ray wagons had been used by the US Army [8.14] and driven by Marie Curie [2.9] during World War I. The trolley, as advertised in the October 1912 issue of the *Journal of the Röntgen Society*, was designed as a mobile X-ray unit for use within hospitals.

Motor Car X-Ray Unit.
12″ Portable Induction Coil with Switchboard, Interrupter and simple type Tube Stand, with cases to hold accessories and with small generator driven direct off the fly-wheel of the car generating 100 v. 10 amps.   As supplied to several Radiologists for use in patients' houses.
Estimates will be gladly furnished for the above groups on request.

X-ray therapy in Philadelphia, 1907: lupus, cancer and epilepsy

[3.24] The most detailed review of early treatment applications with X-rays was given by Kassabian[25] in a chapter entitled 'Therapeutic Value in Disease' in his 1907 textbook. Giving references from both Europe and the USA, he listed the range of ailments thought, at that time, to be suitable for X-ray treatment under the following headings:

Cutaneous affections:
  Lupus erythematosus
  Lupus vulgaris
  Naevus
  Alopecia areata
  Hypertrichosis
  Favus and tinea tonsurans
  Eczema
  Acne
  Sycosis
  Pruritis ani and pruritis vulvae
  Xeroderma pigmentosum
  Psorias
  Senile leg ulcers
  Varicose veins
  Hyperidrosis
  Kraurosis vulvae
  Leprosy

Malignant growths:
  Epithelioma
  Carcinoma
    Breast
    Sternum
    Oesophagus
    Larynx
    Stomach and bowels
    Uterus
  Sarcoma

Constitutional diseases:
  Tuberculosis
  Leukaemia

Miscellaneous affections:
  Trachoma
  Keloid
  Exophthalmic goitre
  Hypertrophied prostate
  Analgesic actions of the rays
  Neuralgia
  Migraine
  Epilepsy

[3.25] Kassabian's X-ray therapy review[25] is unique for its detailed records on the treatment of epilepsy. His experience with treatment of 12 cases of epilepsy from the Insane Department of the Philadelphia Hospital, began in 1903 on patients between 6 and 60 years of age with attacks varying from very mild to severe seizures. His follow-up statistics for 10 patients are given below. Two other young patients died during and immediately after the treatment with autopsies showing congestion of the brain. His general conclusions from this data was that the results were encouraging and that he 'cannot think that any harm can be done, except the alopecia that may be produced'.

Case	Number of attacks			Comments (follow-up to December 1904)
	in 1903	in 1904	decrease	
1	68	41	27	No attacks Oct–Dec
2	845	412	433	Patient died two months later: no new attacks
3	59	14	45	Died five months after treatment started
4	85	80	5	No attacks Oct–Dec
5	144	120	24	No attacks Oct–Dec
6	209	191	18	
7	69	4	65	
8	6	0	6	
9	148	164	(16)	⎰Increased attacks during treatment
10	61	77	(16)	⎱period but no attacks Oct–Dec

Treatment was given in February–April 1904 three times per week, 5 minutes X-ray exposure, alopecia produced after two months treatment. The upper and center photographs show alopecia produced during the three-month treatment period: apart from Case 8 who was not affected. The lower photograph was taken six months later in October 1904 and shows the regrowth of hair. Case 7 was previously bald, but after X-ray treatment the growth of hair reappeared.

Early X-ray patents

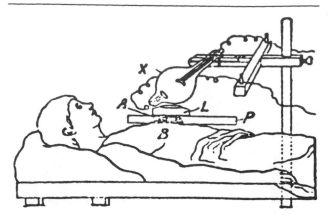

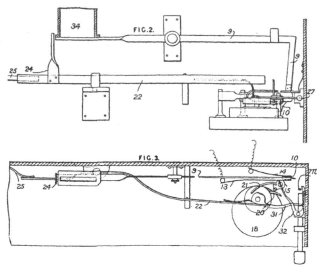

[3.26] It has already been mentioned in [1.20] that Röntgen refused to take any patents, believing that his discovery should be freely used for the benefit of society. Many others did take out patents involving X-rays and the British ones in particular have been reviewed by Ramsey[56] and Hope[57] and include some rather unusual claims. This 1900 patent, by a Charles Johnson in 1900, incorporates an unusual X-ray filter. In this inventor's drawing L is the light filter, P is a metal sheet containing a hole H, and A is a piece of sponge or other absorbent material saturated with medicines. The claim was that the medicines would thus be carried out by the filtered rays into the body of the patient.

[3.28] This ambitious claim, which appeared in the May 6th 1896 issue of the *Electrical Engineer* (New York), was not patented but possibly one of the most interesting series of patents, because of the name of Siemens, commenced in 1918 when a Spanish doctor of medicine by the name of German Perez obtained a patent for an electrical system that converted mercury into gold. Then in 1924–25 the Siemens & Halske company took out five patents which also apparently changed mercury into gold. They used a high frequency discharge of 150 kV and actually obtained a small amount of gold. Unfortunately the original mercury was previously contaminated with gold since it had been part of a gold amalgam used to separate gold from its ore!

[3.27] Other British patents which bordered on the fantastic included designs for X-ray apparatus such as a portable wind-up X-ray machine and several coin operated devices. One such example was patented in 1897 by M. Vidal. The patent commences with 'The circuit of the generating apparatus is completed for a definite time after the insertion of a coin and the operation of a pusher which winds up a clock and raises a shutter in the viewing box. The coin is dropped down a shoot 34 above one end of a beam 9, tilting it until a metallic piece 10 at the other end registers with two strap terminals 13, 14' and continues in this vein until the coin finally rolls off when the X-ray exposure time has been completed.

THE PHILOSOPHER'S STONE AT LAST.

A special dispatch from Cedar Rapids, Ia., of April 20 says: George Johnson, a young farmer residing in Jefferson County, a graduate of Columbia College, who has been experimenting with the X-rays, thinks he has made a discovery that will startle the world. By means of what he called the X-rays he is enabled to change in three hours' time a cheap piece of metal worth about 13 cents to $153 worth of gold. The metal so transformed has been tested and is pronounced pure gold.

Tavern of the Dead: 1896

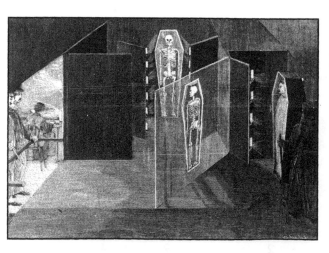

[3.29] The Cabaret du Neant or Tavern of the Dead (top) is a genuine illusion based on the Pepper's Ghost optical illusion which is reproduced from the 'World of Wonder Encyclopaedia' edited by Charles Ray in the 1930s. This illusion was imported to America from Paris and described in the Scientific American of March 1896. The principal interest centred on a stage conversion of a living man into a skeleton (bottom) which was achieved using a series of mirrors and Argand burners, a coffin and some blocks of wood. Two notices were displayed 'Rest in Peace' and 'No Smoking' and the audience was informed that the latter was necessary because a perfectly clear atmosphere was required for the illusion.

A different illusion, based on the scientific principle of diffraction, is described as the 'X-ray gogs' of [23.7].

Biological experiments: Freiburg, 1918 and Paris, 1933

[3.30] Comparative investigations by Krönig and Friedrich[38] in 1918 concerning the strength of the biological effect of gamma rays of radium or mesothorium filtered by 1.5 mm brass plus 5 mm celluloid and X-rays filtered by 1 mm copper. The dosage was measured using an aluminium chamber dosemeter.

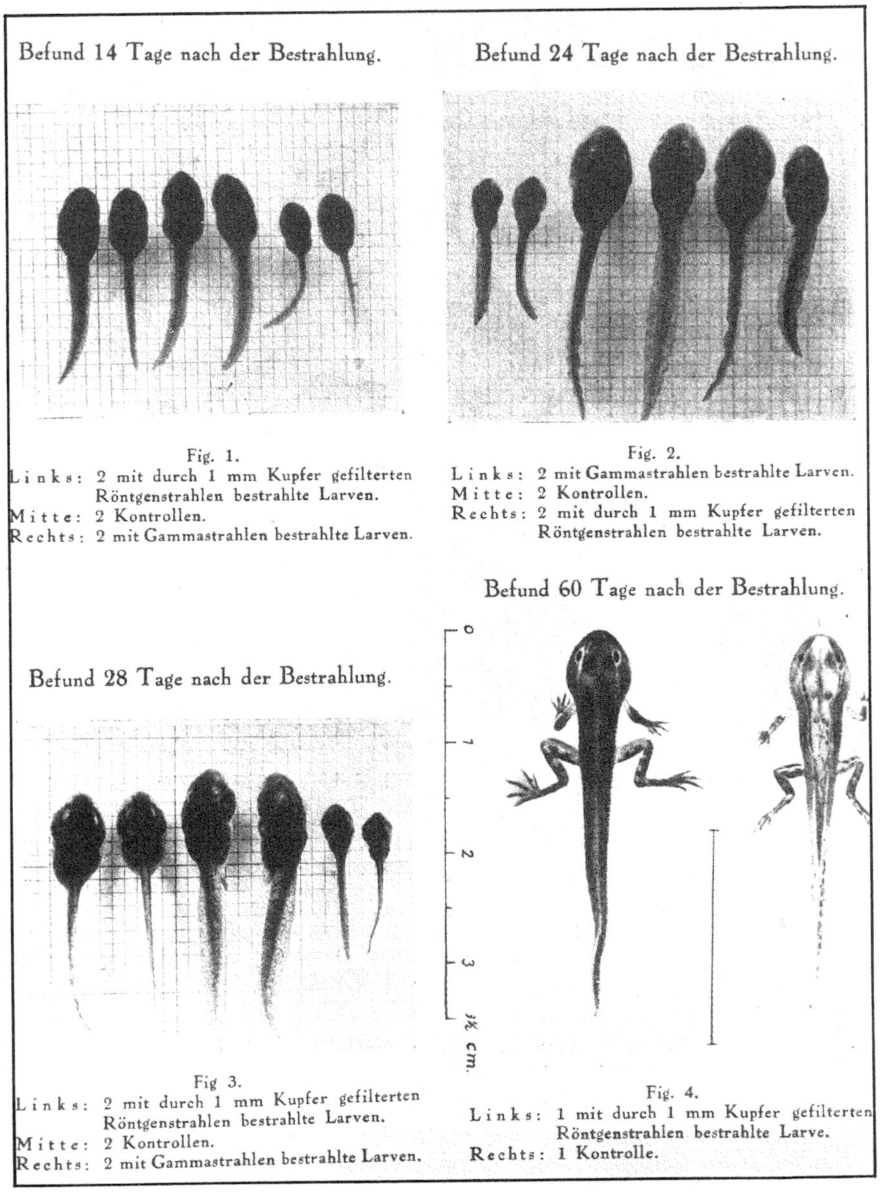

Befund 14 Tage nach der Bestrahlung.

Fig. 1.
L i n k s : 2 mit durch 1 mm Kupfer gefilterten Röntgenstrahlen bestrahlte Larven.
M i t t e : 2 Kontrollen.
R e c h t s : 2 mit Gammastrahlen bestrahlte Larven.

Befund 24 Tage nach der Bestrahlung.

Fig. 2.
L i n k s : 2 mit Gammastrahlen bestrahlte Larven.
M i t t e : 2 Kontrollen.
R e c h t s : 2 mit durch 1 mm Kupfer gefilterten Röntgenstrahlen bestrahlte Larven.

Befund 28 Tage nach der Bestrahlung.

Fig 3.
L i n k s : 2 mit durch 1 mm Kupfer gefilterten Röntgenstrahlen bestrahlte Larven.
M i t t e : 2 Kontrollen.
R e c h t s : 2 mit Gammastrahlen bestrahlte Larven.

Befund 60 Tage nach der Bestrahlung.

Fig. 4.
L i n k s : 1 mit durch 1 mm Kupfer gefilterten Röntgenstrahlen bestrahlte Larve.
R e c h t s : 1 Kontrolle.

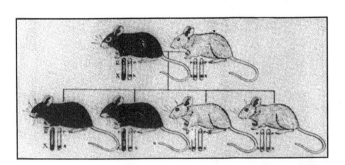

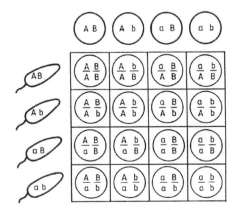

[3.31] Diagram illustrating the genetic strains of mice used for radiobiological experimentation. It is dated April 24th 1933 and was found in the archives of the Institut du Radium in Paris where Claudius Regaud was the first director of the Pavillon Pasteur in 1914. Marie Curie was then director of the Pavillon Curie.

Endoscopic positioning: Paris, 1898

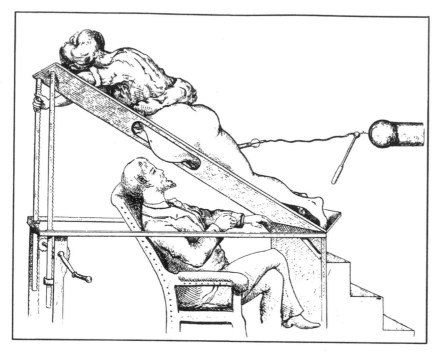

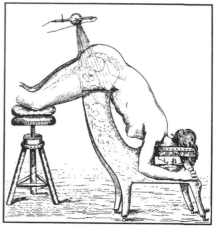

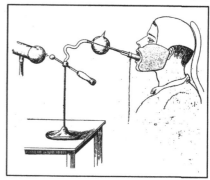

[3.32] These illustrations are from an 1898 M.D. thesis from the University of Paris[55] which is probably the earliest university thesis dealing with patient positioning in X-ray diagnosis as well as one of the earliest on the clinical application of X-rays. They show positions for radiography of the pelvic region (upper right), and for endodiascopy of the vagina and rectum (examination of the anterior region) and of the buccal cavity. The two 'endodiascopy' illustrations were also reproduced under the title *Endodiascopie* in an 1899 issue of the *Archives d'Electricité Medicale Experimentales et Cliniques*.

X-ray injuries and X-ray protection

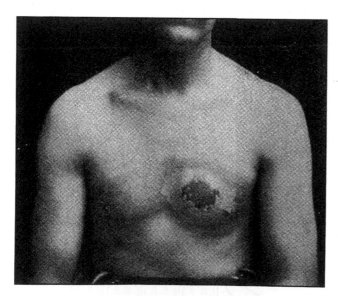

[3.33] Radiation injury[46], 1904, described as 'Röntgen light burn of the second degree'. Sir James Mackenzie Davidson claimed some improvement in X-ray burns by the application of radium and this practice was also described as late as 1926 by Riddell[54], who nevertheless recommended that it be discontinued. For further illustrations of radiation injuries, as they developed over a period of time, see [23.1].

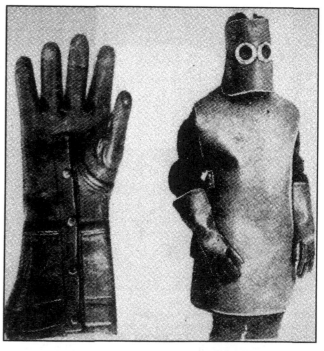

[3.34] Radiation protection in 1910.

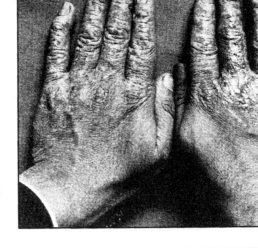

[3.35] Most of the early radiation injuries were to the hands of the pioneer physicians and technicians. The early stage of chronic X-ray dermatitis, described by Beck[46] as 'wrinkled and shrivelled Röntgen hands of physicians' is seen (top right) in the hands of the radiologist Mihran Kassabian of Philadelphia, photographed in 1903 (before amputation of some of his fingers) and published in his 1907 text-book[25]. Kassabian commenced work with X-rays in 1899 and died in 1910, one of the early American X-ray martyrs. Another prominent physician who used his own injuries to warn others was John Hall-Edwards of Birmingham who, in 1908, published radiographs of the bone damage of his amputated left hand and right fingers[60].

A later stage of progression is shown (centre left) by the hands of a London Hospital technician photographed in 1906, ten years after commencing X-ray work. The same technician is also shown (centre right) using his unprotected apparatus in 1903. His left hand is nearest the X-ray tube which, if that was his usual practice, would account for the carcinoma of the dorsum seen on his left hand. The technician died of carcinomatosis in 1934. A much later stage of X-ray damage is seen in the advanced carcinoma of the dorsum (bottom left) and loss of some fingers of a man aged 57 who had stated work in the manufacture of X-ray apparatus some 27 years previously.

A sequential progression from dermatitis to amputation above the wrist is illustrated in [23.1].

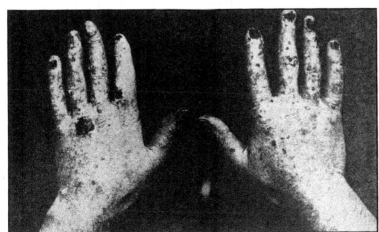

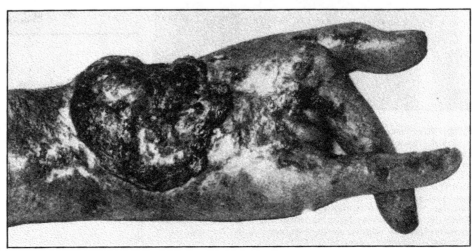

An 1897 catalogue: Erlangen

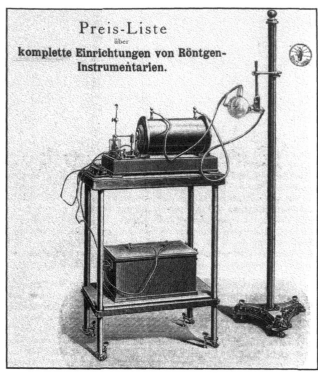

[3.36] The world's first catalogue offering an extensive, illustrated range of X-ray equipment was issued in 1897 by Reiniger, Gebbert & Schall, one of the forerunners (another being Siemens & Halske) of the Siemens company of today. The title page (left) boasts 450 illustrations and the complete set of Röntgen apparatus, reproduced here, cost 827 marks in 1897.

[3.37] The 1897 catalogue included these fine radiographs of the head and the chest, which are the earliest in the Siemen's archive. Reiniger, Gebbert & Schall supplied vacuum tubes to Röntgen so it is reasonable to suppose that these radiographs were obtained using tubes similar to those used by Röntgen [5.9]. They were described as X-ray photographs of living persons taken using the firm's X-ray tubes and spark generators, with 15 cm spark length and exposure times of 11 and 12 minutes, respectively. The chest is reproduced from the catalogue but the head is photographed from the original radiograph in the Siemens archive; it was recently damaged by flood water but the quality of the image can still be assessed.

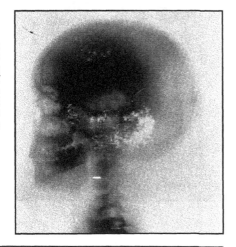

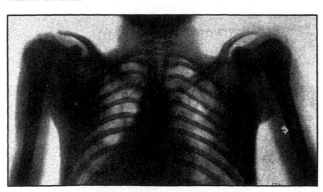

[3.38] The Siemens archive also includes Röntgen's correspondence with Reiniger, Gebbert & Schall. In the letter reproduced here, dated November 3rd 1896, Röntgen writes: 'I forward to you my sincere appreciation for sending me the very beautiful photograph of a head. In response I would like to order on account of the local Institute of Physics two vacuum tubes of the constructive type (with user instructions) and request you to send them to me as soon as possible.' The head to which Röntgen refers is that illustrated in [3.37]. In a subsequent letter, dated 27 November 1896, Röntgen acknowledges the quality of the company's tubes but asks for a substantial reduction in price.

Chapter 4

Archives of Clinical Skiagraphy

The First Radiological Journal: 1896–1899

The *Archives of Clinical Skiagraphy* is the world's oldest radiological journal with its first part appearing in May 1896. Parts numbered 2 and 3 appeared in June and December 1896, but for part number 4 (April 1897) of the first volume the name of the journal was altered to *Archives of Skiagraphy*, removing the word 'Clinical' [4.11].

However, this title was not kept very long either, and the next part (Vol. II, No. 1, July 1897) became the *Archives of the Roentgen Ray*. It was described as the 'only journal in which the transactions of the Röntgen Society of London are officially reported'.

Because of its unique place in radiological literature, this chapter is devoted to a selection of the *Archives*. Most of the *Archives'* illustrations were of clinical skiagraphs (radiographs) but in April 1897, 24 radiographs were included on 'Skiagraphy in zoology' by Norris Wolfenden, who also wrote a special Supplement[1] to the 1897/1898 *Archives of the Roentgen Ray* on 'Radiography in Marine Biology'. Some of these fascinating zoological radiographs have recently been republished[2] in the *Journal of Laryngology and Otology* which was founded in 1887 by Wolfenden and Morell Mackenzie.

A section was also included in the *Archives* called 'Answers to Correspondents', which published only the answers to what must be the first ever X-ray Agony Column! The correspondents are referred to only by their initials or by a nom-de-plume, of which there were three: Rand, Kathode and Ignoramus.

The answer to Ignoramus is the most interesting. 'You may know your focus tube is working correctly when one-half is fluorescing a yellowish green and the other half remains comparatively dark. After prolonged use, and especially if the platinum tube has been overheated, the vacuum is apt to increase. In this case the tube may be restored by heating over a spirit lamp until as hot as the hand can bear. Be careful to keep the tube revolving when heating, so as to equalise the temperature.'

The three issues of the *Archives of Clinical Skiagraphy* (1896) took up only 32 pages with the single issue for the *Archives of Skiagraphy* (1897) being an additional five pages of text plus 24 pages for the skiagrams. For Volume II (1897/98) of the *Archives of the Roentgen Ray* the size had increased to 97 pages (with in addition the Supplement on 'Radiography in Marine Biology') and for Volume III (1898/99) there was a further increase to 117 pages.

As might be expected, as time progressed, the proportion of contents devoted solely to radiographic imaging, the 'Plates', became less, as more technical information became available on, for instance, design of X-ray tubes, radiographic and fluoroscopic technique, and uses of X-rays outside the field of medicine. Much of the latter information was printed in the form of 'Notes' sections such as that in 1897 entitled 'Roentgen Rays and French Customs' (reprinted from *The Globe* newspaper of July 16th 1897), which is reproduced in Chapter 17 where early industrial radiography uses are discussed.

The 'Notes' later became 'Extracts from Medical & Scientific Journals', including newspapers, and various interesting applications for the X-rays were suggested. These included the measurement of potato density (*The Globe*, October 22nd 1897) and a method of printing using X-rays in which five million copies of a newspaper could be produced by a single process (*Daily News*, March 9th and *Manchester Courier*, March 11th 1899). The Manchester reporter was rather cynical about inventors, though, and concluded that 'In the meantime we may breathe freely, for the French submarine boat and the Manchester sewage problems are yet unsolved'.

A new apparatus called 'The Seehear' (*Engineer*, March 10th 1899) was reported as the latest development in Roentgen radiography. It was a device by which a person may observe the action of his own heart, 'said a New York electrical contemporary', and was a combination of a fluorescent screen and a stethoscope.

Under the heading 'Blind to Roentgen Rays' (*British Colonial Druggist*, March 31st 1899) it was stated that 'It has been computed that one person in every 800 is blind to the X-rays; that is to say, when looking through the fluoroscope, they are utterly unable to observe the bones of the body, coins or any other object which is clearly distinguishable by the ordinary observer.'

There was also a lively correspondence on the proposal for the 'new' name radiography, which correspondent called barbarous when compared with skiagraphy. A German suggestion that the term should be pyknography (from the Greek root, dense or obscure) was given in November 1897. Part of the reason for this was that 'if X-rays proved to have therapeutic value' it would be difficult to accept skiatherapeutics or the shadow-cure. By February 1898 the terminology problem was still not settled and diagraphy/diascopy, actinography, skiametry and radiometry were also proposed.

The skeleton

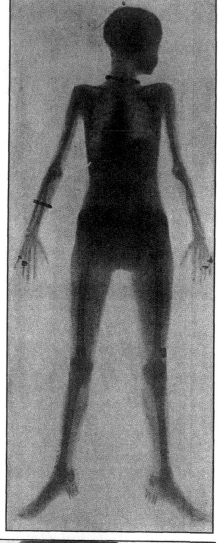

[4.1] The July 1897 issue contains what is probably the most frequently published historical radiograph. It is the first whole body radiograph of a living person taken at a single exposure by William Morton of New York, who was also author of the 1896 textbook[3] *The X-ray*. Hatpin, necklace, bracelet, rings and high button boots with nailed-on heels can clearly be seen. What is not quite so clearly seen is a whalebone corset—pointed out to me when I asked for comments during a lecture to nurses.

The apparatus used by Morton included a 12 inch induction coil whose primary was supplied from 117 volt Edison current of the New York street mains and an ordinary Crookes' tube with a commencing vacuum corresponding to a spark of 2 inches which gradually rose to 8 inches. The distance of the tube to the X-ray plate was 54 inches and the time taken, including stoppages, was 30 minutes.

[4.2] The novelty of a whole body radiograph evidently did not appeal to the fiancée mentioned in the abstract (right) from the 1907 *Archives*.

Marriage and X Rays.—The *Archives d'Électricité Médicale*, August 25, 1906, reports that a case has occurred in New York of breach of promise of marriage, with a claim for 25,000 dollars. The young lady absolutely refused to be examined by an X-ray specialist at the request, or rather command, of her fiancé. There was nothing the matter with her, but the young man wished to be on the safe side before marriage. On her refusal, he broke off the engagement. This breach of promise case is the result, and it is suggested that the judge is in favour of the plaintiff on the grounds that by law no one can reasonably claim an X-ray examination before marriage and on its being refused break the engagement.—C. THURSTAN HOLLAND.

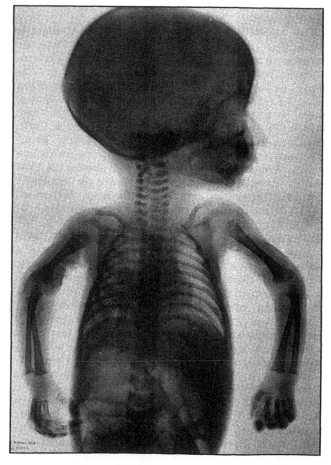

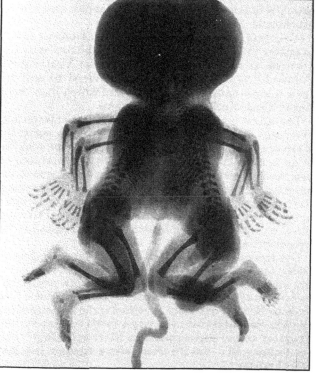

[4.3] The skiagrams in the early *Archives* were given Plate Numbers of which I and II are a double-plate spread entitled 'Child-skiagram of skeleton of full grown child aged three months', 1896.

[4.4] Skiagram of a 'double monster' foetal abnormality, 1896.

Tuberculosis . . . or a coin in the throat?

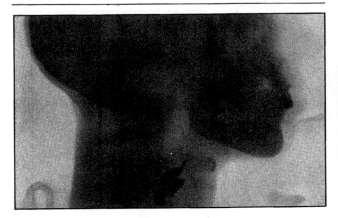

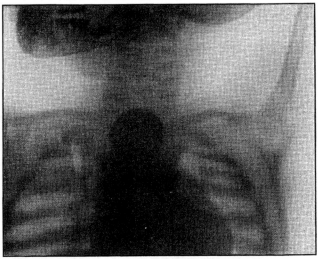

[4.5] The majority of the early clinical radiographs were of bony extremities, normal and abnormal. However, these began to lose their interest to authors and editors when the first improvements in X-ray apparatus design made possible good quality (for that era) chest radiographs and then later, abdominal and skull radiographs.

The coin in the throat right from the *Archives* is of a child of three years old. A similar radiograph was also published by Thurston Holland[4] in 1896 of a child who had swallowed a coin before the discovery of X-rays. It had been incorrectly assumed for some two years that the problem was tuberculosis and only after radiography was the cause of the illness diagnosed as 'a coin'. The radiograph on the (left) reveals a set of false teeth lodged in the pharynx, from the *Archives* of 1903.

However, the most interesting 'swallowing case' was of the Ostrich Man[5] originally reported in the *American X-Ray Journal* of July 1898. He was a vaudeville act travelling the USA and performing on stage by swallowing his watch and chain, wadded cartridges, tacks, money and screws. He stated that as a rule he usually passed metallic bodies within 24 hours. The *Journal* reported that 'A failure in that respect appears to have led to an accumulation and fatal results'.

A similar case was published in the *Johns Hopkins Hospital Reports* of 1900 by the Professor of Surgery, W. S. Halstead. It was a radiogram of a 21 year old juggler, from whose stomach removed 208 foreign bodies and 74 grams of glass. Recovery was complete, in spite of the fact that the items included 20 pieces of a small dog-chain, 54 wire nails, 35 ordinary nails and seven knife blades.

Modern equivalents also exists[6] such as the case of two prisoners in Rhodesia (Zimbabwe) in 1964–66. A 26 year old male was admitted to hospital on seven occasions up to December 1965 having swallowed foreign bodies such as nails, broken razor blades and screws. This led the Governor of the prison to make an order 'to withhold all types of edible foreign materials from the prisoner'. The patient then acquired a taste for bed springs (now displayed in the Westminster Medical School Museum) which he ate in January 1966. After serving sentence in Rhodesia, he was to be repatriated to South Africa, of which he is a national. There he was wanted for gun-running charges which carried a penalty of 25 years hard labour. He hoped by repeated swallowing of foreign bodies that he would be certified insane and therefore unfit to face charges in South Africa.

The second case was his cellmate who had swallowed nine bed springs to keep his friend company.

Cine-radiography

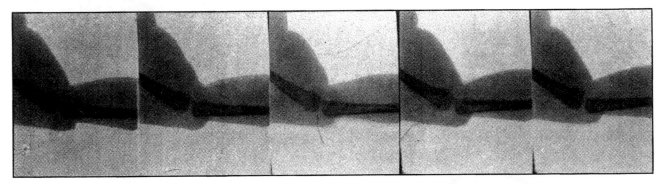

[4.6] This is the world's first example of cine-radiography. The caption stated that John Macintyre of Glasgow had been experimenting for some time on the best methods of obtaining rapid exposures with a view to recording movements of organs within the body. A method using an ordinary camera to photograph images on a fluorescent screen was too slow. The successful method involved the sensitive film passing underneath the aperture in a case of thick lead covering the cinematograph. This opening corresponded to the size of the picture, and was covered with a piece of black paper, upon which the limb of an animal, such as a frog, could be photographed. It was necessary for the movements to be slow and therefore a slow anaesthesia was required. Macintyre was reported at a meeting of the Glasgow Philosophical Society to have moved a film 40 feet in length through the cinematograph; the movements of the frog's leg could clearly be seen when demonstrated on a magic lantern screen by means of the cinematograph.

Another interesting anecdote concerning X-ray cinematography dates back to the late 1920s and early 1930s. James Sibley Watson was an independent avante garde film director whose two most famous films were the *Fall of the House of Usher* (1929) and *Lot in Sodom* (1933). He also experimented with X-ray cinematography using human beings with everyday actions such as shaving, putting on lipstick, swallowing liquids and using a comb. Two examples of his intertitles (text between the pictures) are 'Another way of viewing pictures' and 'At this moment, for example, the microscope is penetrating deeper than ever before into the secrets of living tissue'. (Watson's existing cine films are now in the archives of the International Museum of Photography, George Eastman House, Rochester, New York.)

Treatment of lupus

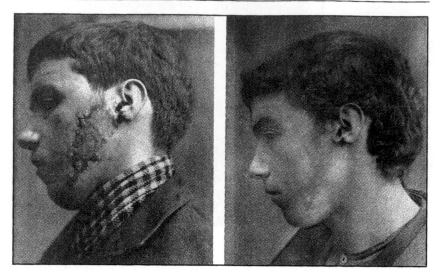

[4.7] The *Archives* of May 1899 contains photographs of a 16-year-old patient before and after X-ray treatment for the non-malignant skin condition, lupus. These illustrations were used by Thurstan Holland in his Röntgen Society lecture of December 19th 1898 on 'Treatment of lupus by X-rays'. A similar report was also published in February 1988 by Albers-Schönberg following a meeting in Dusseldorf of the Deutscher Naturforscher und Arztec. Treatments lasted for some 30 minutes daily. The skin first shows a slight yellow tint, which is soon followed by general redness of the affected part. This redness deepens in colour, and with some patients a slight itching and pricking sensation are felt, together with a feeling of warmth. A sensation of tightness of the skin also follows in some cases, with slight oedema. Excoriation is generally produced, the appearance resembling that of a burn, and extends up to the edges of the area protected by the mask. The skin gradually heals from the edges towards the centre.

Peruvian osteosarcoma aged 600 years

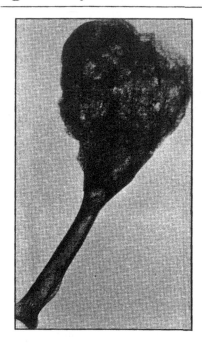

[4.8] Skiagraph of an osteosarcoma. The specimen belonged to the Uhle collection of Peruvian antiquities of the Archeological Museum of the University of Pennsylvania and is at least 600 years old.

Marine biology

[4.9] Skiagrams of a starfish (palimpes placenta), the edible crab (cancer pagurus) and the common lobster (homarus vulgaris). These marine specimens were obtained from dredging in Scapa Flow, Orkney, 'during the autumn months of 1896'. Dr John Macintyre, the pioneer radiologist in Glasgow, provided the radiographic facilities for these marine specimens of Wolfenden. His apparatus was described in the *Archives*. 'The ordinary spring on the induction coil was employed for some, but most were photographed with the mercury interrupter, with exposures varying from one to five minutes, and it was found that with slow interruptions of an average of five per second, the most beautiful results were obtained, with very much shorter exposures than with the ordinary spring. The tubes used were Watson's Penetrator and Palladium, Newton's, and Baird & Tatlock's, all equally good'.

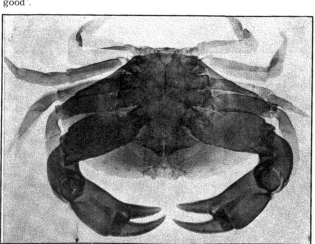

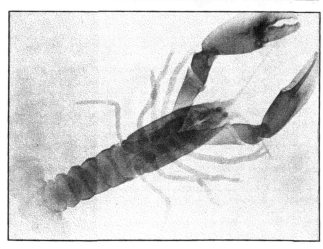

Radiographic positioning

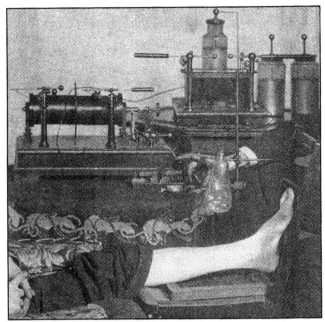

[4.10] The first photograph of radiographic positioning to appear in the 1896 *Archives of Clinical Skiagraphy*. The X-ray tube is of the pear-shaped design used by Röntgen and there is an induction coil and what looks like an accumulator and three Leyden jars. No circuit is given for this experimental arrangement and the caption consists only of 'Photograph of apparatus and method employed for obtaining a skiagram. In this case the leg.'

Other early journals

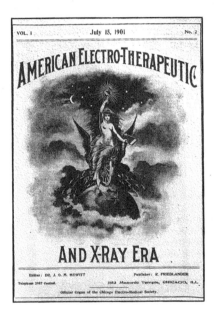

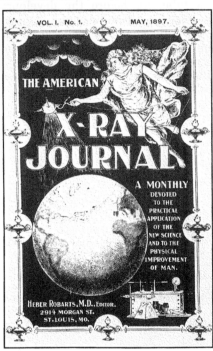

[4.11] First published in May 1896, the *Archives* was only just the first radiological journal. The first issue of the *American X-ray Journal* (above) appeared in May 1897 and the *Fortschritte auf dem Gebiete der Röntgenstrahlen* was founded in Germany in the same year. The *Archives of Skiagraphy* cover shown here is that of the fourth issue, the only issue published under that title, which contrary to the 1898 data on the cover, actually appeared in April 1897. The name changed again, with the July 1897 issue, to *Archives of the Roentgen Ray*[8] with further changes in 1903 to *Archives of the Roentgen Ray and Allied Phenomena* [22.1], and in 1915 to *Archives of Radiology and Electrotherapy (Archives of the Roentgen Ray)*, the bracketed subtitle being discarded in 1918. Finally, in 1928, the two journals which had started as the *Archives* in 1896 and as the *Journal of the Röntgen Society*, in 1904, merged to become the *British Journal of Radiology, New Series* which, with the omission of the wording *New Series*, has continued to the present day.

The *American Electro-Therapeutic and X-ray Era* (above) was another early journal and was associated with the Friedlander company of Chicago, [6.8], as publisher. It eventually merged with the *American X-ray Journal* (above) which was launched by Heber Roberts of St. Louis in May 1897 but passed to new editors in Chicago in 1902. In December 1904 this journal took over the *Archives of Electrology and Radiology*, appearing in January 1905 as the *American Journal of Progressive Therapeutics*. In parallel with these journals, the *Transactions of the American Roentgen Ray Society* was first published in 1902, and continued until 1908 when superseded by the *American Quarterly of Roentgenology* which had been first published in October 1906 and in 1913 became the *American Journal of Roentgenology*. This continues to the present day although in 1923 it was extended to include *Radium Therapy* in its title, and in 1952 to include *Nuclear Medicine*. The other long-standing North American journal is *Radiology* which was first published in 1923 and is the journal of the Radiological Society of North America (RSNA) formed in 1920 having originated as the Western Roentgen Society in 1915[7,9]. The *International Journal of Radiation Oncology Biology and Physics* which first appeared in 1975, and is the journal of the American Society for Therapeutic Radiology and Oncology (ASTRO), is a much more recent addition to the radiological literature, as is the journal *Radiotherapy and Oncology*, of the European Society for Therapeutic Radiology and Oncology (ESTRO).

Chapter 5

Gas Tubes: 1895–1913

When Röntgen discovered X-rays he was investigating the conduction of electricity in gases at low pressures, i.e. gas discharges. He was not alone in making such studies; Sir William Crookes, for example, after whom the Crookes tube was named, was a leading scientist in this field. However, gas discharge is not necessary for the production of X-rays: the requirement is for high speed electrons to be stopped or slowed suddenly. The illustrations in this chapter form a small selection of the wide variety of X-ray tubes available in the gas tube era before the developments by William Coolidge provided the enormous improvement in X-ray production which was made possible by the hot cathode tube (see Chapter 7), which in turn made the gas tubes obsolete. The captions to the illustrations provide details of some of the design developments of gas tubes in Europe and in America.

Sir William Crookes

[5.1] Sir William Crookes (1832–1919) demonstrating the deflection of cathode rays. Later, on January 20th 1896, using a Crookes electrical discharge tube which had been constructed for him in 1879, he demonstrated the production of X-rays. It is well documented that Crookes had previously made the observation that photographic plates stored near his tubes became fogged, and that on one occasion he returned some of the plates to the manufacturer as unsatisfactory. (Courtesy: The Science Museum, London.)

[5.2] Cossor's workshop for producing Crookes tubes in 1896. The photograph was published in the special February 1896 issue of *The Photogram* called *The New Light and the New Photography*, edited by Snowden Ward. The advertisement appeared in 1898.

A. C. COSSOR,

67, Farringdon Rd., London, E.C.

ACTUAL MAKER OF:—

FOCUS TUBES with improved anode, also with movable cathode

TUBES showing different degrees of vacua.

TUBES containing specimens of fluorescent and phosporescent minerals, crystals, salts, rubies, etc.

Fluorescent Screens, Mercury Pumps, Induction Coils, Transformers, Accumulators.

First X-ray picture: Philadelphia, 1890

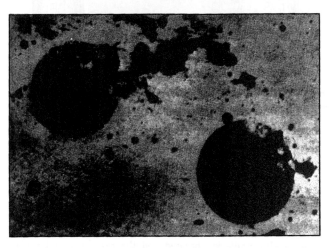

[5.3] Priority for the first 'X-ray picture', albeit unrecognised as such at the time it was taken in 1890, goes to Arthur W. Goodspeed of the University of Pennsylvania in Philadelphia[1,2] when he was photographing electric sparks and brush discharges and using Crookes tubes. The two round discs in the photographic image could not be explained at the time, but six years later in February 1896, Goodspeed repeated the exposure under similar experimental conditions and found the same results.

Pear-shaped X-ray tubes: 1896

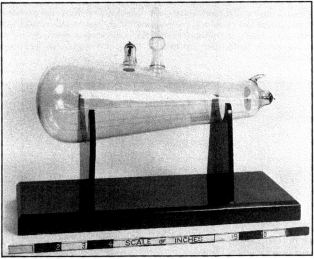

[5.4] Pear-shaped X-ray tube. At the far right is seen a flat aluminium cathode, and on the top of the tube, towards the left of the photograph, is the anode, and next to the anode the taller 'empty' glass tube which was termed an evacuating tube. When the high tension is applied between anode and cathode, X-rays are produced because of the following process.

1. The electric discharge causes ionisation of the gas atoms.
2. Positive ions are driven towards the negative electrode (the cathode) by the electrical potential across the tube.
3. This positive ion bombardment causes the emission of electrons (also known in the 1890s as cathode rays).
4. The stream of electrons when they hit the target (which for a pear-shaped tube is the wide end of the glass bulb) produce X-rays. The target was often termed the anti-cathode.

(Courtesy: The Science Museum, London.)

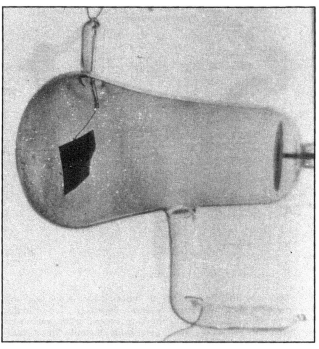

[5.5] The two most important improvements in X-ray gas tube design were by Campbell Swinton, who introduced a sheet of platinum into the tube to serve as a metal target, and by Herbert Jackson, in May 1896, who used a concave cathode to focus the electrons onto a small area of the metal target. The X-ray tube shown here belonged to Campbell Swinton and is from the historical collection of the Röntgen Society.

X-ray tube technology: 1896–1902

[5.6] Two early focus tubes designed by Herbert Jackson of King's College, London. This focus tube improvement ensured that the effective X-ray source was as small as possible, so that the X-ray picture was more sharply defined than could be obtained with the pear-shaped tubes where the X-rays were emitted from a large area of the glass anticathode. (Courtesy: The Science Museum, London.)

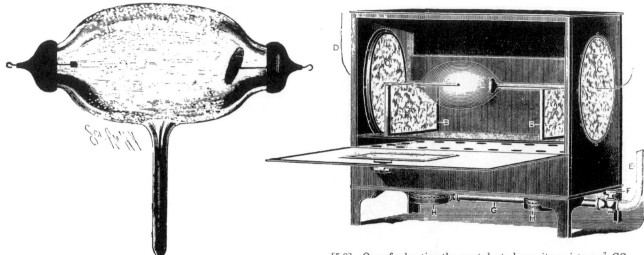

[5.7] This was described in 1896[3] as 'Edison's new X-ray lamp' and was originally published in the *Scientific American*. 'It is a highly exhausted oblong glass bulb having wires sealed in the ends, each wire being provided with a small plate inside the bulb. One of these plates is inclined to cause a distribution of the rays over the side of the lamp. The inner surface of the lamp is covered with a granular mineral substance [originally calcium tungstate but later a more efficient fluorescent material, which was not named] which is fused on the glass and is highly fluorescent. When the lamp is connected to an induction coil the material becomes luminous'. The efficiency of the lamp was quoted as 'light production at the rate of 0.3 watt per candle power'.

The results of manufacturing such lamps were disastrous in terms of radiation hazards, and Thomas Edison soon noted that his assistant Clarence Dally suffered from loss of hair and skin ulcers, and abandoned his fluorescent lamp. Dally died in 1904 at the age of 39 years from the effects of radiation.

[5.8] Oven for heating the gas tube to lower its resistance[7]. CC are asbestos ends through which pass the terminals DD. The X-ray tube is heated by burners HH and is supported by asbestos slabs AA. The front door to the oven is transparent. When the tube is heated, gas from its glass walls is liberated and the resistance is lowered. However, as stated by Williams (1902), 'all methods of lowering the resistance are temporary.'

X-ray tubes used by Röntgen: 1895–1896

[5.9] Tubes of different shapes and targets of various metals were tried throughout 1896 but by the end of that year it was established that tube shape was unimportant. The X-ray tubes shown are some of those used by Röntgen in 1895–96. (Courtesy: Deutsches Museum, Munich).

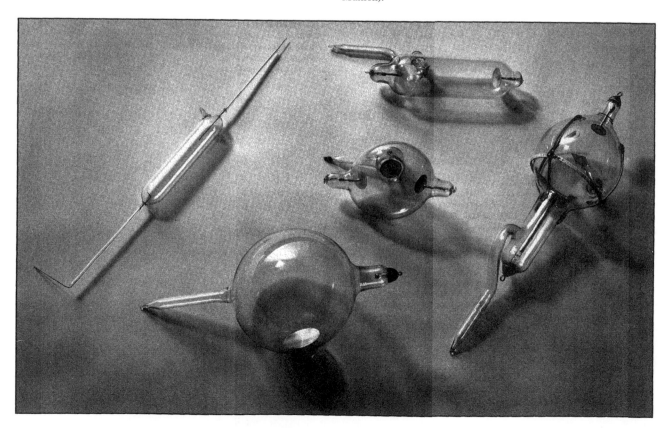

Self-regulating X-ray tubes: 1896–1903

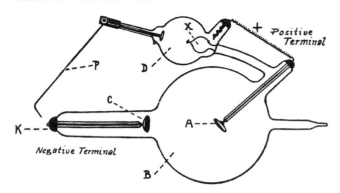

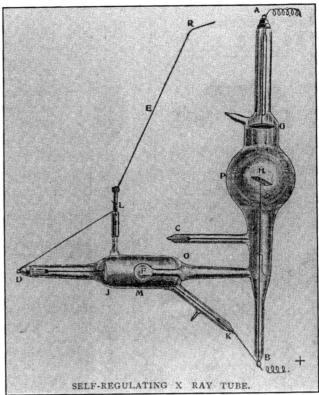

SELF-REGULATING X RAY TUBE.

[5.10] The penetrating power modifications could be affected by the action of gases derived from absorbent substances. Tube design was thus based on the principle that such substances as caustic potash, palladium and potassium permanganate, when placed in an auxiliary glass bulb of low vacuum, liberate gases when heated and reabsorb the gases upon cooling. An example from England is seen in Figure [5.11] and an American example is given above.

This self-regulating design was by L. T. Sayen in 1896 and the tubes were sold by Queen & Company, who probably obtained the best ever collection of X-ray tube testimonials. Those from an 1898 advertisement are given below.

'Especially ingenious': Prof. Dr W. C. Roentgen

'Most satisfactory': Lord Kelvin

'The best I have yet seen': Dr A. W. Goodspeed: see also Figure [5.3]

'Operator can make it do exactly as he desires': Dr C. L. Leonard

'Perfectly satisfied in every respect': Elliott Woods, Washington.

'It can take care of itself': Dr H. P. Bowditch, Harvard.

The description of the operation of this self-regulating tube is as follows. 'The large bulb B contains the main cathode and is connected to the regulating bulb D. The curved cathode C is of hammered aluminium. A is the anode made of platinum. B is exhausted to high vacuum so that initially no electric discharge will pass through it. Bulb B is exhausted to a low Crookes vacuum. Within D is a small pear shaped bulb X, in communication with B and containing a chemical capable of giving off vapour when heated and reabsorbing it when cooled. A small cathode in D is arranged so that the discharge will heat the bulb X and attached to this cathode is an adjustable spark-point P. The end of P can be swung to any desired distance from the terminal of cathode C.

'When put into operation, the high potential secondary current will not initially pass through B (because if its high vacuum) but chooses a path from K to P through bulb D and heats the chemical in the small bulb X. Vapour is given off, reducing the vacuum in B until the vacuum is finally lowered sufficiently so that discharge passes through the main bulb B and X-rays are produced from the anode A.'

When the tube ran properly, the main bulb B was filled with a brilliant green light, with a sharp cut zone through the plane of the platinum anode A. The upper section was more brilliant than the lower section[4].

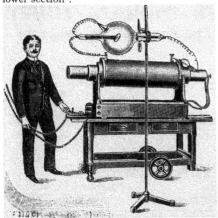

[5.11] The X-ray tubes shown in the previous photographs are of a design known as 'stationary vacuum tubes' in that they were gas tubes in which the vacuum could not be altered during use. They were rather unreliable as the degree of the vacuum varied with the use or misuse of the tube and there was always a danger of permanent damage.

If the tube was too 'hard' (i.e. the vacuum too high) there was a danger of puncturing the tube and making it useless for X-ray production. If alternatively the tube was too 'soft' (i.e. the vacuum too low) then the X-rays would lack the penetrative power to obtain a good contrast radiograph.

The 'self-regulating and regenerative tube' was a great improvement over the stationary vacuum design and the vacuum could, depending on the design, be changed automatically or by the X-ray operator: who was sometimes called an X-ray driver!

A self-regulating tube of November 1897 from Watson & Sons, London is shown. Its use was described in the Archives of the Roentgen Ray in the following terms. 'Bulb A contains a chemical which gives off vapour when heated, and re-absorbs it when it cools. The arm B is adjustable, so that the gap G can be varied. The anode of the auxiliary tube is connected to the anode of the main tube. When put in action the resistance of the main tube, being high, causes a current to pass through the auxiliary tube and across the gap G. The discharge from the cathode of the auxiliary tube heats the bulb and drives out the vapour, thus lowering the vacuum. This action continues for a few seconds until a sufficient amount of vapour has been driven into the main tube to permit the current to go through it. After this only an occasional spark will jump across the gap to maintain the tube at constant vacuum'.

[5.12] A regulating X-ray tube used by one of the early German radiological pioneers, Albers-Schönberg of Hamburg[5] before 1903. It looks to be an early mobile unit from the wheels on the trolley. Anode design was important and in 1904 Beck[6] of New York, noted that 'if currents of a very high intensity are used the platinum disk of almost all tubes becomes white hot after a short time, often after a few seconds, and if kept glowing a little longer, the platinum melts. To obviate this most embarrassing occurrence, tubes have recently been constructed in such a manner that the metallic parts were made very thick and resistant. Such tubes permit a current of maximum intensity for about one minute. The very marked outlines of the picture become less distinct, the tube filling with blue light at the same time, which indicates that it is overheated'. Regulating devices allowed longer exposures, such as that of Albers-Schönberg.

Beck further gave his opinion of how to differentiate between good and bad new tubes when first used. 'The best tubes are those which show a red-hot focus at the platinum disk while a low current is employed. New tubes that show fluorescence only when a high current is used should be rejected.'

Advertising: 1896–1913

X RAY TUBES
15s. each.
A selection sent anywhere on application.
AS SUPPLIED TO H.M. GOVERNMENT & SOME OF THE LEADING HOSPITALS.
COILS, SCREEN, HOLDERS & ALL APPARATUS.

A. & G. SMITH, Opticians, Aberdeen.

BEACON LAMP CO.,
MANUFACTURERS OF
HIGHEST GRADE
Incandescent
Lamps,
Ranging from ⅛ to 300 c. p.

ALSO, A FULL LINE OF

Unexcelled X-Ray Tubes. EVERY TUBE GUARANTEED.
SPECIAL TUBES AND LAMPS TO ORDER.
For Prices and Discounts write
BEACON LAMP CO., Harcourt St., Boston, Mass.
After Oct. 1st, 1896, address New Brunswick, N.J.

[5.13] Very early suppliers of X-ray tubes included electric lamp companies in 1896, and a Scottish optician advertising in 1898.

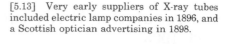
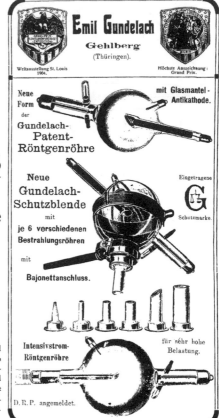

Emil Gundelach Gehlberg (Thüringen).

Weltausstellung St. Louis 1904.   Höchste Auszeichnung: Grand Prix.

Neue Form der **Gundelach-Patent-Röntgenröhre** mit Glasmantel-Antikathode.

Neue Gundelach-Schutzblende mit je 6 verschiedenen Bestrahlungsröhren mit **Bajonettanschluss.**

Eingetragene Schutzmarke.

Intensivstrom-Röntgenröhre

für sehr hohe Belastung.

D.R.P. angemeldet.

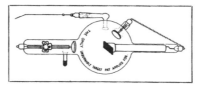

ORIGINAL "MÜLLER" X-RAY TUBES.

No. 12. For General Work.   No. 13. For Heavy Currents.

No. 14 (Watercooled). For Very Strong Currents and any Break.

All Tubes are provided with the "MÜLLER" Regulation of the Vacuum, indispensable for the exact and even work required by modern science.

TO BE HAD OF THE PRINCIPAL DEALERS.

Clover Leaf Tubes

NOTE THE NEW UNBURNABLE TARGET

Are You Using This Unburnable Target Tube?

It is better than the platinum iridium faced target, and you will notice that the prices have not been advanced. This target is our own invention, and we have a patent pending. Our quality is too well and favorably known to need comment here. The new target, however, is the most important invention yet made on a tube. There is nothing just as good. Ask for the Clover Leaf Tubes and see that you get them. They are the standard. Made only by

GREEN & BAUER
HARTFORD, CONN.

[5.15] An American tube design with a platinum–iridium target which carried the claim, in this 1907 advertisement in the *Transactions* of the American Roentgen Ray Society's 8th Annual Meeting, in Cincinnati, that it had 'the only unburnable target'.

[5.14] A water-cooled X-ray tube made by Müller in Germany is featured in this 1905 advertisement which appeared in the *Journal of the Röntgen Society*.

[5.17] The Scheidel exhibit at a convention of the American Roentgen Ray Society held in Chicago, December 1902.

[5.16] The tube at the top of this 1908 advertisement is an osmosis regulating tube. It worked on the principle that heated platinum has the property of being penetrable by hydrogen. A closed bulb of platinum is sealed into the glass bulb and when the vacuum requires lowering the projecting platinum tip is heated to redness in a Bunsen flame. The heated platinum permits the passage of gas through its pores, into the tube, and thus lowers the vacuum. A low vacuum is shown by a bluish tint which indicates that the heating should then stop.

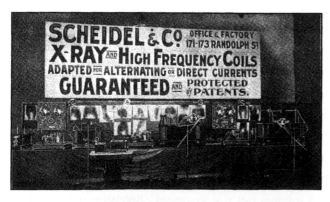

[5.18] A tungsten target tube advertised in the 1913 *Archives of the Roentgen Ray*. This marked the limit of advances in gas tube design, because it was in 1913 that William Coolidge reported the development of the 'hot cathode' X-ray tube and when these soon became widely available, the gas tube was relegated to obsolescence.

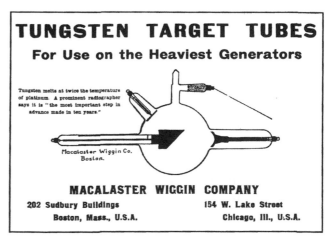

TUNGSTEN TARGET TUBES
For Use on the Heaviest Generators

Tungsten melts at twice the temperature of platinum. A prominent radiographer says it is "the most important step in advance made in ten years."

Macalaster Wiggin Co. Boston.

MACALASTER WIGGIN COMPANY
202 Sudbury Buildings **154 W. Lake Street**
Boston, Mass., U.S.A. **Chicago, Ill., U.S.A.**

Chapter 6

Spark Coils and Interrupters

Until the late 1920s a spark coil (induction coil) was an essential piece of apparatus for all except a few of the earliest X-ray experimentalists who, as an alternative, used electrostatic machines such as the Wimshurst, which were generally unreliable. They were, though, less expensive than induction coils and Snowden Ward[1] in his 1896 textbook *Practical Radiography* commenced his chapter on 'The Apparatus' with 'the whole essentials for radiography' are (a) a source of electricity, (b) an induction coil and (c) a specially made vacuum tube.'

He also went on to say that 'If using a Wimshurst machine as the source of electricity, the induction coil, by far the most expensive part of the apparatus, is done away with. The fact that under these circumstances, the Wimshurst machine is but little used for the purpose, is explained by the fact that many persons who have tried to use it have utterly failed. It has the distinct disadvantage of being exceedingly sensitive to climatic conditions and dust and damp ruin its performance.'

One of the earliest descriptions of the induction coil was given by William Morton[2] of New York, using the electrical circuit shown in Figure [6.1].

'When the switch is closed the electricity flows from S to the circuit breaker B, then through PC and back to S. When the current passes through PC it makes a magnet of the iron core I which attracts the iron head of the spring of B, making a break in the circuit between spring and contact screw. There is a spark at this point which is minimised by the action of C. When the circuit is broken by the attraction of the spring away from the contact screw, the iron core I loses its magnetism because no current is flowing around it.

'The iron head on the spring is no longer attracted and flies back against the contact screw. The instant the screw is touched the electricity again flows through the coil PC and the magnet I acts again. So long as the switch is closed there will be a rapid vibration of B, and consequently a rapid making and breaking of the circuit through coil PC. (An alternating-current dynamo may be used as S without a circuit breaker but it is not nearly so effective for X-ray work because AC does not flow at full strength for an instant in one direction and then suddenly turns around and flows at full strength in the opposite direction).

'The intermittent primary currents in PC induce alternating currents of high potential in the secondary coil SC, and if the terminals T_1 and T_2 are brought a proper distance apart, a brilliant, thin, snappy spark will pass from one to the other. If the terminals are separated so the spark will not jump the gap, the induced current will be available in the Crookes Tube for the production of X-rays.'

Induction coils: 1896 and 1901

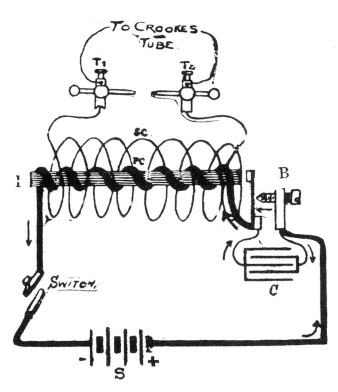

[6.1] Electrical circuit used by Morton[2] in 1896 to describe the induction coil.

I = Induction coil core made of a soft bundle of iron wires.

S = The source of electricity which may be either a primary battery, a secondary battery or a continuous-current dynamo-electric machine.

PC = Primary coil: of small resistance, a few turns of large wire.

B = Circuit breaker. A piece of iron to which is fastened a flat spring held firmly at one end; the end carrying the iron 'head' is directly in front of the iron core and is free to vibrate. Against the side of the flat spring furthest from the iron core presses an adjustable screw. The end of the screw and spot it touches on the spring are both of platinum. When the spring vibrates it flies away from the screw tip and then flies back again, thus breaking and completing the circuit.

C = Condenser connected to the spring and the contact screw of B. It acts as a temporary reservoir for the excessive current flowing when the circuit is broken. This is to ensure that the 'break' in the circuit is very sudden and there is as little sparking as possible.

SC = Secondary coil, wound of great length and with very many turns of very fine wire.

T_1 and T_2 = Terminals which are adjustable so as to regulate the (spark) gap space.

[6.2] 1901 advertisement for a Marconi 10″ induction coil. These coils were also sometimes called Ruhmkorff coils.

Influence machines: 1887–1900

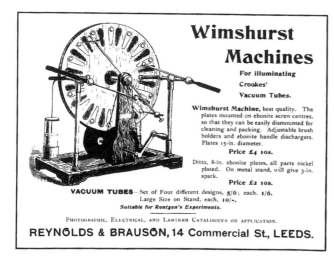

[6.3] This advertisement for a Wimshurst machine appeared in February 1896 in *The New Light and the New Photography*, the special issue of *The Photogram* featuring the English translation of Röntgen's first X-ray paper.

The early circuit breakers such as described above and in Figure [6.1] soon gave way to more sophisticated technology for interrupting the electrical circuit. There were various designs of these 'interrupters' but the most commonly encountered was the mercury interrupter. These were famous for providing dirty work for a service engineer! From the 1985 *Journal of X-Ray Technology* comes the following anecdote from the 1914–1918 war[3].

'The war had been raging for a few months and the reservist Royal Marine responsible for work in the plating shop in the small factory near the present BBC building in Portland Place, London, was already on his way to France earning his Mons Star. X-ray units would be wanted for the Forces and 10″ and 12″ spark coils crackled away on the test benches, whilst the writer was engaged in the delectable task of cleaning mercury 'breaks' or interrupters, surely the dirtiest and most smelly job in the whole of Christendom. The spy scare was on, neighbours' alert ears 'knew' that spies were rampant in the factory in the heart of residential West End London. Could they not hear the Morse Code going out over the air?

'A burly policeman entered the factory, and suspiciously looked round. These crackling induction coils under test could well be used to signal Zeppelins, dropping their load of bombs in the still night air. He was not convinced when matters were explained. More visits from policemen, each time the status and number rose. Only the War Office could placate them, and placate them they did.

'At last we were left to our noisy spark coils and open gas tubes suspended in the middle of a small workshop un-X-ray-protected!! Far more potent than Zeppelins, did we innocents but know it. Well, yes, sometimes they were laid in a lead glass shield (for radiation protection) on test, but who cared, there was a war on?'

Mercury interrupters and spark coils were eventually superseded for the production of high tension by transformers in which the laminated iron core formed a closed loop, and was not a straight core as in the spark coils. These were initially called the 'interrupterless transformer' and the first practical model for X-ray work was designed by H. C. Snook of Philadelphia.

The following description was given in a 1919 textbook on *Roentgenotherapy*[4]. 'Most of the commercial power plants in American cities furnish alternating current to which the interrupterless transformer is especially adapted. The principle of this type of apparatus is the fact that there is a synchronous motor attached to a large revolving disc. The disc is used to rectify the current so that only the peak of the wave is used. The disc revolves with the synchronous motor and is so timed that the current is cut in when the peak of the wave is reached, this motor running in synchronism with the motor at the city power plant. When one wishes to use the current through the X-ray tube, another switch is thrown which throws the 220 volt alternating current directly into the transformer, the rectifying switch cutting off the peak of the load so that it all flows in one direction as it passes from the transformer to the X-ray tube'.

Transformers are still a necessity today for high voltage supply but are of course greatly altered in design and capability from the early Snook transformers and, for example, the usual practice now is to place the transformer in a separate oil-filled box.

[6.4] By 1899 there was considerable controversy on the 'influence machine (of which the Wimshurst was an example) versus induction coil' issue. The argument against the use of a coil centred on its troublesome batteries and contact breakers, whereas its supporters emphasised that machines such as the Wimshurst must be used 'in a room which is large, dry and free from furniture; the machine should be in a case as nearly airtight as possible, and 10 pounds of calcium chloride should be dried, covered by cheesecloth and placed inside the case.' The Wimshurst machine illustrated was constructed in 1887 with twelve 30″ (76 cm) discs and was still in reliable use in 1900, according to Wimshurst himself when writing in the *Archives of the Roentgen Ray*.

Mercury interrupters: 1906 and 1932

[6.6] An American-manufactured mercury interrupter of 1906.

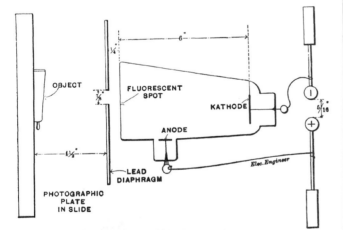

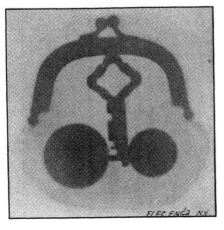

[6.5] An early experiment in which a Wimshurst machine was used[5]. It was undertaken by E. Wilbur Rice Jr and published in the *Electrical Engineer* (New York) on April 22nd 1896. The 'skiagraph' is of a purse containing coins and a key. The Wimshurst is seen on the far right of the upper diagram. Rice's reason for publishing this work was his introduction of a lead diaphragm containing a small central opening of 0.875″ diameter, opposite the fluorescent spot. The X-ray images took 60 minutes with this set-up, but only 30 minutes if the diaphragm was removed. However, the diaphragm made the 'shadows' sharper.

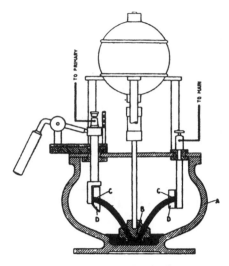

[6.7] A 1932 description[6] of a mercury interrupter. A is a cast iron pot, on the lid of which is mounted an electric motor with vertical shaft. This drives a turbine B in the vessel, which pumps mercury up and sprays it out through two nozzles, CC, onto two copper contacts, DD, which hang down from the lid. While CC are opposite DD which are insulated from the lid, current can pass from the main supply to the primary of the coil when the interrupter is suitably connected. When the nozzles have been rotated further and are spraying mercury against the wall of the pot, no connection exists between DD and no current can flow from main to primary. Two gas taps are fitted to the lid so that coal gas can be introduced into the pot and kept there while the interrupter is working.

Spark gap: 1896–1932

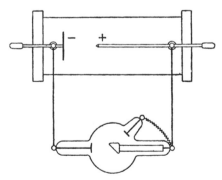

[6.9] The oldest method of specifying the penetrating power of X-rays was by a statement of the length of the spark gap, such that 'kilovoltage = equivalent spark in air'. Schall[6] in 1932 describes the technique as follows. 'A point and a plate are attached to the two discharge columns of the spark coil and placed in parallel with the X-ray tube. At a certain distance between point and plate, depending on the resistance of the tube, the current would prefer to discharge through the spark gap. This distance was known as the equivalent spark gap of the tube, and tubes were rated so many inches.' However, the technique was inaccurate on account of the corona discharge from the point, its variability with air humidity, and ionisation of the air by X-rays. An improvement was to replace point and plate by polished metallic spheres.

Electrolytic interrupter: 1902

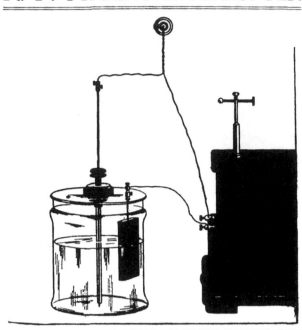

[6.8] An alternative to the mercury interrupter was the electrolytic interrupter, such as that advertised by Friedlander of Chicago in 1902. Its operation was described in Kassabian's 1907 textbook[7]. 'The electrolyte is composed of a 10 per cent solution of magnesium sulphate and the anode is made of German-silver wire. The operator can control the current for the work in hand by simply turning the thumb-screw. It operates by either the direct or alternating currents.'

Interrupterless transformer: 1919

[6.10] The first interrupterless transformer used for medical X-rays. It was installed at the University of Pennsylvania Hospital, Philadelphia and was still in working use in 1919[4]. By this time, in America (though not in Europe), induction coils had been almost entirely replaced by these transformers which make use of the alternating current instead of the direct current necessary for a coil. The first 'Snook apparatus' in England was made in 1907 by Newton & Wright of London, and described as consisting essentially of three parts: the motor converter, the high tension transformer and the high tension rectifier or commutator.

Chapter 7

Hot Cathode X-ray Tubes

William Coolidge and Thermionic Emission

The gas tube (Chapter 5) was unstable and unreliable since its X-ray production depended on its gas content and this was a very variable factor. The so-called Coolidge tube, developed by W. D. Coolidge of the General Electric Company's research laboratories in Schenectady, New York and published[1] in the *Physical Review* of December 1913, was a watershed in the design of X-ray tubes.

It was made possible by Richardson's discovery in 1902 of thermionic emission, which enabled Coolidge to obtain his source of electrons from an incandescent cathode, the 'hot cathode', which could reach temperatures of 2300°C. What at the time was another novelty, was that the gas pressure was so low, with a very high vacuum quoted by various authors as 20 or 100 times greater than that of a gas tube. Thus, unless the cathode was heated there were no X-rays. There was also no requirement for an additional anode, as with gas tubes, and the Coolidge tube contained only a single anode/target/anti-cathode.

In the earliest production design of Coolidge tube the cathode was a small flat spiral of tungsten, surrounded by and electrically connected to a small molybdenum tube which served to focus the electrons onto the tungsten target. The focal spot did not, as in gas tubes, wander or vary in size, and now the penetrating power of the X-rays depended only on the applied voltage, and the intensity varied with the temperature of the cathode filament.

When the cathode is heated to a particular temperature and the high tension is applied to the X-ray tube, the current passing through the tube rises rapidly to a maximum. Beyond this maximum there is no increase in tube current no matter how much the tube voltage is raised. What is known as the saturation current has been reached, see Figure [7.2]. It is due to the fact that all available electrons emitted by the cathode are now impinging onto the target.

If, alternatively, the voltage across the tube is kept constant and the cathode filament temperature is increased, the current through the tube rises very rapidly. Thus if the Coolidge tube is operated at saturation current, the penetrating power of the tube can be varied by the tube voltage and the intensity of the X-rays by varying the cathode filament temperature. These two variable factors are independent of each other and this makes for an excellent control mechanism of the radiation.

The early forms of Coolidge tube, such as in Figures [7.1] and [7.4], operated satisfactorily without a rectifier if the focal spot in the target was at a temperature below that at which it gave off an appreciable number of electrons. It follows that part of the problem of eliminating the rectifier was to keep the target cool.

An improvement was the development of the radiator type[2] of tube, see Figure [7.7], where the target is a tungsten button set in a heavy copper backing which is continuous with a large copper rod extending out of the tube neck. To this are attached a series of discs acting as radiators. In the 1918 *United States Army X-Ray Manual* they were described[3] as being 'designed for 10 milliamps at a 5″ (spark) gap for radiographic work, and for 5 milliamps at the same gap for continuous duty in fluoroscopy'.

With the advent of the Coolidge tube, studies on the performance of X-ray tubes were put on a better scientific basis than during the earlier years of the gas tube era. There were many investigators, including Ernest Rutherford[4] and his co-workers Barnes and Richardson who in 1915 stated the following.

'With a cold cathode the perfection of the vacuum in a Coolidge tube under experiment was such that it withstood 175,000 volts without breaking down . . . and . . . when the cathode was heated, one could not detect (by an ionisation method) any radiation with a voltage less than 10,000 volts on the tube . . . although . . . as the voltage was increased the penetrating power of the radiation increased rapidly and regularly.'

Rutherford's comments on radiation protection were: 'With 175,000 volts it was found that the intensity observable through 3 mm lead was less than 1/10,000th of the initial value . . . this lead thickness may be regarded as affording fairly adequate protection for the workers for voltages under 200,000 volts.'

With a Coolidge tube, unlike with a gas tube, the tube focus could be expressed in a more meaningful manner and in the United States Army manual of 1918 it was stated[3] that although 'not precise', it is customary to speak of broad, medium and fine foci, see Figure [7.9], and that:

Extra fine focus is anything below 3 mm;
Fine focus is 3–4 mm;
Medium focus is 4–7 mm;
Broad focus is anything above 7 mm.

This manual ended with the warning that 'fine focus tubes are not needed in gastro-intestinal work, and should be used in other work with such care that the target does not become pitted.'

X-ray tubes have of course improved since 1918 and tubes with focal spots of less than 1 mm are now used. However, factors governing the sharpness of an X-ray image are not only the size of the effective focus, but also the focus-to-film distance and the object-to-film distance. Sharpness increases with increasing focus-to-film distance and with decreasing object-to-film distance.

It should also be noted that manufacturers usually quote a nominal focal spot size at very low milliamps and that this often understates the effective focal spot size. Thus a nominal 0.3 mm may be 0.45 mm, a nominal 0.6 mm may be 0.9 mm and a nominal 1.0 mm may be 1.4 mm.

Another advance in X-ray tube design was the rotating anode/target, Figures [7.10] and [7.11], first suggested by Elihu Thomson in 1896 but not put into practice until 1915 by Coolidge[5] who achieved 'target rotation of 750 revolutions per second with the focal spot describing a circle 0.75" (19 mm) in diameter. 2.5 times as much energy for the size of the focal spot is obtained when compared with the stationary target.'

A much later type of rotating anode tube is described in 1969 by van der Plaats[6] in the following terms. 'Most rotating anodes at present in medical use have a diameter of 50–125 mm and a minimum speed of 3000 revolutions per minute.'

Indeed, there is now an entire anode disc technology with compound anode discs, discs with slits, with grooves, and materials used such as titanium and zirconium. A review of this technology together with that of insulation and cooling technology is given by Seram[7] in the 1985 publication *X-ray Imaging Equipment*, but most of this type of information is to be found in diagnostic radiology and radiotherapy equipment specification sheets, and in books[6,8] and journals from manufacturers such as Siemens (*Electromedica Journal*) and Philips (*Medica Mundi Journal*).

Coolidge tubes: 1913–1918

[7.1] One of the original Coolidge tubes manufactured in 1913. (Courtesy: The Science Museum, London.)

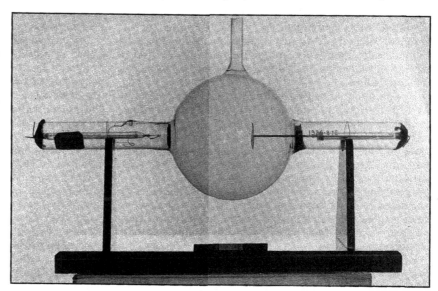

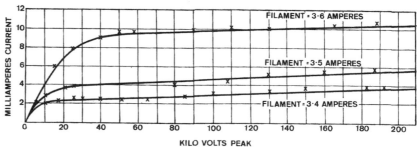

[7.2] Voltage–ampere characteristics of a hot cathode high vacuum X-ray tubes, from a publication[9] by Coolidge & Charlton.

[7.3] 1918 advertisement from the *Journal of the Röntgen Society*, London.

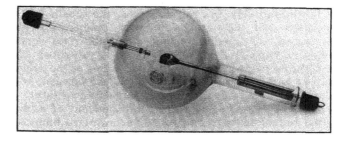

[7.4] A Coolidge tube such as that described in the 1918 advertisement.

Glass blowing technology: 1914

[7.5]   X-ray tube glass blowing technology in 1914 from the catalogue of Reiniger, Gebbert & Schall. (Courtesy: Siemens AG, Erlangen.)

[7.6]   Cuthbert Andrews was a well known supplier of X-ray apparatus to the medical profession for many years and this cartoon entitled 'Gas bulb surrounded by profanity' was one of many amusing advertisements which he issued.

Phoque Rouge Model.

(Suggested by Prof. Jean Lemarcheur).

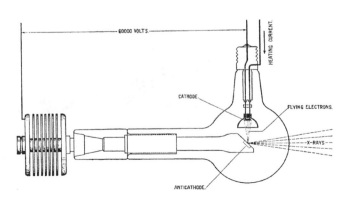

The chief feature of this superb machine is the ease by which it may be removed from the stand. and used in any position (as illustrated).   The only precaution to be observed is that the operator should be always at a slightly lower level than the tube.

Price, with supply of Water	....	....	£14	17	3
Do. do. Fine Old Hungarian Sherry			£17	13	4
Do. do. Oxo	....	....	£3	14	7
Special. The patent Whisky & Water-cooled					
pattern	....	....	....	£314	7	0
Do. with Patient-proof Safe		....	£3147	14	3

The Department of Half-Seas Over Trade writes: "It is an hic. straordinary tube."

Radiator-type Coolidge tubes: 1920s

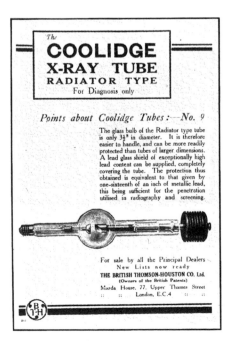

The
COOLIDGE
X-RAY TUBE
RADIATOR TYPE
For Diagnosis only

Points about Coolidge Tubes :— No. 9

The glass bulb of the Radiator type tube is only 3½" in diameter. It is therefore easier to handle, and can be more readily protected than tubes of larger dimensions. A lead glass shield of exceptionally high lead content can be supplied, completely covering the tube. The protection thus obtained is equivalent to that given by one-sixteenth of an inch of metallic lead, this being sufficient for the penetration utilised in radiography and screening.

For sale by all the Principal Dealers
New Lists now ready
THE BRITISH THOMSON-HOUSTON CO. Ltd.
(Owners of the British Patents)
Mazda House, 77, Upper Thames Street
::   ::   London, E.C.4   ::   ::

[7.7]   By the time of this 1922 advertisement from the *Journal of the Röntgen Society*, the radiator type of Coolidge tube had replaced the earlier design, see Figures [7.3] and [7.4]. The glass bulb is stated to be only 3.5″ diameter, whereas the earlier model had a glass bulb of 7″ diameter.

[7.8]   Radiator-type Coolidge tube for dental X-ray work.

Image sharpness: 1918 U.S. Army manual

Rotating anode tubes: 1936 and 1989

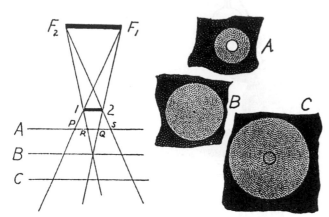

[7.9] United States Army manual[3] illustration of 1918 showing 'variation of size of shadows of small objects when a wide focus is used close to the X-ray plate'. The following explanation was given. 'The size of the focus is found by the use of a pin-hole camera. Its size is important in two ways. First, in relation to the sharpness of the image on plate or screen; second, as fixing the power that may be used without damage to the target. When an electron stream is maintained at high velocity against the target, there is a rapid rise in temperature which may result in vaporisation or fusion of the metal. The rate of removal of heat by conduction is increased by broadening the focal spot, and the amount of metal suffering extreme rise in temperature is increased, so that for two reasons there is less danger of target damage.

'The effect on sharpness of image is shown by using an exaggerated diagram. F_1F_2 are the boundaries of the focal spot and 1–2 is the object. With the plate in plane A, had the only source been a point F_1 a sharp shadow PQ would result. Had F_2 been the only source then RS would result. The only portion entirely shaded is RQ, and if the object is round we have a central white spot with a variable shading out to a diameter PS. If the focal spot were very wide and the object very small a plane B could be found beyond which there would be no white image.

'The ring PR and QS is narrower the closer the object to the plate, the smaller the focal spot and the greater the target-to-plate distance. The apparent size of the shadow will vary somewhat with exposure, as regions partly shaded may be underexposed when the exposure is brief and the true shadow may not appear at all'.

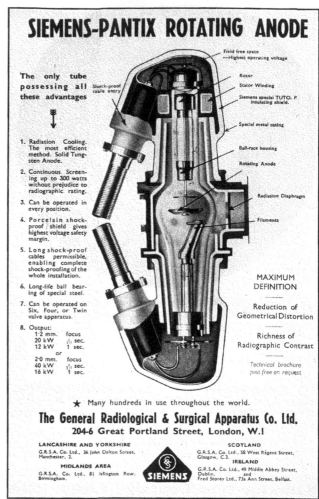

[7.10] 1936 *British Journal of Radiology* advertisement for a rotating anode X-ray tube.

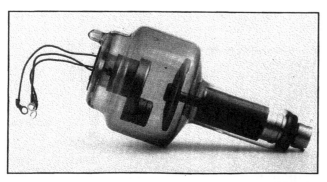

[7.11] Rotating anode X-ray tube used in 1989 for a Philips diagnostic 150 kV machine. The Austrian stamp, illustrating this equipment, commemorates the European Congress of Radiology held in Vienna in 1991.

Chapter 8

Military Radiography

Military uses of X-rays were considered early in 1896 with the Munich *Medizinische Wochenschrift* reporting on February 4th of that year that the Prussian War Ministry in Berlin was undertaking experiments to see if X-rays could be useful to sick and wounded soldiers. The successful evaluation report was published a few months later with the title 'Experiments to determine the possibility of using Röntgen's rays for medical–surgical purposes'. However, it was an Italian who was the first to actually practice military radiography. The Italian army had suffered a crushing defeat by the Ethiopians at the battle of Adowa on March 1st 1896. Two of the soldiers wounded in the forearm were radiographed to locate the bullets by a Lieutenant-Colonel Giuseppe Alvaro at the Military Hospital in Naples, because all other attempts at localisation had failed.

Alvaro described[1] his technique as follows. 'One takes a prepared photographic plate, places it in several layers of black paper, then puts it in a cardboard or wooden cassette, or on a small taboret in such a way that the impressed gelatinous surface is toward and underneath the part of the body of which the shadow is to be taken, it being fixed in this position with gauze. Above is placed the Crookes tube at a distance of 20 to 30 cm, the current being generated by a Ruhmkorff coil. After 20 minutes, or a good half-hour or longer, according to the potential of the current and the nature, thickness and density of the part, one obtains a negative with a relative white shadow on a black base.'

The next military campaign to make use of this new diagnostic aid was the Graeco–Turkish war of 1897. Germany supported the Turks with the German Red Cross providing a hospital unit in Constantinople, while England, France and Russia aided the Greeks. The contribution by the British Red Cross was described in the *Daily Chronicle* newspaper of May 4th 1897. 'The apparatus forwarded will consist of an absolutely complete outfit in itself, similar in every detail to the apparatus in daily use at St. Thomas' Hospital, London. The secondary winding of the induction coil is over 13 miles in length and will give a heavy discharge over 10″ of air. It is hoped that it will be possible to use the fluorescent screen to the exclusion of the photographic method, as the position of the bullet or the seat of the injury may be viewed in many positions rapidly, and the time required to develop a dry plate (although much shortened by the use of Eastman's new X-ray paper) constitutes a serious delay to a busy surgeon.'

It was reported from this 1897 war that a serious obstacle to field radiography was a lack of a reliable source of electrical power. Reliance had to be placed on the Royal Navy warship, *HMS Rodney*, which was used to recharge the wet batteries of the X-ray apparatus. Other novel approaches to solve the problem of obtaining a supply of electric current included: 'pedal power' from a tandem bicycle in the Sudan in 1898; 'hand-operated Roentgen cabinets' in Manchuria during the Russo–Japanese war of 1905; the use of petrol engine driven dynamos from either the truck that carried the X-ray apparatus or from a separate portable single-cylinder air-cooled engine in World War I in 1918; or, also in 1918, from a biplane. The latter was a proposal to the French Ministry of Inventions and called the Escadrille Pozzi, Figure [8.13], but it is not known if it ever flew! However, using automobile engines to provide electrical supply [3.23] was not limited to warfare (this was described for general use in the 1913 *Archives of the Roentgen Ray*) and even horses were recommended for, use with a dynamo[2], Figure [8.9].

The Tirah campaign on the North-West Frontier between India and Afghanistan, near the Khyber Pass, was the first large scale conflict involving the British Army (8,000 British and 30,000 Indian troops) to which was attached a Surgeon-Major (W. C. Beevor) with an X-ray apparatus. It was manufactured by A. E. Dean of London and was supplied with three X-ray tubes by A. C. Cossor. All tubes survived unbroken! It was, though, only after the next major military operation by the British Army, in the Sudan in 1893 culminating in the Battle of Omdurman, that the military use of X-rays became well publicised, Figures [8.1–8.3]. The Surgeon-Major (J. Battersby) with this campaign recorded his experiences in the *Archives of the Roentgen Ray*[3] in 1899 and presented his photographs and radiographs in a lantern slide lecture to the Röntgen Society of London as 'The present position of the Roentgen rays in military surgery'.

His presentation included the following technical details of how to successfully take radiographs in extreme heat in the desert. 'Beevor demonstrated that successful results can be obtained in the cold and mountainous regions of Northern India . . . in the Sudan the temperature varied from 100° to 120° Fahrenheit in the shade. Before leaving Cairo I took the precaution of having very thick felt covers made to surround the outer boxes containing the coils and storage batteries, and by keeping these constantly wet, the internal temperature was considerably reduced. My apparatus had to travel for two days and a night in an open truck, exposed during the day to the blazing sun. The felt was kept wet every two hours. We later proved that the temperature in the centre of the induction coil did not exceed 85°F.'

Battersby's conclusions were that 'radiography can boast its most brilliant results in obscure injuries to bone, especially when the injured parts are too swollen to admit of careful examination by ordinary methods, or when such examination cannot be borne by the patient.' Also, that 'after the Battle of Omdurman, 121 British wounded were conveyed to the surgical hospital at Adadieh. Of this number there were 21 cases in which we could not find the bullet by ordinary methods. In 20 of these an accurate diagnosis was arrived at with the help of the rays. The remaining case, a bullet in the lung, was too ill to examine.'

American experience in military radiology commenced with the Spanish–American war of 1898, which resulted in the annexation of Puerto Rico and the Philippines and the setting up of Cuba as an independent republic. Relatively few soldiers were wounded in this conflict and the principal cause of war dead was typhoid fever. However, a full report was written for the United States House of Representatives by the Medical Department of the U.S. Army[4] providing a series of case histories as well as details of X-ray apparatus and technique. The section on 'lodged missiles' was subdivided into Mauser bullets deformed by ricochet, undeformed Mauser bullets, shrapnel bullets and brass-jacketed bullets. Figure [8.4] is a photograph of a Dr Gray in the X-ray room of the hospital ship *Relief* during the Spanish–American war, Figure [8.5] is of a private soldier in the First Nebraska Volunteers who had a Mauser bullet lodged in the brain and [8.6] and [8.7] show X-ray burns on soldiers.

It was, though, not until some 20 years later, in World War I in 1914–1918, that American military radiology was practised on or near the battlefield. By this time, fleets of U.S. Army X-ray camions (wagons) were available [8.14, 8.15]. Marie Curie, who wrote a treatise[5] on *La radiologie et la guerre*, served in the radiographic service of France and Belgium from 1915 and taught pupils of the American Expeditionary Force in 1919, also drove one of these X-ray wagons [2.9].

The Russo–Japanese war was fought in 1905 and after the May battle of Tshushima Straits there were 83 wounded sailors on the Russian cruiser *Aurora*, 40 of whom were radiographed in the ship's infirmary. (The *Aurora*'s claim to fame is more than merely radiological! It was the same cruiser that on October 24th 1917 shelled the Winter Palace in Petrograd (which city in 1991 has now reverted to its original name of St. Petersburg and discarded the name Leningrad) and signalled the fall of Kerensky's provisional government and the Bolshevik takeover.)

An earlier 20th century war that received much more international attention than the battles between Russia and Japan was the Boer War of 1899–1902 [8.10–8.12]. British Army X-ray apparatus was provided as standard issue for all general hospitals and the *British Medical Journal* of 1899 lists the standard X-ray kit as follows. '10″ Apps–Newton induction coil with condenser, spring-hammer interrupter, rods and electrical cables; 2 six-cell lithanode accumulators; 6 Cox 'Record' focus tubes with tube stand; 1 Mackenzie Davidson cross-thread localiser with stand [8.13, 8.14]; 108 Edwards cathodal XXX plates, photographic paper and chemicals'. Later, a 2.5 horse-power motorcycle engine and dynamo was added to the kit and this was used in practice by fixing engine and dynamo to an army bed frame.

To bring military radiography up to date it is reported[6] that during the Gulf war of 1991 the United States army set up two CT scanners in field hospitals in Saudi Arabia and that there were also two CT scanners aboard U.S. Navy hospital ships. This is a change from what had previously happened in warfare, and certainly during the Vietnam war the only imaging modality near the front lines for triage (determining which casualties need immediate attention and thus avoiding waste of time on unnecessary surgery, especially laparotomies) was conventional X-ray apparatus. The first demonstration of the value of CT in combat triage was by Israel during the fighting in Lebanon in 1982. CT scanners were used instead of the usual skull films and abdominal CT scans instead of plain films. With such CT available, it dramatically changed the ability to diagnose penetrating injuries caused by blunt objects.

Sudan and the Battle of Omdurman: 1898

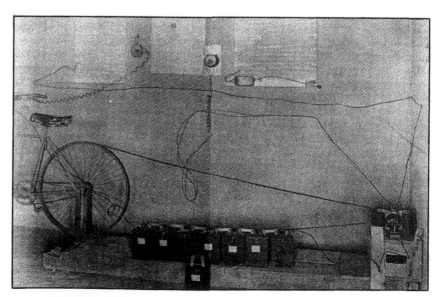

[8.1] Method by which electricity was generated for charging storage batteries at the Sudan campaign base hospital in 1898 by the Nile at Abadieh, which was 1,250 miles upstream from Cairo[3].

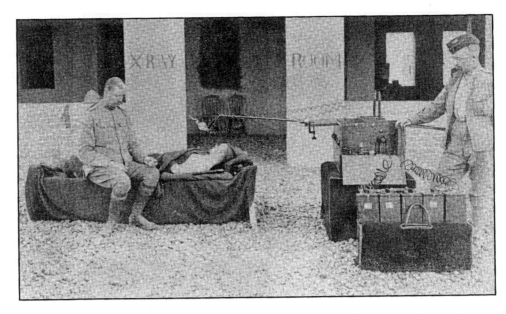

[8.2] Surgeon-Major Battersby and his orderly taking a radiograph in 1898. The X-ray tube was described as 'a modified Crookes tube suspended by means of an ingenious holder'. They used a 10″ induction coil made by A. E. Dean of London[3].

[8.3] Bullet wound in the left thigh of a private soldier in the Lincolnshire Regiment, Abadieh 1898[3]. At the top right of the radiograph, Sudanese dust is visible, which could not be prevented from blowing onto the X-ray plate. Developing work in the Sudan was undertaken at 3 a.m. in the morning when the temperature in the mud brick dark room varied from over 90°F to 110°F. This was the coolest time available. Constant dust storms also proved a problem and one night the roof of the mud hut blew off and 11 X-ray plates were destroyed.

The Spanish–American War: 1898

[8.4] X-ray room on the hospital ship *Relief*.

(Courtesy: Otis Archives of the Armed Forces Institute of Pathology, Walter Reed Medical Center, Washington)

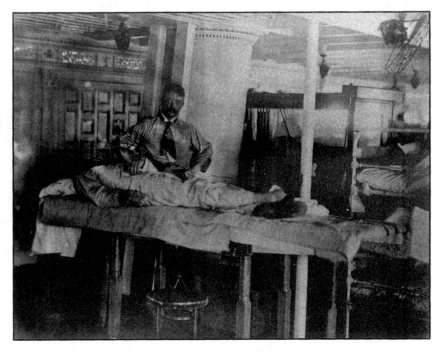

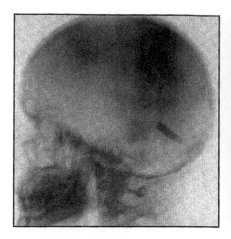

[8.5] Radiograph of a bullet (left) lodged in the left occipital lobe of the brain of Private John Gretzer, taken in August 1899 on his return to San Francisco having been injured at long range in the Philippines in March of that year. He entered the U.S. Mail Service afterwards and returned to Manila on duty. The photograph (right) was taken five months after the injury; the scar of the entrance wound is above the left eye[4].

[8.7] Röntgen ray burn on the abdomen of a private in the U.S. Hospital Corps. This was caused when 25 minute X-ray exposures were made every other day for three days in December 1899 to investigate a possible calculus in the left kidney. Ulceration occurred but disappeared in about 10 days[4]. This and other experiences in the Spanish-American war led the author of the report to state that the factors which influence the production of Röntgen ray burns are fourfold: the length of the exposure; the nearness of the tube to the surface of the body; the patient's physical condition; and individual idiosyncracy.

[8.6] Röntgen ray burn of the right breast of a former private of the 6th United States Infantry who had received a gunshot fracture of the upper third of the right humerus for which an excision of the upper part of the humerus was made. In December 1898 using a 20 minute exposure at 10″ from the shoulder an attempt was made at radiography, but the result was so unsuccessful that second and third attempts were made, also unsuccessfully. Six days after the last exposure a slight erythema appeared, later forming small coalescing ulcers and tissue necrosis. The burn showed no sign of healing for four months and was not entirely healed until 11 months after the Röntgen ray exposure[4].

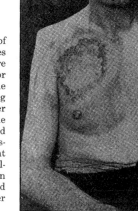

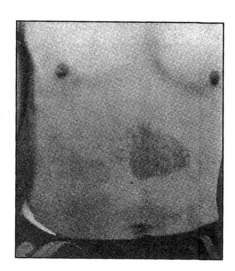

Mill power, 1901 and horsepower, 1909

[8.8] In the *Archives of the Roentgen Ray* of 1901 Lieutenant F. Bruce described his experiences during the Boer War with the X-ray installation at Ladysmith during the siege of that city. One of his biggest problems was the necessity of getting batteries charged and in order to achieve this he visited a flour mill which was in operation 24 hours a day. 'I went to interview the manager, with the object of asking his assistance to get driving power for the dynamo. He was most obliging but had no suitable engine to offer me. I asked to have the dynamo driven from the mill shafting. He readily agreed. The number of revolutions of this shaft having been ascertained, also the diameter of the pulley of the dynamo, it remained a matter of calculation to find out what the diameter of the pulley should be in order to obtain a speed of 2000 revolutions per minute on the dynamo. This installation worked remarkably well and as well as supplying the X-ray apparatus, it also supplied electric light to the operating room at night'.

[8.9] From a 1909 textbook[2]: 'A portable X-ray outfit may be designed for driving by horsepower. Where men are readily available, such as in the service, a pedal gear may be arranged similar to the driving gear of a bicycle. In emergency, a serviceable drive may be obtained by supporting an actual bicycle frame, and connecting the dynamo by belt to the back wheel'.

The Boer War: 1899–1902

[8.10] Boer War radiography of the foot[9], 1899–1902. The X-ray apparatus shows no X-ray protective shielding. The portrait in the background is of Queen Victoria.

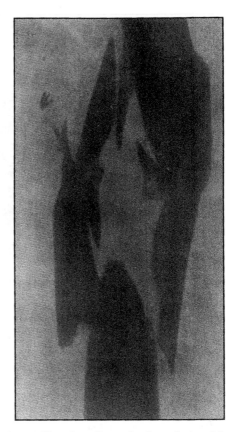

[8.12] One of Lieutenant Bruce's radiographs from Ladysmith showing a gun shot wound of the right thigh. The patient was an officer in the Imperial Light Horse at the battle of Elandslaagte. No amputation was necessary and the officer returned to duty.

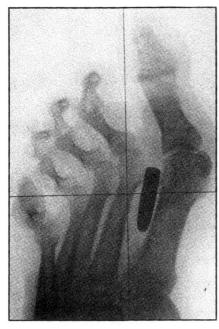

[8.11] Radiograph of a Mauser bullet in the foot, taken using the X-ray apparatus in [8.10].

The Escadrille Pozzi: 1918

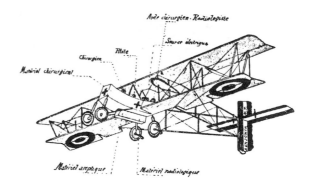

[8.13] The Escadrille Pozzi of 1918 was a project proposed by a Russian and a Frenchman, Nemirovsky and Tilmant, to the French Ministry of Inventions. It was a single-engine biplane which carried a pilot, a surgeon and a radiologist, together with the equipment for emergency surgery and X-ray apparatus. The electrical power for the apparatus was provided by the aircraft engine. (Courtesy: Bibliotheque Medicale, Centre Rene Huguenin, Paris.)

X-ray wagons: 1914–1918 war

[8.14]   A United States Army X-ray camion (*American Journal of Roentgenology*, 1919). Colonel A. C. Christie who was in charge of radiology on the Western Front compared apparatus of the French, English and American armies in the following words[7]. 'The advantage of the English and French camions over the American is that the two former have the body of the car arranged for a dark room, while the latter is not. Experience in this war has shown that the greater part of the Roentgen ray work in hospitals in the forward areas, including evacuation hospitals, is fluoroscopic. An elaborate dark room is therefore unnecessary. The few plates that will be made can conveniently be developed in the small portable dark room furnished with the American apparatus'. Christie also commented on the automobile engine: 'the engine of the American camion is not used to generate power for Roentgen rays or lights, and can therefore be overhauled and placed in order between trips'. He saw this as one advantage for the United States design, others being its much lighter weight, a demountable apparatus and the availability of spare parts.

[8.15]   A Siemens X-ray motor wagon (*Journal of the Röntgen Society*, 1918). The view shown was entitled 'The table arranged for use, but without operating tent erected'. The Siemens London advertisement (the British Government took over the Siemens operation in London at the start of the war) claimed that radiography, radioscopy and therapeutic treatment could be undertaken in the car. The generator specification was 20 amps at 150 volts; the accumulators were 30 amps at 150 volts; and the generator and accumulators could be used in parallel.

Projectile localisation and fluoroscopy: 1914–1918 war

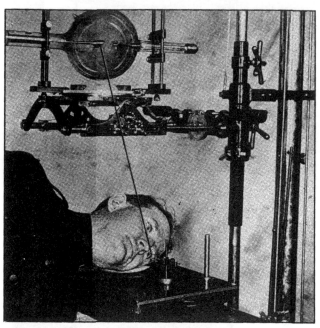

[8.16]   From the *United States Army X-ray Manual*[8] of 1918 showing part of the technique for the localisation of 'projectiles in the eye'. A tube shift method was used with two radiographs taken.

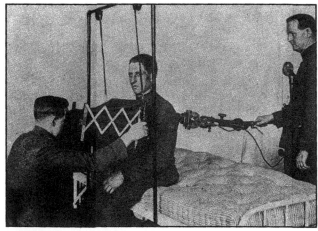

[8.17]   From the *United States Army X-ray Manual*[8] of 1918 showing 'a bedside unit with simple vertical fluoroscope for chest examination at the bedside'. It was stated to give 'good average results when two conditions are met: (1) a tube current of 5 mA; (2) a proper low tension voltage applied to the transformer terminals'.

Chapter 9

Animal Radiographs

Right from the start of world knowledge in January 1896 of Röntgen's discovery, small animals were used as test objects to illustrate 'the new light' and from the literature it would appear that the frog was the most popular choice [9.1–9.3, 4.6]. Indeed, it was a foot of a very small frog that was used by the Röntgen Society as one of their test objects when in 1900 the then President, Dr John Macintyre (the same who demonstrated cine-radiography [4.6]) decided to offer a gold medal to the maker 'of the best practical X-ray tube for both photographic and screen work', Figure [9.2].

The gold medal winning X-ray tube is illustrated in the 1909 *Journal of the Röntgen Society*[1–3] in the review entitled 'The historical collection of X-ray tubes (of the Röntgen Society)'. It was the only such competition ever held by the Society, perhaps because of the furore it caused[1]. 'A committee of experts was formed to act as judges, and some 28 tubes were sent in by manufacturers, both at home and abroad. There was a good deal of grumbling when the award was made to Mr C. H. F. Müller of Hamburg' and 'it was a pity that the very elaborate tests through which the tubes were passed was not made public'. The X-ray plate of the frog's foot (the hooked nail at the tip of each toe was only one-tenth of a millimetre width) was the final test that decided the award.

The 'radiographic zoo' in this chapter, the majority taken within five years of the discovery of X-rays, encompasses a wide variety of different animals, fish and birds which have been used as test objects or to illustrate radiographic publications. Elsewhere, the lobster, the crab and the starfish appear in [4.9], a (cine-radiographic) frog's leg in [4.6] and a 'February 1896 frog' in [3.3].

Frogs

[9.1] An American frog (*Archives of the Roentgen Ray*, 1898) exhibited at a conversazione of the Röntgen Society, having been sent by Miss E. Fleischman of San Francisco, a pioneer X-ray technologist who died in 1905.

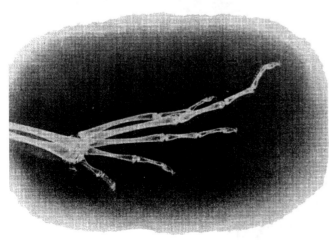

[9.2] The frog's foot[1] used as a radiographic imaging test object by the Röntgen Society in 1900 for the award of a gold medal for 'the best practical X-ray tube'.

Birds and bat

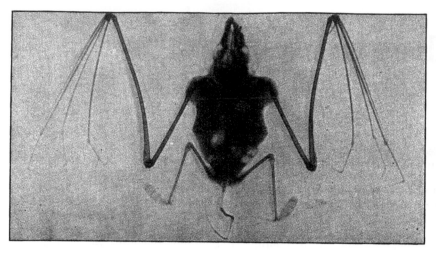

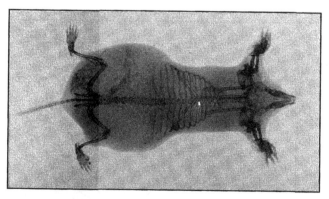

[9.3] A bat of 1896. This was from one of the popular magazines of the day, the *Strand Magazine*[6], and other animal radiographs in the article were a plaice, dog's paw, rabbit's paw, chicken's foot, newt, frog, pigeon and puppy. No other 1896 publication had such a large 'zoo'. Although it is recorded in the *Journal of the Röntgen Society* of 1920 that 'A collection of prints of reptiles (slow worm, grass snake, adder, frog, lizard and toad) were shown to the Röntgen Society in November 1897 by J. H. Gardiner' and were shown for a second time in January 1920 at the only joint exhibition ever held between the Röntgen Society and the Royal Photographic Society, when some 200 radiographs were exhibited.

[9.4] A canary published in an 1896 text-book[4] by Edward Trevert of Lynn, Massachusetts. The radiograph was taken by Fritz Giesel of Braunschweig. Trevert stated that it was taken immediately after death and 'The rays passed unimpeded through the feathers, nothing of which shows, the fleshy parts are slightly outlined, but the impenetrable bones have come out distinctly.

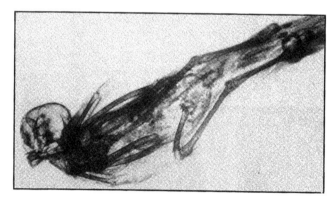

[9.5] A mummified bird from an Egyptian tomb. A friend of Thurstan Holland[5] brought him a large red block, hard and smooth, the shape of and the look of a rather large red brick and he was told that it had come from an Egyptian tomb and was at least 2000 years old. The owner wondered if the X-rays could tell what was inside the brick. The radiograph was taken in October 1896.

Cat, dog, mole, rabbit and chameleon

[9.6] The common mole (*Journal of the Röntgen Society*, 1904) taken by George Rodman who also had in his collection of radiographs those of a hedgehog, monkey, chameleon and a lizard with a double tail.

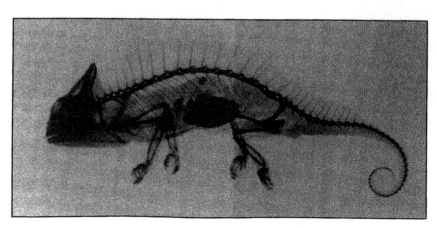

[9.7] Radiograph of a chameleon from the portfolio of Eder and Valenta[8] which was presented to Röntgen in 1896. (Courtesy: Fachhochschule Würzburg-Schweinfurt.)

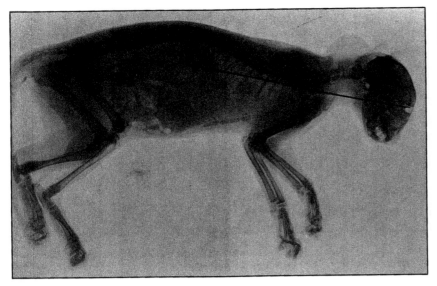

[9.8] A cat which had swallowed a hat-pin (*Archives of the Roentgen Ray*, 1900), which was published with the following case history. 'This plate is from a radiograph sent to us by Messrs Watson and Sons (the X-ray apparatus manufacturer), who inform us that the cat, which is nearly fully grown, swallowed the hat-pin when playing. The owner did not know this at the time, and had the cat poisoned; at the same time the point of the pin made its appearance through the eye. This illustrates the possible value of the X-rays and the fluorescent screen to the veterinary surgeon to assist in the diagnosis of some of the mysterious diseases of cats and kittens. The radiograph is also interesting as showing the intestine very clearly.'

[9.9] A dog from a French textbook[7] of 1898 by A. Londe who also published a radiograph of a guinea pig.

[9.10] A rabbit which was killed with shot (*Archives of the Roentgen Ray*, 1899) taken by W. R. Cooksen who also published a radiograph of a partridge. A rabbit was also mentioned in the *Archives of the Roentgen Ray* of August 1898 when a Röntgen Society conversazione was reported: 'In a large screen-room, darkened for the purpose, were given demonstrations by Mr Watson Baker, who brought a living rabbit in a leather bag, and a bird in a cage for examination with the rays'.

Fish

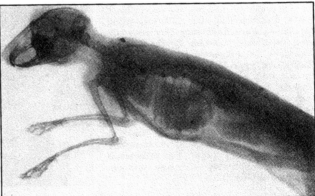

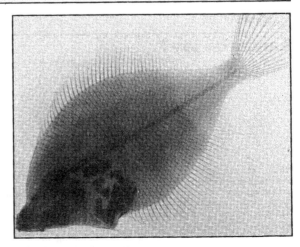

[9.11] Radiographs of *Zanclus cornutus* and *Acanthurus nigros* from the portfolio of Eder and Valenta[8] which was presented to Röntgen in 1896. (Courtesy: Fachhochschule Würzburg-Schweinfurt.)

[9.12] Plaice with a small crawfish in its guts: taken by A. W. Isenthal[9] in 1898.

Snake and crocodile

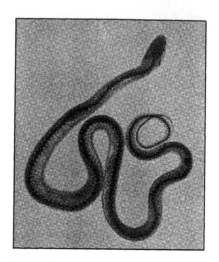

[9.14] A young crocodile killed with a shotgun; the pellets are clearly visible in this radiograph. (Courtesy: Deutsches Museum, Munich.)

Elephant and beetle

[9.13] This snake was chosen to demonstrate the production of a radiograph using a short exposure (10 seconds) and a small spark gap (1″) from an induction coil. It was stated by L. Wright in the 1897 book[10] that the snake 'was of course chloroformed'.

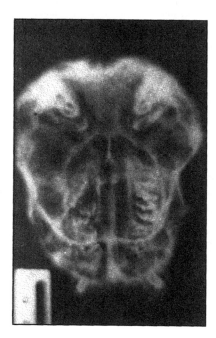

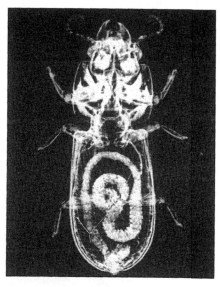

[9.15] Foetal elephant skull. The date when the radiograph was taken is not known but it is from the historical slide collection of the British Institute of Radiology. Elephants only have one further mention by the Röntgen Society: in 1930 a report under the heading of 'Radiographing an elephant'. This related to an animal on a rubber estate in Ceylon. The elephant had become dangerous owing to what appeared to be a foreign body lodged in its head and accordingly it was marched to Colombo General Hospital. There was a nervous moment when the elephant 'tried to touch the delicate Philips X-ray apparatus with the tip of his trunk'. The radiograph revealed a small bullet which was later successfully removed.

[9.16] A Peruvian wood-boring beetle taken in the 1960s. It is about 2.5 cm in length and the internal anatomy is clearly shown. (Courtesy: Eastman Kodak.)

Kangaroo jaw?

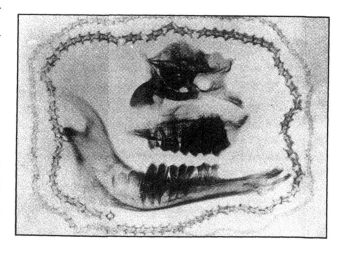

[9.17] This composite radiograph 'Parts of animals from Australia' was published in the 1903 *Archives of the Roentgen Ray* with an invitation as follows: 'This skiagram was sent to us sometime ago by Dr J. G. Beckett of Alma, Northcote, Melbourne, Australia, unfortunately without any description, and in spite of our efforts to get some accounts of it, we have so far been unsuccessful. It was requested that the plate should appear, and, failing any description from the sender, we invite suggestions as to its meaning from our readers.' There were, however, no replies.

Chapter 10

Diagnostic Radiology: I

Röntgen's first communication made public the fact that X-rays excited fluorescent substances and the world's first fluoroscope quickly followed in January 1896. It is attributed[1] to the Italian scientist E. Salvioni who demonstrated his cryptoscope, Figure [10.1], before the Perugia Medical Society. It was of a very simple design: a tube having at one end a pasteboard cover coated with fine crystals of platino-barium cyanide and at the other end an eyepiece.

The shape of the variously called cryptoscopes/skiascopes/fluoroscopes/lorgnette humaine for fluoroscope/radioscopy was soon changed to that of a pyramid, Figure [10.9], and this type of design remained in use for many years and it was still the recommended model for use in 1915 by the United States Army[2]. An alternative was to use a dark cloth over the head[3], Figure [10.5], in conjunction with a large fluorescent screen such as shown in Figure [10.6].

There were many fluoroscopic experiments in 1896 and Julius Mount Bleyer of Naples produced the first ever photofluoroscope by combining a camera and a skiascope, Figure [10.2]. It was initially demonstrated[1] on April 7th 1896.

However, the most prolific investigator in the field of fluoroscopic design was Thomas Edison, who by April 1896 had examined 1800 chemicals to detect and to compare their X-ray fluorescent properties, if any. Of these, calcium tungstate was found to exhibit six times the luminosity of platino-barium cyanide. Of other tungstates, strontium had a luminosity of half that of calcium, whereas lead and barium scarcely fluoresced.

Almost a century later in the *British Journal of Radiology*[4] of 1989, calcium tungstate was still a source of discussion. The topic then concerned the facts that radiation exposure in diagnostic radiology is less for rare earth screens than for calcium tungstate screens and that Sir William Crookes was in 1881 the discoverer of rare earth fluorescence with cathode rays.

The early fluoroscopic screens could be easily damaged and advice was often given on how to protect them[5]. One such example by Knox of the Royal Cancer Hospital, London, in 1923 was as follows. 'Before placing the screen in the holder it should be carefully dusted with a wide camel-hair brush. The film side of the plate is brought into contact with the fluorescent coating on the screen, care being taken to avoid rubbing the surfaces together. When the screen is not in use it should be placed in a position such that it cannot be damaged or splashed with chemicals. When a large number of plates are taken with an intensifying screen, it is useful to fix the screen permanently in the cassette. A sheet of fine paper should always be placed over the screen and the cassette closed when it is not in use. A sheet of clean glass may take the place of the paper.'

Other problems of the early days of intensifying screens are taken from the 1976 review by L. J. Ramsey[6] who for many years worked for Ilford Ltd. one of the then major manufacturers of radiographic film in the United Kingdom. 'Campbell Swinton [3.2] is claimed as having experimented as early as January 1896 by including a fluorescent substance in emulsions, and this idea of incorporating the fluorescent salt within the X-ray plate was later taken up on a commercial scale. Substances such as powdered fluorspar or calcium tungstate were used but the results were not very satisfactory due to the granularity.

'Much later though, in 1921, the idea was revived and X-ray plates were made with an extra surface coating of gelatine, impregnated with the fluorescent material, which had to be renewed before the plate was developed. The recommendation was to place the plate in hot water at 43°C for at least three minutes, gently wipe with damp cotton wool and follow this by thorough rinsing in cold water. Normal processing could then begin.

'Since glass plates were coated with emulsion on one surface only, the possibility existed of using a screen either as the front or as the back screen. Although naturally with the screen surface always in contact with the emulsion side of the plate. Both methods were used but the problem with a front screen was that the lack of uniformity of coating thickness led to image variations, whereas with a back screen the image density depended also on the X-ray absorption by the glass of the plate. This factor was usually ignored by X-ray plate manufacturers.'

However, Morgan and Lewis[7] writing in 1945 on 'The use of protective glass in photofluorographic equipment' showed that in 35 mm photofluorography, removal of the glass improved the clarity of the image by 33%, although for 70 mm the results were less striking. They came to the conclusion that the need for protection of the screen against dust and accident does not appear to be very significant and that the benefits accruing from the use of protective glass are either non-existent or may be obtained equally well by other means. They also commented on the current developments in 1945 concerning screens which could be covered by a thin transparent acetate film and which require no further protection.

An interesting exercise is to compare two standard textbooks[3,5] of the years 1902 and 1923 and see how the diagnostic radiology (radiography and fluoroscopy) workload altered over the first quarter of the century. By taking the total number of pages of each of these two weighty textbooks, from Boston City Hospital in

1902 and from the Royal Cancer Hospital, London in 1923, the following distribution (which excludes therapy and physics) was obtained.

Application	Workload (%)	
	1902	1923
Thorax	58	10
Skeleton	28	34
Alimentary and urinary tract	10	42
Foreign body localisation	5	15

For the thorax in 1902 a very large number of applications was quoted: pulmonary tuberculosis, pneumonia, emphysema and bronchitis, pleurisy and empyema, hydrothorax and pneumothorax. Each rated a large section and it is sometimes forgotten in the 1990s that the term 'screening' meaning fluoroscopy was derived from its use in screening the chest for pulmonary TB and for silicosis.

The 1902 section on foreign body localisation contained a variety of techniques, with the Mackenzie Davidson localiser always prominent [10.13, 10.14]. The extra coverage in 1923 was in large part due to World War I, 1914–1918, and this is also mirrored in journals such as that of the Röntgen Society.

Screening using just a fluorescent screen as in [10.19] has long been an obsolete technology and modern apparatus makes it possible for the quality of the fluoroscopic image to be greatly increased by the use of an image intensifier tube. Such an increase in image brightness makes possible a large reduction in the radiation exposure necessary when compared with that required in the earlier years. It also obviates the requirement for the technique to be performed in total darkness.

Figures [10.17] and [10.22] show rooms of the early 1970s and late 1980s for fluoroscopic work. The image intensifiers and the TV monitors are clearly seen in each photograph. Figure [10.23] shows a barium meal taken in the X-ray room for general fluoroscopic work in [10.22]. The quality of this image can be compared with that of [11.4] which was taken in 1923.

Digital technology has revolutionised diagnostic X-ray imaging, and digital subtraction angiography (DSA) is mentioned in Chapter 13, but another feature of digital advances is digital spot imaging (DSI). This offers real time image reviewing using random access memory (RAM) storage so that, for example, images can be automatically enhanced and displayed in either single or 16-image format for immediate evaluation. Using equipment such as the Philips Diagnost 76 DSI, spot radiography can be achieved with 1024×1024 pixels and hundreds of grey shades.

Digital image processing opportunities are improving with each new generation of equipment as, for example, with the Siemens Digiscan system, Figure [10.24]. This is a cassette-based digital acquisition unit which includes an interactive workstation for display and archiving on optical disc, laser camera and processor for hard copy output and a laser reader/cassette unloader and loader. This removes the phosphor screen from the exposed cassette, reads the image by means of a laser scanner, regenerates the phosphor screen, loads a new (regenerated) screen into the empty cassette and transfers the digital image data to the interactive workstation.

Diagnostic radiology started with X-ray plates and screens of a very basic nature and the plates were relatively quickly replaced by X-ray film, but the future for the start of the second century of X-rays is the filmless X-ray Department and a computer network with workstations and links to digital archives and radiological information systems and Imaging Departments which not only have X-ray (radiography, fluoroscopy, CT) but also magnetic resonance (MR), ultrasound and nuclear medicine imaging facilities within an umbrella framework [10.25]. The computer network technology will enable efficient management and communication facilities to be used for image (and associated data) storage, retrieval and analysis which would have seemed as impossible to Röntgen in the early 1900s as sending man to the moon seemed impossible in the 1950s. All we can be sure of is that technological achievement will continue, even if we cannot guess in exactly what direction it will progress over the next century.

Fluoroscopy: 1896–1902

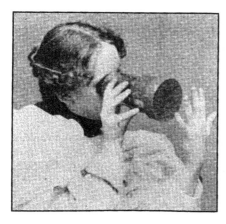

[10.2]   The Bleyer photofluoroscope of 1896.

[10.1]   The earliest design of a fluoroscope: that of Salvioni of Perugia, Italy: 1896.

[10.3] Thomas Edison and his skiascope[1], which he devised in February 1896 using platino-barium cyanide. He later called the instrument a cryptoscope and finally, when he used calcium tungstate, it was termed the Edison fluoroscope. It was demonstrated at the New York Electrical Exposition of the Electric Light Association in 1896 in what was called 'Thomas Edison's beneficient X-ray exhibit'. Edison and his X-ray lamp have already been referred to in Figure [5.7], as has Clarence Dally, the first X-ray martyr in America. It is Dally in the background of this photograph, using his hand as a test object.

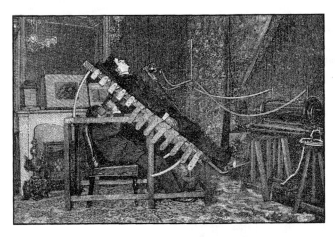

[10.4] The first tilting table specially designed for fluoroscopy[8]. It was invented by de Bourgade of Paris.

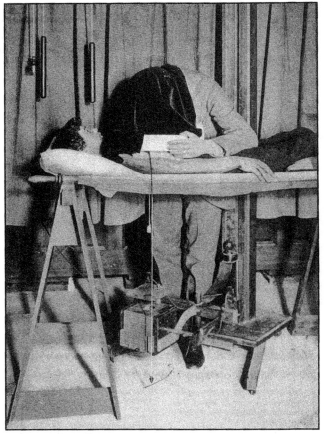

[10.5] A 1902 Boston City Hospital method of examining the whole thorax with a large open screen 30 cm × 35 cm placed in a shallow box, the side of which is shown under the hand. The dark cloth was drawn aside for the purposes of the photograph but would normally be covered[3].

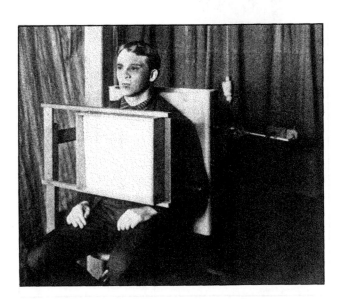

[10.6] The fluorescent screen is the same as that used in Figure [10.5]. The X-ray tube was used in a lead-lined box and was 71 cm from the screen, directly behind the patient[3].

Intensifying screens: 1896–1935

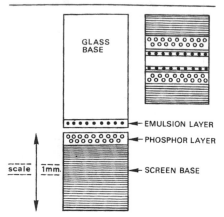

[10.7] Diagram after L. J. Ramsey[6] of the cross-section of an intensifying screen with a glass plate and a screen–film combination of 1976. Overall thicknesses are to scale, but for greater clarity, the thin layers are not to scale. Michael Pupin of New York is credited[11] as being the first to use a 'reinforcing' screen, in February 1896, to shorten X-ray exposure times.

Fluoroscopy in the U.S. Army: 1918

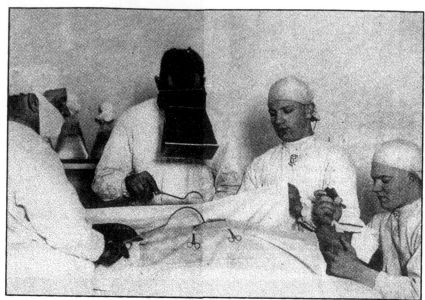

[10.9] Fluoroscopic assistance during an operation in 1915. The United States Army description[2] was 'intermittent control: surgeon and roentgenologist working simultaneously'.

X-ray plates and films: 1896 and 1920s

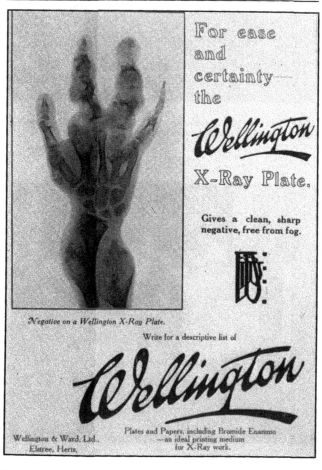

[10.10] Advertisement for X-ray plates in a 1923 issue of the *Journal of the Röntgen Society*.

[10.8] Advertisement for intensifying screens in a 1935 issue of *Acta Radiologica*.

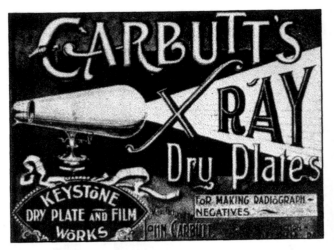

[10.11] John Carbutt of Wayne Junction, Philadelphia owned a company famous for its X-ray plates. Working with Goodspeed [5.3], a special Roentgen X-ray plate was developed[11] which permitted an exposure of 20 minutes where previously one or more hours were required. It was first tested in February 1896 at the Maternity Hospital in Philadelphia and a report appeared in the September 26th 1896 issue of the *Western Electrician*. This described radiography of a mummified hand encased in gold leaf which proved by a 2 minute exposure that the plates were indeed quite sensitive.

[10.12] Advertisement for X-ray film in a 1927 issue of the *British Journal of Radiology*.

Mackenzie Davidson portable localiser: 1897

[10.13] The Mackenzie Davidson portable localiser as advertised in the *Journal of the Röntgen Society* in 1910. Localising equipment such as this was essential in the wars in the Sudan, South Africa and in World War I.

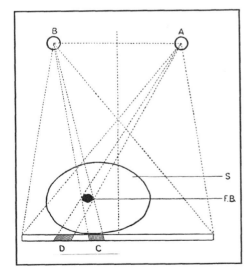

[10.14] This diagram shows the geometrical essentials of the Mackenzie Davidson method and was first described in 1897[9]. A is the first position of the X-ray tube; B is the second position of the X-ray tube; D is the shadow thrown on the plate by tube in position A; C is the shadow thrown on the plate by tube in position B; S is the skull or limb; F.B. is the foreign body. What is termed the 'central ray' is located by means of cross-wires or a plumb line running on pulleys and moving with the X-ray tube. Thus whatever the position of the tube, the plumb line always indicates the central ray. The anode–couch-top distance is constant. The distance from the X-ray plate to the couch top is also taken into consideration so that the anode–plate distance is known.

Stereoscopy: 1896 and 1901

[10.15] Stereoscopic radiography became of interest from early 1896. For example, James Mackenzie Davidson published in the *British Medical Journal* of December 3rd 1896 a paper 'Stereoscopy in clinical photography and skiagraphy'. His examples were for photography of a case of smallpox, and for skiagraphy the bullet in the leg shown here. The patient was age 14 years and was radiographed on June 6th 1896 making this stereo pair the earliest taken in England.

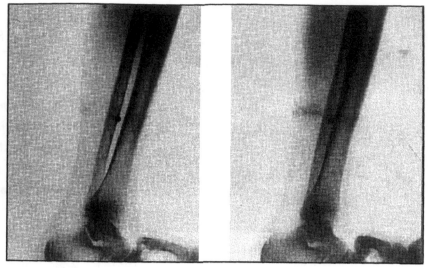

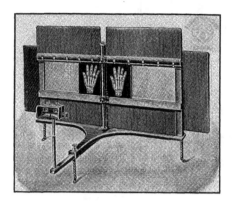

[10.16] In Germany in 1901 from the Hamburg-Eppendorf Hospital, a series of stereoscopic images on cardboard were produced for test purposes[10], three sets of which can still be seen at the Deutsches-Röntgen Museum in Remscheid-Lennep. These are I: Das Arteriensystem des Menschen (1901) as illustrated right, III: Die kongenitalen Hüftgelenksluxationen (1902) and VI: Deformitaten und Missbildungen (1903). Also shown (above) is the prism stereoscope of Bernhard Walter of Hamburg, as advertised by Reiniger, Gebbert & Schall of Erlangen prior to 1907.

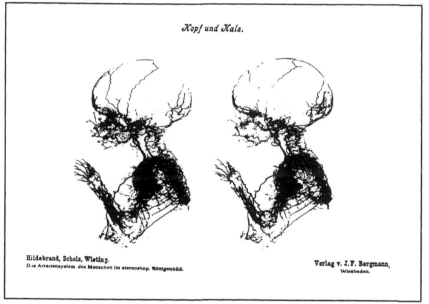

Kopf und Hals.

Hildebrand, Scholz, Wistinz.
Das Arteriensystem des Menschen im stereoskop. Röntgenbild.

Verlag v. J. F. Bergmann,
Wiesbaden.

These stereo pairs may be visualised in three dimensions by viewing them cross-eyed until the two images coincide. Another stereo pair, showing a heart with injected coronary arteries, is reproduced in [13.4].

Image intensifier: 1972

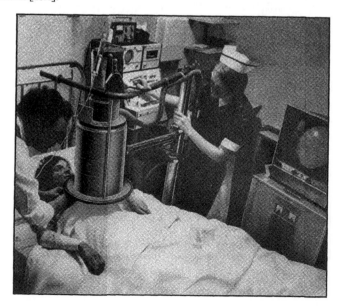

[10.17] Image intensifiers date from the early 1960s but that shown here is a later model from 1972, the Watson Cardiovision, described as a 'new concept of radiology in the intensive care unit': a mobile image intensifier with TV monitoring. It was developed with the Edinburgh Royal Infirmary.

Fluoroscopy: 1923 and 1956

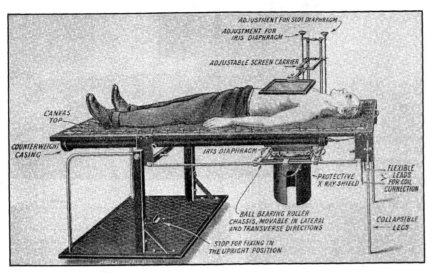

[10.18] Combined fluoroscopy stand and X-ray table as sold by the Medical Supply Association of London in 1923.

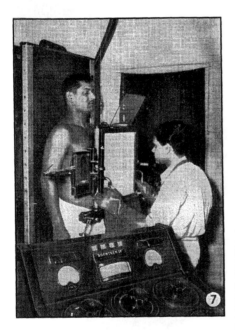

[10.20] Fluoroscopic equipment in use at the Port Commissioner's Hospital, Calcutta, from a photograph in the *Watson & Sons X-ray News* of Spring 1956. The generator is a Roentgen IV which was introduced in 1939 and was still commercially available 10 years later.

[10.19] Upright fluoroscopy stand manufactured by Newton and Wright of London[5] in the early 1920s and being described as suitable for taking radiographs of the thorax, and, when a rapid plate-changing device is added, also for stereoscopic work.

Fluoroscopy advertisement: Paris, 1897

[10.21] Arthur Radiguet was one of the early manufacturers of apparatus in Paris. After many months of experiments he suffered painful and chronic dermatitis of the hands and died in 1905 aged 55 years. His 1897 advertisements[13] states that his X-ray apparatus was awarded a Gold Medal and a Diploma of Honour at the Rouen exhibition of 1896.

A translation of his French advertisement text is: 'A special laboratory has been placed at the disposal of the doctors for using a Radioscope to instantly examine the interior of the human body. The Radiguet Company can provide in their laboratory or in a home, at a moderate price, the radiographs necessary for examination and verification of surgical operations.'

Barium enema and barium meal: 1988

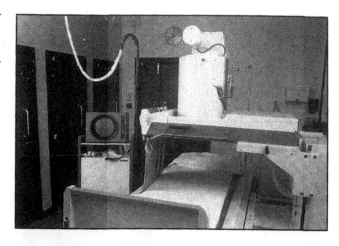

[10.22] An example of an X-ray room installation in 1988 (Westminster Hospital, London) for general fluoroscopic work where body movements can be studied, such as with barium meals for viewing of swallowing mechanisms. Many other diagnostic investigations are also possible using this equipment, such as of the GI tract, deep vein studies of the lower leg, myelography, renography, proctograms and crystography. The equipment shown is a Shimadzu model YSF 20050 with TV 900A monitor, YSF serial changer and diagnostic table type YSF-200-90.

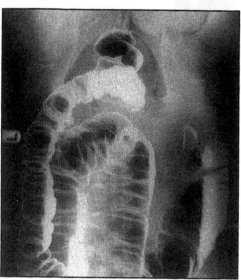

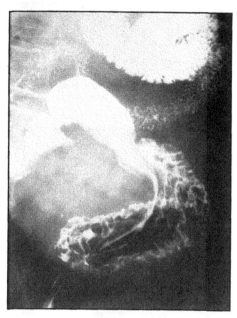

[10.23] Barium meal (left) and barium enema (right) studies using the equipment in [10.22].

Image processing: 1988

[10.24] Digiscan processed images, 1988. Negative zoom image (below) of a wrist. Chest radiography (right) demonstrating edge enhancement.

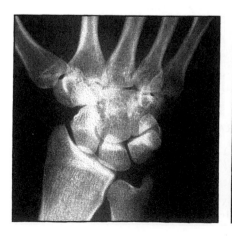

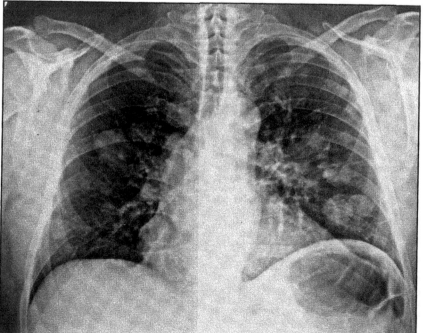

Diagnostic imaging modalities: 1993

[10.25]   As an example of diagnostic imaging possibilities for the 1990s, bladder cancer has been chosen to demonstrate this spectrum and is from a COMETT Educational Project[12] on interactive media. For bladder cancer, *urography* should be performed before cystoscopy and can provide information about bladder wall irregularities in the upper urinary tract. A *CT scan* is helpful in the detection of enlarged lymph nodes in the pelvis and the para-aortic region. The chest *radiograph* can evaluate distant spread to the lungs or the presence of a primary lung tumour: the radiograph shows bilateral metastases. *MR imaging* is very good for assessment of extravesical spread. An *ultrasound scan* can detect liver metastases and is mandatory in the case of an abnormal liver function test. A *bone scan* can detect bone metastases. (Courtesy: Prof. J. J. Battermann, University of Utrecht, and Prof. B. G. Szabó, University of Groningen.)

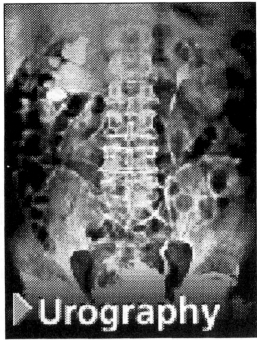

Urography

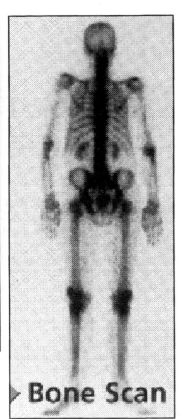

Bone Scan

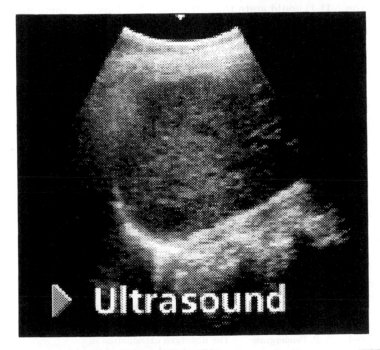

Ultrasound

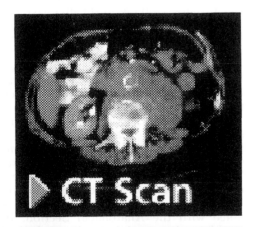

CT Scan

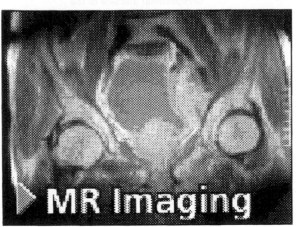

MR Imaging

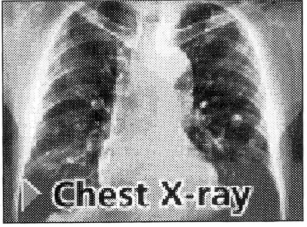

Chest X-ray

Chapter 11

Diagnostic Radiology: II

The first published record of an investigation of the gastrointestinal tract using X-rays was by H. Strauss[1] of Berlin in 1896 who used gelatin capsules (similar to those used to administer castor oil) filled with reduced iron oxide and bismuth subnitrate. This was not very successful and in the same year, J. C. Hemmeter[2] suggested the use of plumbic acetate in a gutta percha bag which could be swallowed, and later withdrawn.

Animals were used for experimental fluoroscopy (their use in experimental radiography is described in Chapter 9) and W. B. Cannon of Harvard published in the 1897 issue of the *American Journal of Physiology* a paper on 'Movements of the stomach by means of the roentgen rays'. His first experiments related to the movement of food in the oesophagus of a goose. The head and neck were surrounded by a tall pasteboard collar which allowed free movement of the head of the goose, without restricting its neck. For the stomach, his observations were made using a cat.

Contrast materials in these early years included bismuth-filled capsules covered with celluloid to prevent them from dissolving in the digestive tract, gas or air introduced into the colon, a spiral wire introduced into the oesophagus, an oesophageal sound filled with mercury or lead shot, and bismuth subnitrate in a suspension or mixed with food. However, by about 1910, barium sulphate had, at least in America, replaced bismuth preparations as a contrast material for the GI tract.

In a standard British textbook[3] of the 1920s it was stated that 'the examination of the stomach and intestinal canal has become one of the most important spheres of radiographic work' and related the following. 'At the meeting of the Electro-therapeutic Section of the British Medical Association, held in Liverpool in 1912, a joint discussion on the normal stomach took place with the Anatomical Section. It was almost unanimously agreed that radiography had modified the opinion held formerly as to the position of the normal stomach.'

There were considerable differences of opinion on the establishment of a standard opaque meal and a routine method of preparation and examination, as demonstrated by the following recommendations. (1) The standard meal should consist of either bread and milk, or porridge. (2) The total bulk of the meal should be about half a pint. (3) The meal should be mixed with 2 ounces of barium sulphate or 2 ounces of bismuth oxychloride. (4) The meal should be taken as nearly as possible on an empty stomach. (5) No aperient or other medicine should be taken within 36 hours of the first examination, and if the bowels are not opened naturally an enema should be given on the morning of the examination.

The various stages of a bismuth meal from immediately after ingestion to 48 hours later are shown in [11.1], whereas [11.2] shows variations in filling phenomena. Both are taken from the 1923 textbook[3] by Robert Knox of the Royal Cancer Hospital, London. Figure [11.3] is from a much earlier book, 1908, from the Medical School of Rouen, and shows the radiographic appearance of a tumour following a bismuth meal[4].

However, by 1945, the time of the 50th anniversary of the discovery of X-rays, James T. Case[5] writing in the *American Journal of Roentgenology* (this series of 50th anniversary papers in the *AJR* are given in Table 11.1) could state the following. 'No longer is there an effort to fill the stomach or bowel with a large opaque meal; now one seeks rather to visualise the mucosal pattern and to complete the study with a minimum of contrast material. Nor is the main effort carried out fluoroscopically in preference to the roentgenograms. The screen studies serve for observation of peristaltic activities and for gross details, but for the fine details which are needed for sure diagnosis, one must have recourse to roentgenograms, including the "spot films" made with compression when needed.'

However, bismuth and barium salts were not the only media used for contrast enhancement in fluoroscopy and radiography and, for example, as early as 1919,

Table 11.1. Series of papers published in the *American Journal of Roentgenology* Volume 54 (1945) on the 50th anniversary of the discovery of X-rays.

Author	Title of paper	Page
G. Failla	Protection against high energy Roentgen rays	553
W. D. Coolidge	Experiences with the Roentgen-ray tube	583
O. Glasser	Early American Roentgenograms	590
E. P. Pendergrass	The Roentgen examination in occupational diseases of the lungs	595
J. T. Case	50 years of Roentgen rays in gastroenterology	605
B. H. Nichols	Roentgen diagnosis in urologic disorders	626
B. R. Kirklin	Background and beginning of cholecystography	637
L. M. Davidoff	The development of modern neuroentgenology	640
R. Spillman	Early history of Roentgenology of the sinuses	643
H. Roesler	History of the Roentgen ray in the study of the heart	647
L. Reynolds	The history of the use of the Roentgen ray in warfare	649
A. D. de Lorimer and M. Dauer	The Army Roentgen-ray equipment problem	673
E. H. Quimby	The history of dosimetry in Roentgen therapy	688

workers in Sweden, as a result of injecting air into the intrathecal space as a treatment for tuberculosis of the spine, discovered the enhancement properties of air as a contrast media. There was also considerable interest in the 1920s in heavy elements, such as calcium, strontium, rubidium and bromine. It was, however, sodium iodide that emerged as the preferred inorganic salt for the cardiovascular and urinary system.

Thorium dioxide in colloidal suspension (Thorotrast) was used from 1929 until the mid-1950s/early 1960s in various countries with disastrous effects, because it was found to have carcinogenic properties. Its use as a diagnostic contrast media was then abandoned and extensive follow-up studies commenced on some of the populations exposed to Thorotrast. The main countries which were involved were Denmark (where, for example, 140 epileptics were given Thorotrast during cerebral angiography), Germany, Japan and Portugal. Hunter[6] in his textbook on occupational diseases states that it was used for hepatolenography, retrograde pylography and arteriography and that it was retained indefinitely in the liver, renal pelvis or subarachnoid space in these diagnostic procedures. The most recent review of the data is given in a 1989 British Institute of Radiology *Report*[7]. Taking an example from this report for the German Thorotrast study (most of the patients were injected intravascularly during the period 1937–1947) of 2326 Thorotrast exposure cases and 1890 controls, diseases with high excess mortality rates are liver cancer, liver cirrhosis, myeloid leukaemia and bone marrow failure.

It was the custom in the early 1920s to treat syphilis with high doses of sodium iodide. The urine in the bladder was observed to become radio-opaque during this treatment and further studies showed that the iodine content was responsible. However, sodium iodide was too toxic for satisfactory intravenous use and less toxic iodinated compounds had to be found. The early iodinated contrast agents were ionic and it was not until the early 1970s that the first non-ionic, low osmolar contrast agents, which are now preferred, became available. There are numerous radiological examinations involving contrast media and some examples from 1990 are given below.

Oesophagus. Barium swallow, designed to coat the lumen (and is therefore thicker than the barium meal) to study the swallowing mechanism to look for an obstruction.
Stomach. Barium meal which is very liquid, as this coats the stomach lining best, and gas (carbon dioxide) tablets to produce some stomach distension. When the outflow of barium is poor, this is an indicator of stenosis and of duodenal/gastric ulcers. This method provides a good picture of the stomach surface. In earlier years when a thick barium meal was used, the patient had to be turned round and ulcer craters could only be viewed in profile.
Large bowel. Double contrast techniques are also useful for this organ when diagnosing cancer or ulcerative colitis.
Small bowel. Barium follow-throughs and small bowel enemas are used. A tube is passed down the stomach and duodenum with this technique which is useful for the diagnosis of Crohn's disease.
Venograms. Iodinated contrast media is used for deep vein thrombosis of the leg.

For other earlier examples of the range of applications of contrast media, one should refer to the two-volume 1964 work *Classic descriptions in diagnostic radiology* edited by A. J. Bruwer[8]; and to the 1969 work *The Rays: a history of radiology in the United States & Canada* by Ruth and Edward Brecher[9], which was published due to the initiative of the American College of Radiology Foundation. The section titles in the Brechers' chapter on 'Progress in X-ray diagnosis' summarise the areas which were considered to have made significant progress by the late 1960s:

Air as a contrast medium
  Ventriculography
  Pneumoencephalography
Soluble iodine compounds
  Contrast media in the veins and arteries
  Gall bladder diagnosis
Non-absorbable contrast media: Lipiodol
  Lipiodol myelography
  Lipiodol bronchography
Contrast-medium catheterisation
Mammography

Image intensifiers [11.6] were mentioned in the previous chapter, [10.17], and their introduction into the armamentarium of diagnostic radiological equipment was a turning point in diagnosis. (In the Brechers' 1969 historical treatise[9], it is interesting to note that only two major headings were given for their chapter on 'Improvement in diagnostic equipment'. One was body-section radiography and the second was image intensification.) With such a device and a suitable contrast media, dramatic advances were made in the techniques for studying, for example, the GI tract and the swallowing mechanism in the oesophagus.

Modern image intensifier systems are coupled to TV monitors, video systems and 100 mm film cameras. Such technology has not only improved this field of diagnostic imaging out of all recognition from the pre-intensifier days, but has also significantly reduced the radiation exposure received by patients, radiographers and radiologists.

However, all equipment has to be checked routinely and the following is a summary of those fluoroscopic equipment parameters which require checks. The list is taken from the World Health Organisation 1982 guidelines on *Quality assurance in diagnostic radiology*.

Tube and generator performance
Automatic exposure rate control
Field size and distortion
Beam collimation and alignment
Conversion factor
Contrast ratio/veiling glare
Grey scale test object for TV monitor
Limiting resolution
Threshold contrast (noise)
Minimum visible detail versus contrast test object
Lag/afterglow.

This is very different to the earliest quality control method when the physician exposed his own hand as a test object, that hazardous practice being superseded by the use of a chiroscope or osteoscope [23.4, 23.5] to ensure both good image quality and good radiation protection.

In conclusion, this chapter closes with the same topic with which it began: contrast media. Ionic and non-ionic contrast media are now available, but

Bismuth meal: 1908, 1919 and 1923

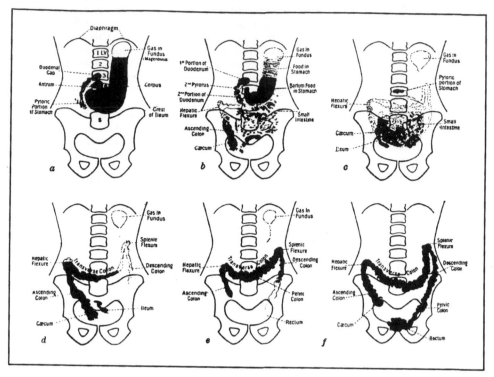

[11.1] Diagrams published in 1923 to illustrate the stages of a bismuth meal. (a) Stomach immediately after ingestion of the meal, (b) 2 hours after the meal was taken, (c) 4 hours after, (d) 6 hours after, (e) 24 hours after, (f) 48 hours after[3].

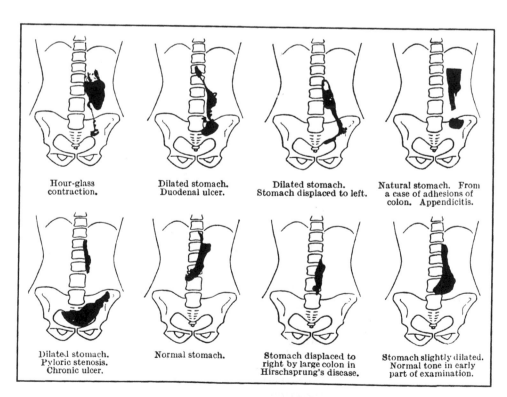

Hour-glass contraction.

Dilated stomach. Duodenal ulcer.

Dilated stomach. Stomach displaced to left.

Natural stomach. From a case of adhesions of colon. Appendicitis.

Dilated stomach. Pyloric stenosis. Chronic ulcer.

Normal stomach.

Stomach displaced to right by large colon in Hirschsprung's disease.

Stomach slightly dilated. Normal tone in early part of examination.

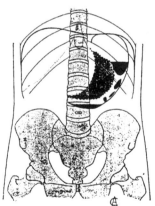

[11.3] Cancer of the lesser curvature ('la petite courbure') of the stomach: Rouen, 1908. The shaded area represents not only the tumour but also the spreading of the barium meal[4].

[11.2] Diagrams to illustrate the variations met with filling phenomena. They were described by Knox[3] as 'very instructive' since 'it is possible to foretell the type of stomach under examination, by an observation made in the first few minutes'.

non-ionic media are generally considered to be preferable to ionic media. Although they are more expensive, they are considered to be less toxic. For example, they are less viscous and therefore easier to inject, they do not attach to haemoglobin (i.e. do not displace oxygen) and are therefore good for patients with asthma, and they are useful if the patient is allergic to iodinated compounds. Some examples of contrast media in current use are Niopam (Iopamidol) from Merck Pharmaceuticals, Urografin from Schering AG, and Gastromiro, also from Merck. The following are some of the manufacturer-recommended applications, which together make a list of the major contrast media diagnostic investigations now possible at the end of the first century of radiology. Niopam: cerebral angiography, peripheral arteriography and venography, angiocardiography, left ventriculography and selective coronary arteriography, aortography, selective visceral angiography, CT enhancement, DSA and arthrography. Urografin: intravenous urography, cystography, hysterosalpingography. Gastromiro: GI tract CT. (Contrast media X-ray images are also given in Chapter 13 when arteriograms from the early years of radiology are shown, together with DSA images.)

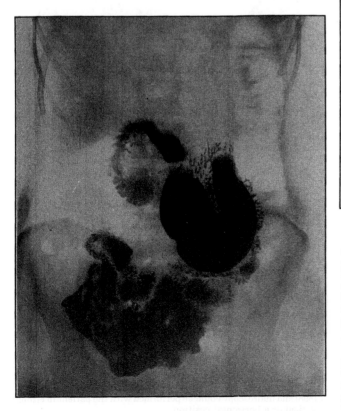

[11.4] Radiogram by Knox[3] of an opaque meal in stomach, duodenum and small intestine. Single flash exposure with an intensifying screen: initially from the *Journal of the Röntgen Society*, 1919.

Private radiology clinic: London, 1897

[11.5] In the immediate period following the discovery of X-rays, radiographs of patients were more often than not taken by physicists and engineers. The circular below was sent by C. E. S. Phillips to local medical men. He received numerous replies saying that the scheme was good, but that it could not even be expected to pay expenses. The scheme was therefore abandoned. (Courtesy: The Royal Marsden Hospital.)

> Castle House,
> Shooters' Hill,
> Kent.
>
> *November , 1897.*
>
> SIR,
>
> *The practical Medical utility of " X " rays has now become well established, and in view of the difficulty experienced I believe by local Medical men in obtaining for a moderate fee an X-ray examination of their patients, it has occurred to me to open a Consulting Room for this purpose in some central part of Woolwich.*
>
> *May I ask if you will be good enough to let me know your views as to the advisability of such a step, and whether you consider that such an institution would fulfil a really useful function in this town.*
>
> *The Scheme would embrace, in addition to a Consulting Room, the attendance when necessary of a qualified operator at the patient's own residence.*
>
> *I am, Sir,*
> *Very faithfully yours,*
>
> CHAS. E. S. PHILLIPS.
>
> *Dr.*

Phillips (1871–1945) was an active physicist a few years before the discovery of X-rays and was from about 1892 England's first hospital physicist, being based at the Royal Cancer Hospital, London. In 1898 he produced the first detailed radiological bibliography[10] and kept a notebook of his X-ray experiments with an album of 'Röntographs' including those he took on behalf of the police, so perhaps he was also the first Metropolitan Police radiologist (one of Phillips' Röntographs was of the knee of the Police Commissioner, Sir Charles Warren). Phillips obtained his first X-ray pictures on February 10th 1896 using a Lenard tube [3.2].

Under the heading 'History of the Tube' he recorded 'At first when I excited the tube it glowed with a blueish colour with flickering whiteish flames here and there. This gave no Rönto-effects with 45 minutes exposure. The tube then turned greenish after two days pretty well continual excitation and Rönto-effects were obtained with one hour exposure. The tube became bathed in green flames internally licking the glass after another day or two and then the best effects were obtained. This best condition lasted about a week and then the resistance of the tube began to increase.'

Image intensifiers: 1962, 1969 and 1975

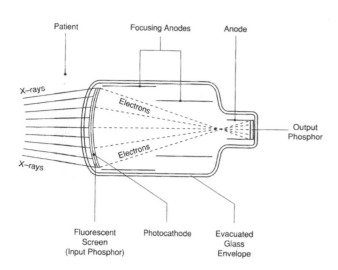

[11.8] The Kompact 6 image intensifier of 1969. Technical developments had significantly reduced the size of the instrument from that shown in [11.7].

[11.6] Schematic diagram of an image intensifier tube. The X-ray beam passes through the patient and the outer casing (not shown) and glass envelope of the image intensifier, and impinges on the fluorescent screen (the input phosphor). Light scintillations are emitted which fall onto the photocathode. Electrons are emitted in proportion to the light photons (which in turn are proportional to the incident X-ray photons) and they are then accelerated and focused (a voltage of 25 kV between photocathode and anode) onto the output phosphor. Compared to the use of only a fluorescent screen, the image intensifier increases the brightness of the image by a factor of some 1000. The output phosphor emits light in proportion to the electrons it absorbs.

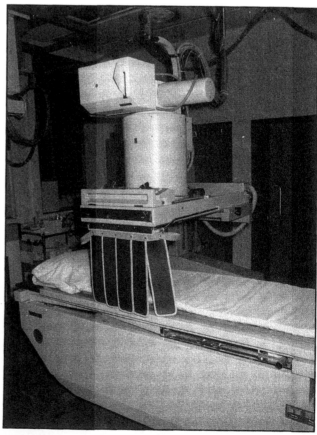

[11.7] The early image intensifiers were bulky as seen from this 12″ Old Delft Cinelix image intensifier with a cine camera. This photograph was published in the *Watson & Sons X-ray News* of Spring 1962 which issue also described the early use of image intensifiers in association with TV cameras. The applications were listed as: heart diagnosis, chest diagnosis, pharynx and oesophagus, stomach and duodenum, small bowel, large bowel, retrograde pyelography and cystography, hysterosalpingography, aortography, and central nervous system (myelography, ventriculography, localisation prior to pallidectomy, angiography). Three sizes of intensifier were available at the time, 7″, 9″ and 12″. It was also reported that extensive use had been made of cinematography for GI tract examinations and during angiocardiography, with excellent results. Those for the stomach were taken at 16 frames/second, 100 kV, 10 mA and an FFD of 75 cm. The dose rate was stated to be 10 roentgens/minute. For angiocardiography, up to 30 frames/second were used and the film was run for 12–15 seconds.

[11.9] Fluoroscopy equipment used in a busy general hospital (St. Stephen's Hospital, Chelsea, London) throughout the 1970s and 1980s. The protective lead rubber strips are seen in the centre of the photograph. The table is a tilting table (see [10.4] for a design of 1898) and fluoroscopy could be performed with the patient in either a vertical or a horizontal position.

Chapter 12

Diagnostic Radiology: III

For the most part, the initial radiographic (as distinct from the fluoroscopic) applications of X-rays were devoted to fractures, dislocations and congenital abnormalities of bony structures, to the thorax, to foreign body localisation of items such as bullets (see Figure [8.11]) and swallowed coins, and calculi. The latter, kidney stones and bladder stones, were a logical follow-on to the use of X-rays in foreign body localisation. Figures [12.1] and [12.2] show radiographs of kidney stones taken in 1899 and 1900.

Figure [12.3] shows a patient in position for a kidney radiograph almost a quarter of a century after [12.1] and [12.2] were taken. The shoulders and the lower extremities are raised to obtain a closer positioning of the spine to the X-ray plate[1]. A certain amount of compression is required and [12.5] shows the compression apparatus of Albers-Schönberg[2] which was one of the more popular designs.

An interesting anecdote of this period concerning the kidney was recorded by Cochrane Shanks[3] in 1973 in a paper describing radiology in the 1920s. 'In 1929 Lichtenberg and Swick introduced uroselection and laid the foundations of modern intravenous urography. I will always remember my first intravenous urography—still known as IVP instead of IVU. In the film I could just make out the feint shadow of a right hydronephrotic kidney, but no shadow on the left. The surgeon, ignoring my one-sided report, decided to remove the right (and only) kidney, and we saw the patient gradually sink into uraemic coma and die a fortnight later.'

An example of an early radiograph of a bladder stone is given in [12.4] taken by Thurstan Holland who also described the advantages of compression, using a pressure tube apparatus, when giving his Röntgen Society Presidential Lecture in 1904. He referred to 'exposure through a pressure tube apparatus such as that suggested by Albers-Schönberg and stated the following. 'I have used this method latterly with a modification of the apparatus designed by myself. Two advantages are obtained by it. Firstly, by cutting off all but a small direct stream of rays, and by cutting off secondary rays, a much sharper picture is obtained. Secondly, by the pressure of the tube I think that the respiratory movements are modified on the side being examined to such an extent as to prevent the up and down movement of the kidney, and thus permitting longer exposures. I would strongly recommend it to all radiographers.'

Turning now to gall stones, prior to 1924, when Graham and Cole published in the *Journal of the American Medical Association* their method of gall-bladder diagnosis using intravenous injection of tetra-bromophenolphthalein, gallbladder studies were by ordinary radiograph, although these were supplemented in some cases by the injection of air or of opaque material into the stomach, duodenum and colon. Kirklin[4] described Graham's 1931 reminiscences in his paper 'Background and beginning of cholecystectomy'.

'One evening in the winter of 1922, the idea occurred to Graham that since Abel and Rowntree in 1909–10 had demonstrated the fact that chlorinated phenol-phthaleins are excreted almost entirely through the bile, it might be possible, by substituting for the chlorine atoms other atoms which would be opaque to the X-ray, to obtain a shadow of the gallbladder. After several months Graham obtained some of the free acid of tetraiodophenolphthalein. To render it more soluble it was converted into the sodium salt and in July 1923 it was injected intravenously into six dogs by Graham's associate, Warren Cole.

'In only one dog was a shadow of the gallbladder obtained and that shadow was faint. On inquiry it was learned that by an oversight this dog had been required to fast, while the other five had been fed as instructed. From the standpoint of the future development of cholecystography, said Graham, we often feel grateful to that one dog which cast a shadow, probably because he was accidentally given no food. If we had failed to get a shadow in all these animals we probably would have abandoned the whole idea as a fruitless one. It is curious how fragile a thread the destiny of some events hangs.'

However, the technique was not without its difficulties as recorded by Cochrane Shanks[3]. 'Intravenous chole-cystectomy, using tetraiodophenolphthalein, a blue dye, gave reasonable pictures of the gall bladder, but was intensely irritating if any escaped outside the vein. I saw two cases in which a minimum escaped and in both the ante-cubital fossa sloughed and had to be skin grafted.'

Figure [12.7] is an advertisement for the Potter–Bucky diaphragm, designed to obtain better radiographs when dealing with the heavier parts of the body such as the spine, kidney and pelvic regions, particularly when the patient is obese[1,5]. The principle for the grid was patented in Berlin in 1913 by Gustav Bucky (who also had a joint United States patent with Albert Einstein on the automatic adjustment of a photographic lens for motion picture cameras by means of a photoelectric cell).

Kidney stones: 1899 and 1900

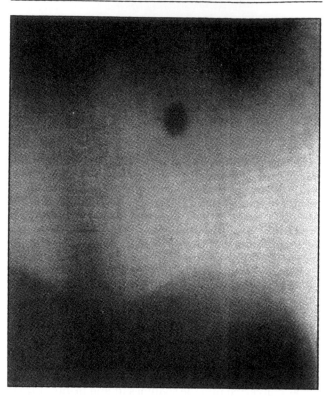

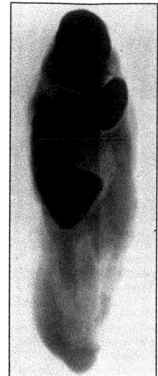

[12.2]   From the January 1900 issue of the *Archives of the Roentgen Ray* and taken at St. Bartholomew's Hospital, London. The stone was radiographed in situ and (as shown in this Figure) when removed from the body. It was stated that 'the case is an interesting one owing to the fact that the diagnosis lay between a renal calculus and a tuberculous kidney. The Roentgen rays cleared up the doubt.'

[12.1]   From the November 1899 issue of the *Archives of the Roentgen Ray* and taken at the London Hospital. The final paragraph of the case history reads as follows. 'This case is of importance because all the ordinary signs of renal calculus were absent; and although the diagnosis was made and the operation arranged for before the radiograph was taken, the evidence afforded by the Röntgen rays changed what would have been an exploratory erasure into an operation definitely undertaken for the removal of a calculus the exact size and position of which were known for certain'.

Bladder stone: 1904

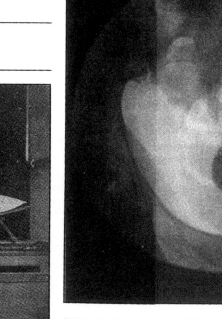

Radiographic positioning, kidney: 1923

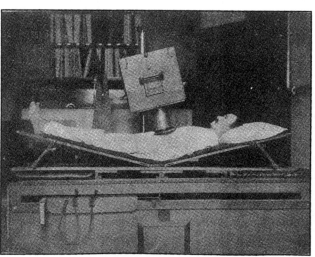

[12.3]   Radiography of the kidney in 1923.

[12.4]   Radiograph of a bladder stone taken through a diaphragm compressor by Thurstan Holland in Liverpool and published in the *Journal of the Röntgen Society*, 1904.

Compression apparatus: 1903 and 1914

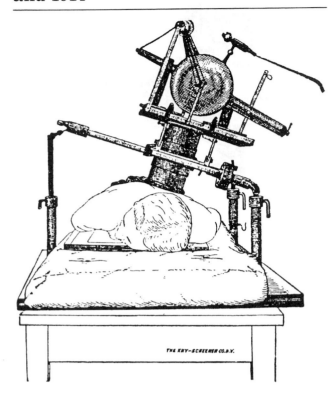

[12.5] The compression apparatus of Albers-Schönberg, 1903.

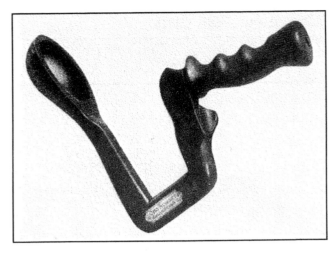

[12.6] A very simple device for compression, known as the Holz-knecht Spoon, was devised to enable the radiologist to keep his hand outside the X-ray beam. This model was shown at the 1914 Radiological Congress in Berlin.

Potter–Bucky grid: 1920s

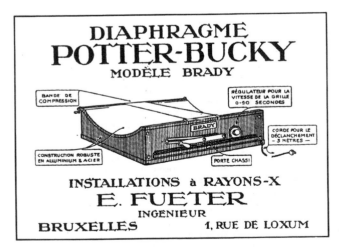

[12.7] Advertisement[6] for a Potter–Bucky diaphragm, for obtaining better radiographs of the heavier parts of the body. The Bucky principle depends on the selective action of a number of lead strips set radially to the X-ray source, forming what is generally known as a 'grid'. It served the purposes of cutting out the secondary and scattered radiation, but the grid image was recorded on the emulsion and disfigured the radiograph. This problem was overcome by Hollis Potter of Chicago who eliminated the grid shadows by ingeniously giving the grid a constant motion during the radiation exposure.

Surgery and handwriting: 1896

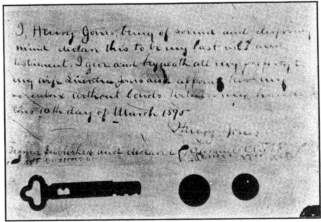

[12.8] The conclusions drawn by William Morton in his 1896 textbook[7] *The X-ray or Photography of the Invisible and its Value in Surgery* were as follows, accompanying the last (above) of his 91 illustrations:

'It is difficult to say what final uses the simple fact discovered by Roentgen that different substances are more or less opaque to the X-ray, according to their density, may be put. We have laid stress mainly upon its uses in surgery but there are several other applications of great interest.

'Flaws and weldings may be discovered in metals; investigation as to the amount of metal in different ores may be made, and it will interest dealers in diamonds to know that the X-ray positively distinguishes between the true diamond and the paste imitation.

'Not only can handwriting in sealed envelopes be photographed with the aid of the X-ray (which purports to be a man's will sealed in an envelope as an experiment) but in a negative not here produced an X-ray photograph of handwriting inside the plateholder was taken even after the X-ray has passed through a human skull.'

Couches: 1898–1918

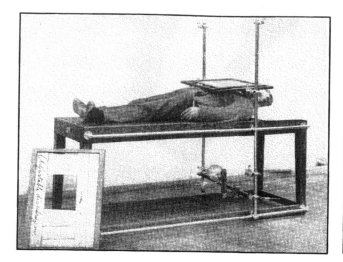

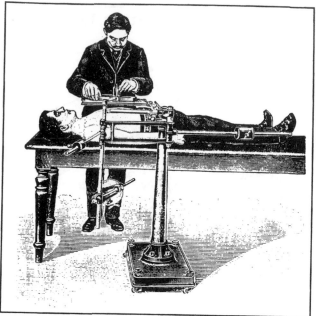

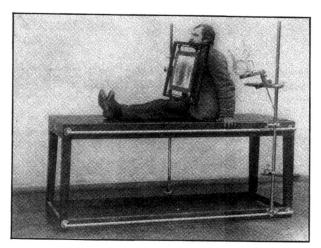

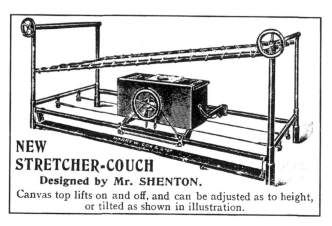

NEW
STRETCHER-COUCH
Designed by Mr. SHENTON.
Canvas top lifts on and off, and can be adjusted as to height,
or tilted as shown in illustration.

[12.9] Two examples of pre-1900 designs of diagnostic couches are seen in [3.32] for endioscopy of the vagina and rectum and in [10.4] for fluoroscopy. The earliest couches were, however, more often than not merely plain tables as in the illustration (top right) of the free-standing orthodiagraph designed by Emil Grunmach (1848–1919).

Other very simple designs such as that of Kassabian[8] incorporated horizontal rails and moveable vertical rods to which the X-ray tube, plate holder of fluoroscope could be attached. Kassabian's 1907 caption for his couch (top left) is: 'Table for skiagraphing the heart and lungs. An adjustable lead cover diaphragm rests against the foot of the table.' The same couch is shown (centre left) with the patient in a sitting position for stereo-radiography.

The stretcher-couch (centre right), having an adjustable canvas top, was designed by a Mr Shenton, the Senior Radiographer at Guy's Hospital, London, and advertised by Harry W. Cox & Co. Ltd in the July 1911 issue of the *Journal of the Röntgen Society*. By that time, however, more sophisticated designs incorporating lead shielding were becoming available, such as that of Ironside Bruce (bottom left), as advertised in the same journal in October 1912 by Watson & Sons, who offered a similar Mackenzie Davidson design in their 1918 catalogue.

Chapter 13

Diagnostic Radiology: IV

Angiography had its inception in the researches of Egaz Moniz in Lisbon[1], with, for example, his papers on 'Arterial encephalography: importance in the localisation of cerebral tumours' in 1927 and 'Angiopneumonography' in 1931, although there were other seminal papers of the period, such as 'Catheterisation of the right heart' by Werner Forssmann in 1929 and 'Experimental cardiac ventriculography' by Henri Reboul and Maurice Racine in 1933.

However, well before Moniz injected 25% solutions of sodium iodide into the internal carotid artery, and later into the common carotid, interest had arisen in arteriograms using cadavers. These produced some remarkable images of the venous system. Nevertheless, these early radiographic images were of limited value as indicated by Tousey[2] in 1915.

Tousey refers to 'radiographs' showing the injected blood vessels of normal human subjects made for him by a Mr Bush of the Metropolitan Hospital in New York. 'The Röntgen ray may be used in studying post mortem anatomy without any limitations as to length and strength of exposure, and with the advantage derived from absolute immobility and the injection of blood vessels and other hollow organs with opaque substances or with transparent gases. Pictures produced in this way show the blood vessels in their natural relations, undisturbed by dissection.'

There were many experiments of this nature in the early days and one of the first was by Franz Exner who injected the hand of a cadaver with Teichmann's mixture (lime, cinnebar and petroleum) via the brachial artery. He showed the results to the Chemical-Physical Society of Vienna on January 18th 1896. This presentation was followed up in Vienna by a physicist, Eduard Haschek, and his physician colleague, Otto Lindenthal, some six days later. Their experiment was described in the following terms[3].

'In an amiable manner Herr Doktor Tandler, a Viennese anatomist, placed at our disposal the hand of a cadaver. We injected through the arteria brachialis, Teichmann's mass, which consists essentially of chalk. We also wrapped copper wire around the index finger to compare the degree of impenetrability of metal, bone and chalk. The exposure was 57 minutes.'

Other reports of the time included the use of sulphate of lime and of gypsum and, in *The Lancet* of December 19th 1896, N. Raw from Manchester published a paper on 'Skiagraphing the arteries' using calcium sulphate. Some early examples are shown in [13.1–13.4].

Figure [13.6] is an early radiograph of a deformed hand and [13.7] is of a typical laboratory around the turn of the century where the early arteriograms, using cadavers, would have been performed. Somewhat different from that X-ray equipment is the versatile

Arteriogram, injection of four pounds of mercury: London, 1899

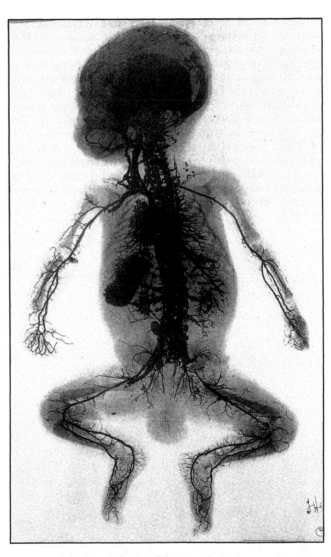

[13.1]   This arteriogram was taken in 1899 at St. Thomas' Hospital, London, and published in the *Archives of the Roentgen Ray* with the title 'An injected infant'. The contrast media was four pounds of mercury into the external circumflex branch of the femoral artery. The exposure was 20 minutes.

angiographic unit of the early 1990s, a Siemen's Multiskop [13.8] which is suitable for digital acquisition in conjunction with the Siemen's Polytron unit. Alternatively, it can be used in an analogue mode with 100 mm film. Figure [13.9] can be compared with [13.2], there being some 85 years difference between these arteriograms of hands.

Figure [13.12] is the first published 'skiagraph of a head' and a comment to the caption was that 'nasal bones appear like eyelashes'. It appeared in 1896[4] and was taken by A. W. Goodspeed of the University of Pennsylvania in Philadelphia, who in 1890 had taken the first documented radiograph [5.3].

Early radiographs of the heads of mummies [13.13] were often used for demonstration purposes in museums and in the early literature a series of radiographs were published of bullets in the head, some of deceased persons and some of skulls as test objects [13.14].

Figure [13.15] is not a skull radiograph but is a very relevant image of the 1990s with which to conclude this chapter. With the advent of magnetic resonance (MR) imaging, this has replaced in part some of the previous applications of X-rays. MR is still an evolving field as more and more information on clinical applications is being gathered. However, it is quite clear that there are some useful MR applications in neurology, myelography and in orthopaedic disorders. For example, MR has been demonstrated to have value in the diagnosis of brain infarction, white matter disease including evaluation of multiple sclerosis and cerebral manifestations of AIDS, posterior fossa and cranio-cervical junction lesions, changes consistent with Alzheimer's disease, and it has a well defined role in orbital imaging.

Arteriograms from Australia: 1904

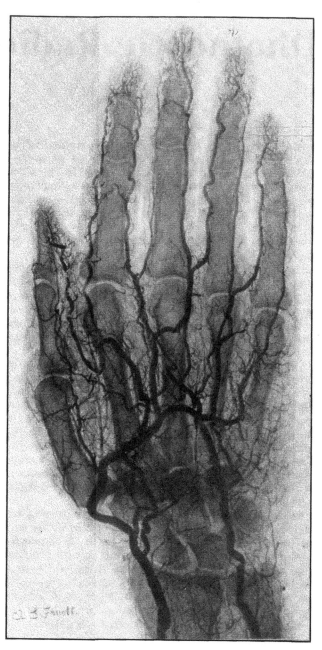

[13.2] In the February 1904 issue of the *Archives of the Roentgen Ray* a 'radiograph of shoulder showing injected arteries' was published. This was taken by Mr Alfred G. Fryatt FRMS of Melbourne, Australia and the *Archives* stated the following. 'This is one of a number of beautiful stereoscopic pictures sent to us by Mr Fryatt. We shall later reproduce some of these in duplicate, and trust to be in a position to inform our readers how these beautiful pictures were produced.' However, the *Archives* never did inform their readers as to the technique, so we can only guess. Nevertheless, an arteriogram similar to that above, made using mercury, was included in the 1898 edition[5] of *Practical Radiography* by Isenthal and Ward.

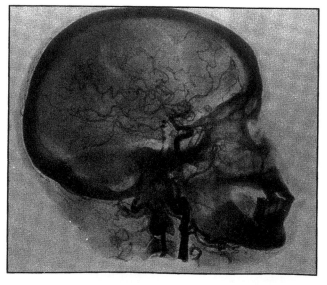

[13.3] 'Radiograph of half a human skull (arteries injected)' by A. G. Fryatt (1904).

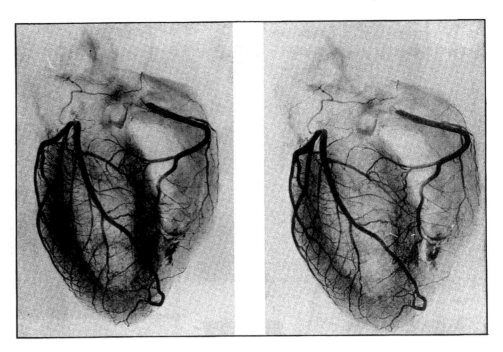

[13.4] 'Stereoscopic plates of heart showing injected coronary arteries' by A. G. Fryatt (1904).

Cardiac image: Glasgow, 1896

[13.5] For comparison with Figure [13.4], this is a 'radiogram of a human heart in situ', by John McIntyre of Glasgow. It was published as the frontispiece in the first edition[6] of *Practical Radiography* by Snowden Ward 1896. Radiography was so new at this time that this cardiac image was published upside down!

Snowden Ward and a double thumb: 1896

Southport Social Photographic Club.

The New Light

AND THE

NEW PHOTOGRAPHY.

LECTURE BY

H. SNOWDEN WARD, Esq., F.R.P.S.,

Assisted by Mr. E. A. ROBINS.

TEMPERANCE INSTITUTE,

TUESDAY, MARCH 24, 1896.

Doors open at 7-30. Commence at 8.

SECOND SEATS, 1s.

Geo. Cross, *Hon. Sec.*,
15, Cambridge Arcade

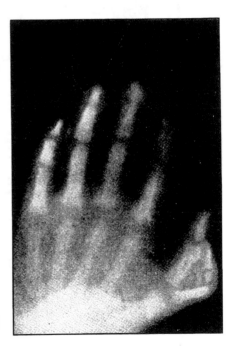

[13.6] Snowden Ward[5,6] was one of the most prolific public speakers on X-rays in the period 1896–1897 in England, touring many cities to give demonstrations [1.22]. One such lecture was in Southport, Lancashire on March 24th 1896 and was reported in the local newspaper, the *Southport Visitor*. The audience was 'composed largely of medical gentlemen, professional and amateur photographers, scientific students and hospital nurses'. The history of the radiogram of the hand with a double thumb was given as follows. 'One of the most successful radiograms of a surgical case had been made that afternoon. The patient was a little boy with a double thumb and the exposure was less than one minute. The hand was moved after 15 seconds and from that fact the image was slightly blurred'. Ward conjectured that the vacuum tube was operating for only about one-quarter or one-third the exposure time. During the evening, he tried to take another radiogram of the boy's hand, but the experiment failed.

A. W. Isenthal's London X-ray laboratory: 1899

[13.7] A. W. Isenthal was the director of the London firm of Isenthal, Potzler & Company which specialized in electro-medical instruments well before 1895. The photograph (right) is of his 'Röntgen' laboratory in 1899[7] and is typical of the period.

Isenthal joined with Snowden Ward in the second (1898) edition of *Practical Radiography*[5], the frontispiece of which (below right) was described as 'Radiographic outfit arranged for a demonstration by A. W. Isenthal before the Royal Photographic Society, February 22nd 1897. The authors as operator and subject.'

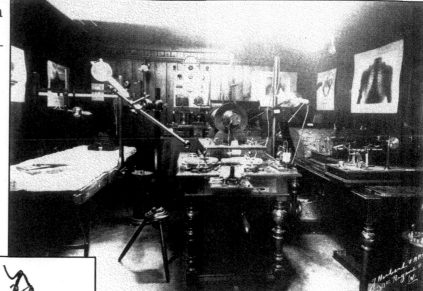

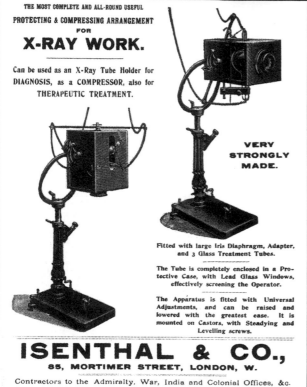

The frontispiece of their third (1901) edition[8] is the radiograph of Röntgen's rifle [17.1] which was presented to Isenthal when he interviewed Röntgen in 1898. That interview is recorded in Chapter 1, including Röntgen's cautionary comments about the harmful effects of X-rays. Radiation protection was evidently not a concern during the 1897 demonstration (below) but by 1906, when advertising (left) in the *Journal of the Röntgen Society*, Isenthal had introduced X-ray tube protection incorporating lead shielding and a lead glass window.

Angiography: 1980s–1990s

[13.8] Siemens Multiskop angiographic unit, 1990. The image intensifiers are 40 cm or 33 cm and the C-arm allows AP/PA, lateral, left and right anterior oblique and caudal/cranial views to be used. Because the ceiling to floor track allows motorised movements of the C-arm and stepping motion for peripheral angiography, the patient can remain stationary throughout the procedures, which is important during interventional radiology.

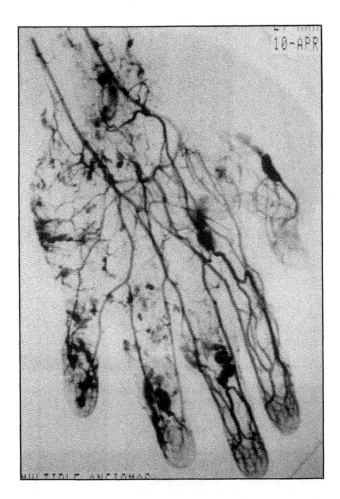

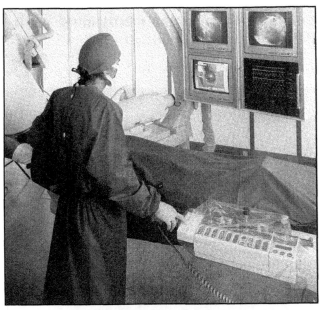

[13.10] Siemens Digitron 3 VACI system for vascular 512 × 512 bit digital acquisition for coronary angiography. The features include up to 50 frames/second, remote control operation from the side of the table, up to 17,000 images on line, bi-plane operation possible and the equivalent of five reels of 60 m cine film can be archived digitally on one video cassette in digital form using a Gigacassette.

[13.9] Modern angiogram of the left hand of a 15 year old female showing multiple angiomas. This was taken using a Siemens Digitron 3 system for vascular and quantitative assessment.

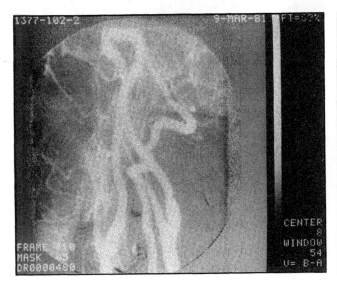

[13.11] Digital subtraction (DSA) images, 1984, of carotid arteries, showing stenosis of the left internal carotid after an intravenous injection to the superior vena cava. On the left is the contrast–mask image and, right, the mask–contrast image. (Courtesy: Technicare Imaging Ltd.)

Skulls: 1896 X-rays compared with 1990 magnetic resonance

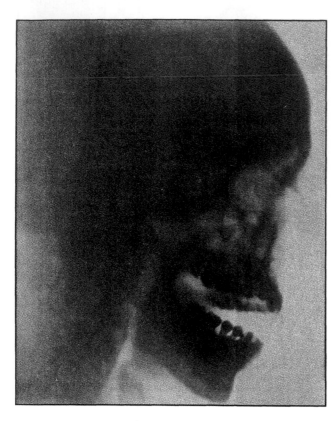

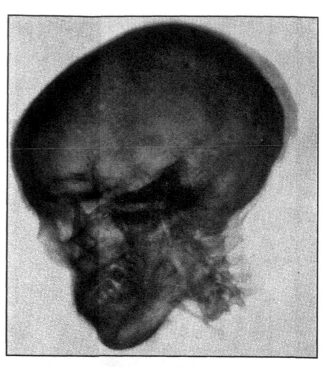

[13.13] 1897 radiograph of the skull of an Egyptian mummy[5], taken by T. Brinkmann of Frankfurt-on-Maine.

[13.12] 1896 'skiagraph of a head' by Arthur W. Goodspeed [5.3] of Philadelphia, first published in the *International Medical Magazine* of June 1986 and reproduced later the same year in the textbook of E. P. Thompson[4] who noted that 'Nasal bones appear like eyelashes and the cervical vertebrae are distinguishable in the original, but barely so in the half-tone. Fillings are located.'

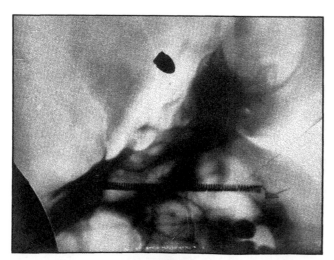

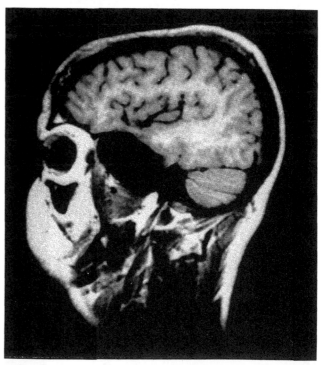

[13.15] Magnetic resonance (MR) sagittal image of a head using a General Electric Medical Systems Sigma MR scanner, 1990.

[13.14] An X-ray of a bullet in the base of the brain, May 2nd 1896. This is thought to be a radiograph of a 'simulated bullet within a skull' taken by Sir Arthur Schuster to determine experimentally the required exposure for a radiograph of Elizabeth Ann Hartley of Nelson, Lancashire, who had been shot in the head by her husband, Hargreaves Hartley, on April 23rd 1896.

Chapter 14

Diagnostic Radiology: V

Dental radiography was one of the very early medical applications of X-rays with William Morton reading a paper on the subject to the New York Odontological Society on April 24th 1896. One of Morton's radiographs [14.1] 'showed the roots, fillings and pulp chambers and location of disease at the roots and in the bones.' It was classified as of a 'non-living object' and Morton sold the life-size reproduction for 60 cents a copy.

In a short chapter on 'The X-ray in dentistry' in his 1896 textbook[1] he has this to say of the new dental diagnostic tool. 'The density of teeth is greater than that of bone, and for that reason, pictures of the living teeth may be taken by the X-ray even of wandering fang or root, however deeply imbedded in its socket. Also children's teeth may be photographed before they have escaped from the gums, and the extent and area and location of metallic fillings may be sharply delineated, even though concealed from outer view. The lost end of a broken drill may be found, and, what is more interesting, the fact that even the central cavity of the tooth may be outlined, so that diseases within the tooth may be detected.'

Six years later in 1902, Francis Williams of Boston was dwelling on the difficulties[2]. 'For the successful use of X-rays in dentistry, sharp definition in negatives is necessary, and differentiation is required between tissues that do not suffer very much in the obstruction they offer to the passage of the X-rays. The roots of the teeth are only a little less permeable to the rays than the surrounding bone, and therefore it is difficult to get a clear picture of their ends.'

However, Williams recommended[2] a particular design for a dental X-ray tube. 'For good differentiation between tissues that are not very different in permeability the tube must have a low resistance.' Of the methods then available for lowering the resistance of a gas tube when it had become too high, Williams recommended heating the tube [5.8]. Figure [14.5] shows Williams's dental apparatus of 1902 and [14.2] one of his early dental radiographs. This can be compared with [14.3] to show the improvements achieved some 70 years later, both in image quality and radiographic technique.

It was very early recognised that glass plates were less useful than film in dental radiography, and William Rollins of Boston devised in 1896 a method which consisted of placing small films, 3.6 cm × 4.2 cm, inside the mouth [14.2]. These films had been previously enclosed in a thin bag of black soft rubber, or waterproof paper, to exclude light and moisture. The X-ray tube was placed at a distance of 28 cm from and opposite to the same jaw as the films. Several films were used with each separated by a thin sheet of foil so that each negative received a different exposure.

To bring dental radiography to a more recent age, [14.4] is an example[3] of a panoramic radiograph, taken with a specialist dental X-ray unit in 1985.

Dental radiographs: 1896, 1902, 1969 and 1985

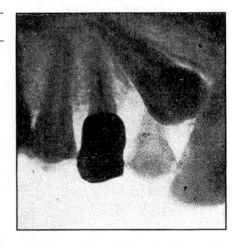

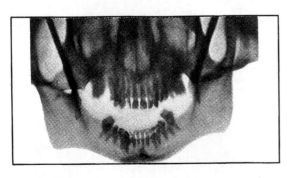

[14.1]   Dental radiograph of 1896 by William Morton of New York[1].

[14.2]   Example of a 1902 dental radiograph[2] from the Boston City Hospital. It shows 'a case of delayed dentition. The girl of 17 still has her temporary cuspid, with no indication that the permanent has ever formed. The radiograph shows it to be imbedded against the central at an angle of about 45°. It is now being placed in its proper position. The gold crown is conspicuously shown on the next tooth.'

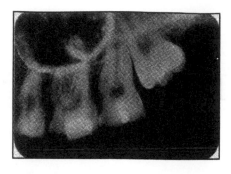

[14.3] Radiographic exposure of approximately 0.5 seconds of the maxillary molar region, 1969[4].

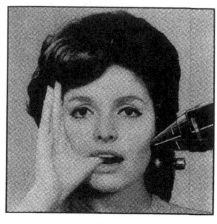

[14.4] Radiograph obtained in 1985 with a Siemens Orthopantomograph 10, a specially designed X-ray unit for panoramic dental radiographs[3].

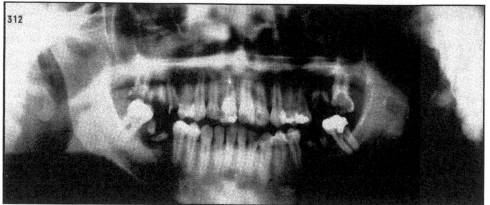

Dental X-ray apparatus: 1902, 1915 and 1920

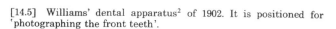

[14.5] Williams' dental apparatus[2] of 1902. It is positioned for 'photographing the front teeth'.

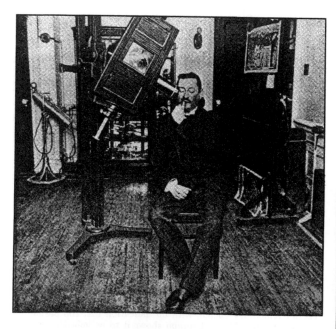

[14.6] 'Radiography[5] in 1915 of the upper teeth upon a small film held against the teeth and the roof of the mouth. Ripperget shield.' This examination of the teeth and maxilla was described as 'one of the most important applications of the X-ray'.

[14.7] Dental X-ray unit advertised in 1920 in the *Journal of the Röntgen Society*.

Exposure chart: 1921

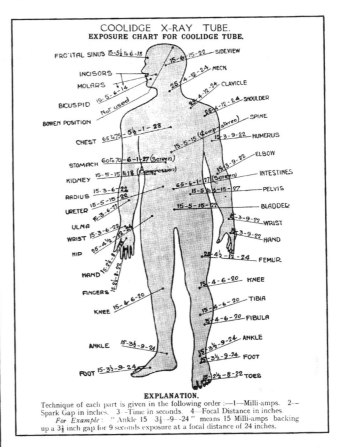

[14.8] Exposure chart for Coolidge X-ray tubes, circa 1921, from the Watson & Sons (Electro-Medical) Ltd 5th edition catalogue. For dental work the recommendations were 15 mA, 5 inch spark gap, 6 inch exposure and a focal distance of 14 inch (36 cm).

Sea shells: London, 1904

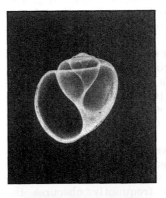

[14.9] In the first decade after the discovery of X-rays, shells were sometimes used as test objects to illustrate how radiology could be used to distinguish intricate detail. These images are from the 1904 *Archives of the Roentgen Ray*.

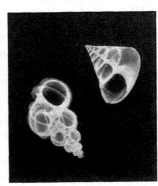
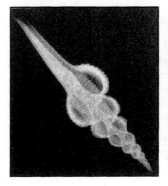
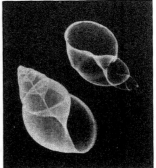

Chapter 15

Diagnostic Radiology: VI

Tomography, as a special type of radiography, originated shortly after the 1914–18 World War and the name is derived from the Greek 'tomos', meaning a section, and was proposed by Grossman[1] of Berlin in 1935 for the apparatus he designed for lung tomography [15.1, 15.2]. In the United States, planigraphy ('a method of roentgenographic projection of plane sections of solid objects') as it was termed by Andrews, was reviewed in detail between 1936 and 1947[2-4]. The early major applications were for chest tomography, but other sites were also considered by 1940, such as the larynx[5], the external auditory canal[6] and the spine in problems which are difficult to solve by conventional radiography. Examples are Pott's disease with extreme kyphosis and rib-crowding when many of the vertebral bodies are collapsed[7].

Bricker[8] in Bruwer's *Classic Descriptions in Diagnostic Roentgenology* describes the basic principle of linear tomography in the following terms. 'The X-ray tube and the film move synchronously in opposite directions during the exposure. They are connected by a bar permitting this about an adjustable fulcrum in the plane of the desired section. The images of all points in a plane through the fulcrum and parallel to the film are clearly recorded as these occupy unchanging positions on the moving film. The images of all other points are blurred by the movement, and the degree of blurring increases with the distance from the selected plane. A thinner section is obtained with a longer tube movement or a shorter focus–film distance, and conversely. The central ray may move perpendicular to the plane of the section throughout or be continuously directed to the centre of this plane. Numerous modifications have been explored, particularly possible movements of the tube focus which includes rectilinear (straight-line or arcuate), circular, spiral and sinusoidal.'

Figure [15.3] shows an example of a linear tomogram taken in 1988 for an intravenous urography study with an IGE Linotome unit. As well as being used for IVU studies where bowel gas frequently obscures the kidneys, the Linotome is also employed for simple linear tomography of other organs such as the chest and larynx.

However, when the word 'tomography' is now used, the connotation is computed transmission tomography, computerised axial tomography (CAT), or more often, just computerised tomography (CT) and the production of CT scans. The start of CT, which led to the 1979 Nobel Prize for Medicine being jointly awarded to Godfrey Hounsfield and A. M. Cormack, is well summarised by Webb[9]. 'The announcement[10] of a machine used to perform X-ray computed tomography in a clinical environment, by Hounsfield at the 1972 British Institute of Radiology annual conference, has been described as the greatest step forward in radiology since Röntgen's discovery.'

The classic papers by Hounsfield and his clinical colleague Ambrose[11,12], of the Atkinson Morley Hospital in Wimbledon, left the scientific community in no doubt as to the importance of this discovery. However, it was made quite clear by Hounsfield that he never claimed to have 'invented CT' and the question as to who really did invent CT has been much debated since. According to Webb 'The original concept is credited to Radon in 1917[13], whilst Oldendorf in 1961[14] is often quoted as having published the first laboratory X-ray CT images of a "head" phantom. What Oldendorf actually did was to rotate a head phantom (comprising a bed of nails) on a gramophone turntable and provide simultaneous translation by having an HO-gauge railway track on the turntable and the phantom on a flat truck, which was pulled slowly through a beam of X-rays falling on a detector. He showed how the internal structures in the phantom gave rise under such conditions to characteristic signals in the projections as the centre of rotation traversed the phantom relative to the fixed beam and detector. He was well aware of the medical implications of his experiment, but he did not actually generate a CT image.'

Webb[9] also noted that a CT scanner was built in the Ukraine in 1958 and Korenblyum and colleagues published the mathematics of reconstruction from projections, together with experimental details. They wrote 'At the present time at Kiev Polytechnic Institute, we are constructing the first experimental apparatus for getting X-ray images of thin sections.' Their analogue reconstruction method was based on a TV detector and a fan-beam source of X-rays. Others were also working in this field of reconstruction tomography and in 1963, Cormack[15] in particular was performing laboratory experiments in CT which would lead to his sharing the Nobel Prize with Hounsfield.

Figure [15.4] is the experimental CT scanner used by Hounsfield at the Central Research Laboratories of EMI Limited at Hayes, United Kingdom. It was the forerunner of the first EMI-Scanner which was commercially available in 1973. This was suitable only for head scans and had a 6.5 minute scan time. A normal scan by Ambrose and Hounsfield is shown in [15.5] and a left frontal meningioma in [15.6]. By 1975 the EMI-Scanner CT1010 [15.7] had been developed and 60 seconds were required for a head scan. However, EMI was no longer the only manufacturer of CT scanners and during the decade 1973–1983 there were 18 different manufacturers, of which only 11 survived to 1983, several of which have since closed their manufacturing operations. EMI itself was taken over by GE in 1979[16].

An elegant analogy in 1978 by General Electric[17] to explain CT in layman's terms made use of an orange. 'Imagine that you are attempting to find the seeds inside an orange, using a very thin wire probe. By piercing the orange at many points along its 'equator' you will hit a seed every once in a while. You would record each point of entry, and whether or not you hit a seed. Eventually, with enough probing and record keeping, you would get a reasonably accurate picture of where the seeds were located. CT works by a similar logic, but is a non-invasive procedure, involving no intrusion into the body. A large number of very fine, low-level, pulsed X-ray beams are used as probes; and a xenon detector on the opposite side of the patient measures how much of each X-ray beam is transmitted through the body. To make an image of a particular slice, you rotate the pulsing X-ray source and detector around the patient. For each of tens of thousands of pulses and positions, a computer records the information received. Finally, the computer compiles all the data and reconstructs an exact image of the organs, bones or tumours, etc. inside. The physicians can then distinguish fat from muscle, healthy tissue from diseased, and at times, benign from malignant tumours.'

There have been several types of CT scanner from the various manufacturers and some examples[16] are: (a) translation–rotation system with one detector per slice; (b) translation–rotation system with a multi-detector array per slice; (c) rotating system with rotation of X-ray tube and detector arc; (d) ring-detector system; and (e) hybrid system with translation/rotation and pure rotation. Figure [15.8] defines two types as of 1976, after Zaklad[18] of Elscint. (For further reading on the principles of CT see the retrospective review of 10 years of CT by Dümmling[16], the 1976 review from the National Institute of Health, Bethesda[19], the review by Pullan[20] in 1979, the book edited by Webb[9] and the 1978 review by Kreel[21] which covers not only head scanners but the first EMI body scanner (1975–76), the CT-5005).

The concluding illustrations in this chapter are body scan images [15.9–15.11] taken in 1989 using a General Electric CT-9800 scanner, the original model of which, introduced in 1983, was the first 2-second CT scanner. A Philips CT scanner of the 1990s is shown in [15.12] and may be compared with the 1976 EMI head scanner in [15.7] which predated CT body scanners.

Linear tomography: 1935 and 1988

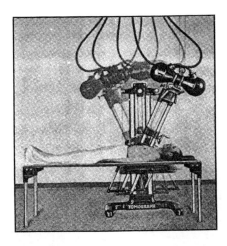

[15.1] Grossman's tomograph of 1935[1].

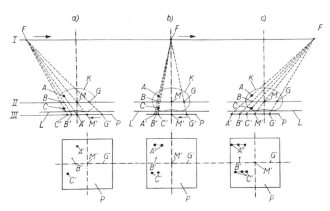

[15.2] Tomographic geometry after Grossman[1]. I is the movement of the X-ray tube focus F; II is the central plane of the tomographic slice; III is the plane of the film, P. The lower three squares represent the film on which is indicated the positions of points A, B, C, M and G within the body, as the focus F moves from position (a) to position (c). The focus F and the film P are moved along straight lines in this diagram.

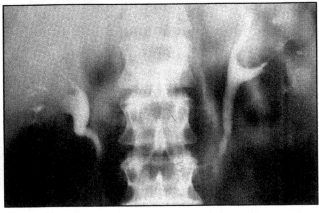

[15.3] Example of a linear tomography installation in 1988 (Westminster Hospital, London). The unit is an IGE Linotome which is dedicated to linear tomography. The X-ray tube and the film are linked. During an exposure the tube moves along the length of the table and the film moves in the reverse direction. Only at the axis point (fulcrum height) will body organs not be blurred. The fulcrum height is adjustable and therefore the required longitudinal body plane can be selected for imaging. The tomograph shown above is for an intravenous urography (IVU) study.

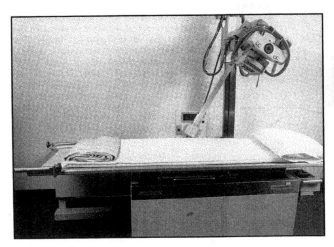

EMI CT scanners: 1972–1976

[15.4] Hounsfield's laboratory lathe-bed X-ray scanner at EMI in 1972 with a portion of a pickled human brain in the specimen position.

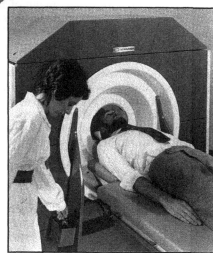

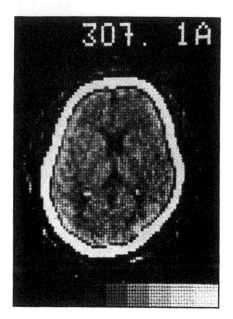

[15.5] Normal scan published in 1973 by Ambrose and Hounsfield[10] using an EMI-Scanner. It is a polaroid picture and the tissues exhibiting the highest density are shown as white areas, whilst tissues of lowest density are shown as black areas. The subarachnoid space, the interhemispherical fissure, the difference between the cortex and white matter, the latter ventricles, septum pellucidum, third ventricle and choroid plexuses are major features which are easily defined. (Courtesy: *British Journal of Radiology*.)

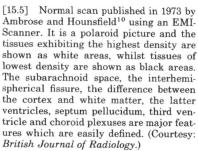

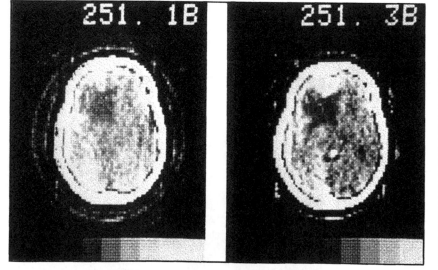

[15.6] Scans of a left frontal meningioma, published in 1973 by Ambrose and Hounsfield[10] using an EMI-Scanner. Scan 1B shows a dense left frontal tumour immediately behind a local thickening of the left frontal bone. The area of low density behind the tumour is oedematous brain. Scan 3B was taken two hours after carotid arteriography and at a slightly lower level. Note the increased density due to retention of contrast medium by tumour tissue. (Courtesy: *British Journal of Radiology*.)

[15.7] An EMI CT-1010 head scanner, 1976. In the first series of commercially available EMI-Scanners, the head was confined in a waterbag and two adjacent slices were scanned simultaneously in 5 minutes using the arrangement of an X-ray source and two radiation detectors. The CT-1010 was the second generation EMI-Scanner and was of a 17-detector design where the patient's head was scanned using a tightly collimated fan-shaped beam of X-rays. Again, two adjacent tomographic slices, each 13 mm thick, were examined simultaneously: using eight sodium iodide detectors to measure X-ray intensity for each slice and a reference detector to measure the X-ray intensity of the primary beam. The CT-1010 was of a translate–rotate design, indexing around the patient's head by 3° rotations between each linear traverse. Absorption values of materials normally encountered in radiology were established for the CT-1010 on an arbitrary scale with water scaled 0, air scaled −1000 and dense bone scaled +1000. This is termed the Hounsfield Scale:

+1000	Dense bone
+80–1000	Bone/calcification
+56–76	Congealed blood
+36–46	Grey matter
+22–32	White matter
0	Water
−100	Fat
−1000	Air

[15.8] As important as short scan times are short reconstruction times since it is the total of these two imaging periods that truly limits patient throughput. One technique (of which the CT-1010 was an example) reducing the total scan time makes use of a fan-shaped segment of the X-ray beam and 10 or more detectors (left). At the end of the linear scan, the gantry tilts 10° or more, instead of the 1° step of the simplest scanner. Yet another method for cutting scan time is to use a beam covering the width of the body, completely eliminating the linear travel of X-ray tube and detector (right). However, a much larger number of detectors, perhaps several hundred, are necessary[18].

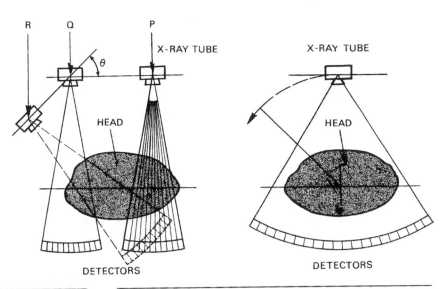

CT scanners: 1990s

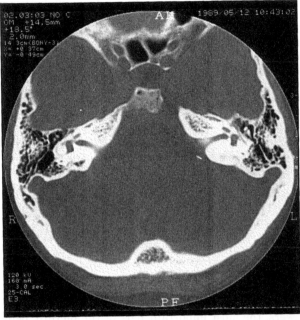

[15.9] General Electric CT-9800 scan in 1989 through the petro-mastoid bones. The slice width is 1 mm (in the CT scanners of circa 1978 the minimum slice width was 5 mm). The sphenoid sinuses, the pituitary fossa and the canal for the acoustic nerve are clearly seen.

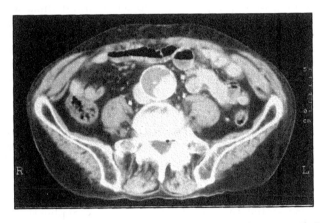

[15.10] CT-9800 scan in 1989 which demonstrates the presence of an aortic aneurysm. The white rim round the aorta is indicative of calcification in the wall of the aorta. On the right of the aortic image a normal flow of blood with IV contrast media is seen, whereas on the left of the aortic image is seen the thrombus. The iliac crests are also seen in this scan.

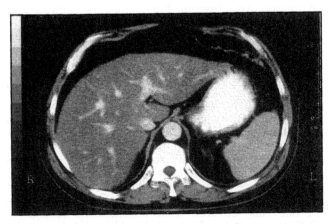

[15.11] CT-9800 scan in 1989 taken with diluted Gastrografin in the stomach and Urografin in the liver. Stomach, spleen, liver and aorta are all clearly seen.

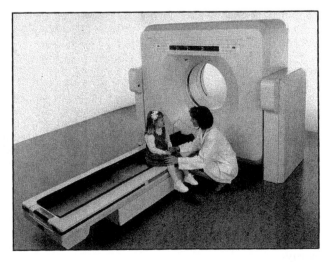

[15.12] A CT scanner of the 1990s: the Philips Tomoscan SR.

Chapter 16

Paintings and Museum Artefacts

Radiography of paintings is a now well established technique in the art world but the earliest investigations were reported in the medical and scientific literature. The first radiograph of a painting is attributed[1] to Walter König of Giessen, a former student of Röntgen, in 1896, but the details of his investigation are not well documented. Many years later König referred[1] to 'a röntgen picture of an oil painting on wood, which was brought to me by a gentleman from Frankfurt.'

The Royal Photographic Society had, from the earliest days, taken a keen interest in X-rays and radiographs [3.1] and developed informal links with the newly founded Röntgen Society. This relationship climaxed in 1920 when for the month between January 6th and February 7th, 192 radiographic prints were 'tastefully hung in the meeting room of the RPS, where, as the Visitor's Book shows, they attracted considerable public attention.' The Hanging Committee of the Röntgen Society listed all 192 radiographs in the April 1920 issue of the *Journal of the Röntgen Society* and reproduced six typical radiographs from the exhibition. Figure [16.1] is one of these examples while others included industrial radiographs from the 1914–1918 World War.

Figure [16.1] was submitted to the Röntgen Society–RPS exhibition by L. G. Heilbron of Amsterdam and it was stated that it 'illustrated a new application of X-rays and showed how radiology can determine whether one picture has been superposed on another.' This was followed up less than a year later in the *Journal of the Röntgen Society*, July 1921, by a paper on 'The radiography of pictures' by Andre Cheron, reprinted from the *Proceedings* of the French Academy. There was, in fact, an earlier radiographic investigation of overpainting by von Faber, in Germany, who not only published in *Zeitschrift für Museumkunde* but also patented, in 1914, his investigative procedures.

Figure [16.2] shows two of Cheron's radiographs, both from the 15th century French school. The 'Royal Child at Prayer' is in the Louvre in Paris and it is interesting to note that as far back as 1921, the curators of the museum had documentary evidence that the original background of the painting had suffered extensive deterioration which had been obliterated a century or so earlier by the uniform black background seen in the 20th century. The radiograph fully confirmed this and revealed the damaged original background beneath the black paint.

In 1930 the *British Journal of Radiology* published an interesting art case history[2] from the Chief Curator of the Royal Museum of Art of Belgium, L. van Puyvelde, which could never have been solved without X-rays.

'On the advice of the Commission d'Art Ancien, the Royal Museum of Art purchased a very fine 17th century painting by Daniel Seghers, entitled 'Guirlande de Fleurs' [16.3]. But the effect of this exquisite garland of flowers was spoilt by a large medallion at the centre bearing a portrait of Helene Fourment, the second wife of Rubens. This portrait was carried out in harsh and flamboyant colours, quite out of keeping with the flowers, and the harmony of the picture suffered greatly as a result. Doubts were entertained concerning both the period and authenticity of this portrait. One rarely finds a 17th century portrait surrounded by a garland of flowers. As a rule the subject depicted in the medallion is a religious one, treated either *en grisaille* or in comparatively restrained tones. It seemed probable that some owner had the religious subject of this picture painted over with a likeness of the young and beautiful Helene Fourment. The picture was therefore submitted to test by X-ray photography. The radiographs revealed beneath the portrait a Nativity, still apparently in a good state of preservation. We did not hesitate to have the portrait removed. This work was carried out with all the necessary precautions and by safe methods. Scarcely had the agents been applied when the cracks on the portrait disappeared. They were only the varnish, and it proved to be an easy matter to remove the portrait without affecting the painting beneath. The medallion in the centre does in fact portray, in subdued tones, the Nativity. The whole picture now presents a beautiful aspect, the garland of flowers, the most important part of the picture, being seen to the best advantage.'

Medical radiological journals have continued occasionally to publish papers on applications of X-rays in art, but only very infrequently, such as papers in 1939 from the National Gallery in London[3], in 1943 from the Worcester Art Museum, Massachusetts[4], and in 1960 from the Washington County Museum of Fine Arts[5]. This is because this non-medical application of X-rays is now an established technique and the radiographic studies of paintings now appear in art books rather than in medical journals.

Figure [16.4] is from a later period, from an Eastman Kodak Company 1960s advertisement for Kodak Ektachrome film which used the title 'Anatomy of a Painting' and the following description. 'Among radiography's positive contributions to art is the discovery of previously unknown works. Often in the past, artists painted over existing works and such was the case in the discovery of the portrait of St. Catherine of Alexandria. The 17th century Italian oil painting was at one time covered with a handsome landscape. Delicate restoration brought the hidden painting to the surface. Radiography also revealed the use of metallic compounds to repair the ground and original paint, the conditions of the canvas, and the rubbing away of some varnish and paint that formed part of the painting's laminated structure.'

One of the most recent descriptions of the use of X-rays in art is given by van der Wetering[6] in the book accompanying the 1992 Rembrandt exhibition, 'The Master and his Workshop', at the Rijksmuseum in Amsterdam. In the mid-1600s the Praelector of the city of Amsterdam Surgeon's Guild gave a public demonstration, only once or twice a year, of the dissection of a human corpse. This event when a Dr Tulp was Praelector was painted by Rembrandt in 1652 and then in 1656 Rembrandt painted a similar event by Tulp's successor, entitled 'The Anatomy Lesson of Dr Joan Deyman'. The painting [16.5], or rather the fragment of it which remained after it was burnt in a fire in 1763, was exhibited in the 1992 exhibition, together with an accompanying radiograph, which is surely the only radiograph of a painting on which so many features (ten) can be demonstrated radiographically [16.5].

Radiography of museum artefacts preceded that for paintings, the first artefacts to be investigated being Egyptian mummies. The earliest report [3.1] in the *British Journal of Photography* of February 28th 1896, refers to radiographic investigation by Dedekind of the Vienna Museum of Natural History; a mummy which appeared to be similar to a human mummy bore inscriptions suggesting that the wrapping enclosed a ceremonial ibis, and the radiograph confirmed the outline of a large bird[1]. Other early radiographs, of a cat mummy and the knees of a child mummy [16.6], were made in March 1896 by Walter König[7] in the Senckenberg Museum in Frankfurt-on-Main. The same museum was also the source of the head of a mummy [13.14] published in 1898 by Isenthal and Ward[8] but attributed to T. Brinkmann. Another mummified bird was found by Thurstan Holland [9.5] when he X-rayed an artefact from an Egyptian tomb in October 1896.

A comparison of whole body mummy radiographs, separated by 60 years, is provided by [16.7], and an 11th century Peruvian mummy is shown with its modern radiograph in [16.8]. These were real mummified objects, but occasional frauds were also discovered with the use of X-rays, such as reported in the *Archives of the Roentgen Ray* in 1900. 'The fraud was detected in the case of a supposed mummy which, when examined with the rays, was found to contain only a block of wood, and no human remains at all.'

The radiograph of another Peruvian antiquity, a 600-year-old osteosarcoma [4.8] was published in an early issue of the *Archives of the Roentgen Ray*.

Since those early days many other museum artefacts have been radiographed, including ceramics, sculpture and wood and metal items: mainly either as an aid to detect forgeries[9] or to investigate in a non-destructive manner, for the benefit of a restorer, or for scientific curiosity, the structure and inner mechanics of the artefact. Examples illustrated in this chapter include the world's first globe [16.14], the bust of Queen Nefretiti [16.12] and an 18th century grand piano [16.13].

Radiographs of paintings: 1920–1930

[16.1] 'The Crucifixion' by Engelbrechts. Left: Natural photograph. Centre: Radiograph showing monk in surplice underlying portrait of Donatrice in right foreground. Right: Natural photograph taken during process of restoration, revealing monk. *Journal of the Röntgen Society*, 1920.

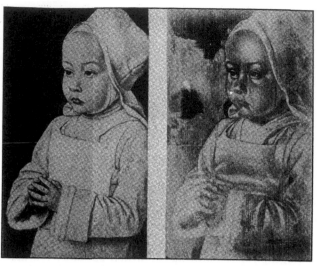

[16.2] Radiographs in 1921 of 15th century French paintings. Left: 'The Virgin' by Stella. Right: 'The Royal Child at Prayer' by Cheron.

[16.3] Guirlande de Fleurs' before and after restoration in 1930[2].

Discovery of a 17th century painting

[16.4] 1960s discovery of a 17th century Italian oil painting of St. Catherine of Alexandria.

The Anatomy Lesson of Dr Joan Deyman

[16.5] 'The Anatomy Lesson of Dr Joan Deyman', by Rembrandt (1620–66), canvas 100 cm × 134 cm, and its radiograph[6] revealing ten different features: (1) painting style; (2) metallic pigments; (3) adjustments by the artist; (4) canvas texture; (5) stretch deformation; (6) damage by fire (1763), and restoration; (7) burnt paint; (8) fillings: traces of restoration; (9) nails and frame; (10) vandalism (1931) by axe.

Egyptian and Peruvian mummies

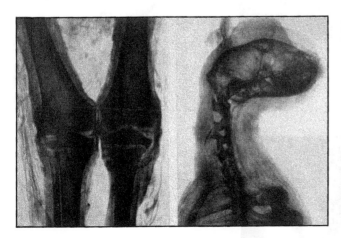

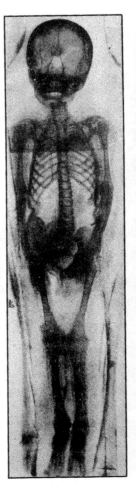

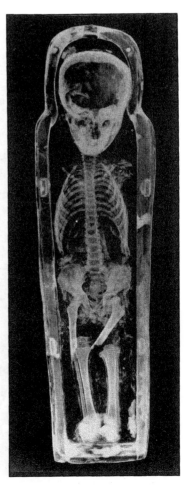

[16.6] Some of the first recorded radiographs[7] of Egyptian mummies were made in March 1896 by Walter König, including the mummified cat and the knees of a child mummy as illustrated here.

[16.7] Whole body radiographs of Egyptian mummies in France in 1898[10] (left) and in the USA in a Kodak exhibition in the 1960s (right). Kodak gave the following history. 'More than 2500 years ago a young boy named Pediamon perished in Egypt. Irregular calcium development in the ends of the long bones revealed by this radiograph indicates that malnutrition was an ancient as well as a modern problem. The coffin available was too short for the boy's body, so the undertaker broke the joints, overlapped the upper and lower leg bones and left out the arms.'

[16.8] Photograph and radiograph of a 900-year old Peruvian mummy of a male aged 40–50 years. (Courtesy: Deutsches Röntgen-Museum, Remscheid-Lennep.)

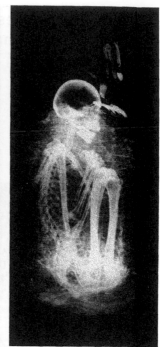

Authentication of museum artefacts

[16.9] X-rays have been useful in authenticating museum artefacts as well as paintings and in this example the handle of the bronze dagger (A), dating from 800 BC to 500 BC reveals (below) that the blade is held in place by solder, the opaque white area, and that the weapon is not completely authentic. In the case of the sword (B), radiography proved that the weapon was a composite of an original hilt with a replacement blade. Solder had again been used and these areas carefully filled and then 'aged' with corrosive chemicals to give a result which appears genuine but in reality is a fake. (Courtesy: Eastman Kodak.)

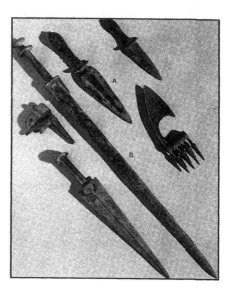

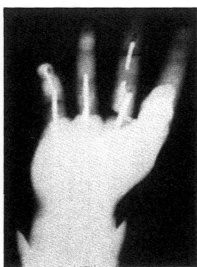

[16.10] Radiography can also be helpful in revealing repairs and alterations previously unrecorded. One example[11] was when Michael-angelo's dense Carrara marble sculpture 'La Pieta' was moved from St. Peter's Basilica in Rome to the Vatican Pavillion at the 1964 World's Fair in New York. An exhaustive X-ray examination was made and revealed that metal pins had at some time been used to rejoin the broken fingers on the hand of the Virgin. It was also found that shallow holes had been drilled to accommodate a halo, later removed. (Courtesy: Eastman Kodak.)

19th century doll

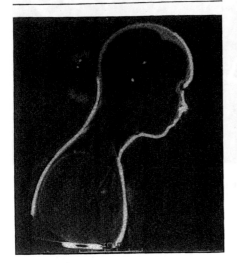

[16.11] This 70 cm doll was manufactured about 1845 and is shown wearing a long white blouse, woollen pants and two lace half-shirts under a bright checked cotton dress together with a scarf made of cotton lace. White cotton stockings and leather shoes decorate the doll's feet. The two radiographs show the structure of the doll's head and chest which are made of papermaché and leather bellows. She also has glass eyes and genuine hair. (Courtesy: Siemens Med. Erlangen and Germanisches Nationalmuseum, Nürnberg; see also [16.13].)

Queen Nefretiti

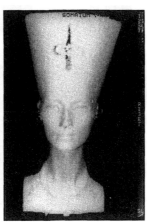

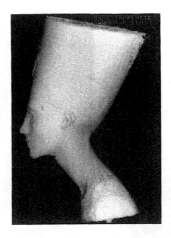

[16.12] The three most beautiful female faces in the world are often said to be those of Helen of Troy, whose face launched a thousand ships and started the Trojan War, of the Swedish actress Greta Garbo, and of the Egyptian Queen Nefretiti who lived during the XVIII dynasty which was between 1364 BC and 1347 BC. Her bust was discovered in 1912 by the German archeologist Ludwig Borchardt at the excavation site of Amarna in Middle Egypt. In 1920 it was donated to the National Museum in Berlin-Dahlem where it is still part of the Egyptian Collection. It had been noted in 1925 when the bust was being copied that several corrections had been made on the porous sandstone core by plaster near the crown and the neck. However, it was not until 1992, when using the 3D reconstruction capabilities of the Siemens Somatom Plus CT scanner at the Radiological Clinic of the Rudolf Virchow Medical Center of the Free University of Berlin[12], that it became apparent that Nefretiti was not quite so perfectly formed as previously thought! The neck and shoulders were quite uneven and plaster 'lifts' of 4 cm to the right shoulder and 2 cm to the left shoulder had been incorporated to reduce considerably what would otherwise have been a rather skinny neck. Also, the support of the reclining crown on the top of her head was only achieved by adding the lighter plaster to the upper and dorsal part of the crown: otherwise a neck fracture of the bust would have been very likely and the attractive position of the crown could not have been maintained. In the radiographs (centre) the soft sandstone is seen as a homogeneous white colour and in the CT axial view of the crown (far right) the plaster is seen as a grey material. (Courtesy: Dr Christian Zwicker.)

Grand piano of 1749

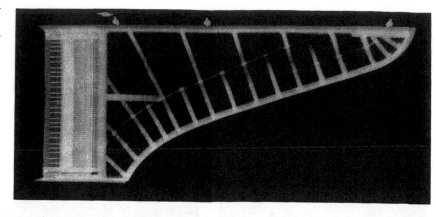

[16.13] Most museum exhibits which have been radiographed are not extremely large, but an exception is this grand piano which is part of the musical instrument collection of the Germanisches Nationalmuseum in Nuremberg. It is the best preserved of the three oldest grand pianos crafted using the hammer mechanism and was built in 1749 in Freiburg/Sachsen by Gottfried Silbermann (1683–1753) who since the year 1730 presented

all his newly developed grand piano models to Johan Sebastian Bach (1685–1750) who, however, only approved the later models. It is therefore quite likely that Bach himself played this particular piano. The radiograph clearly displays the construction principle of the body of the piano, the strings and the 86 different keys of the keyboard. (Courtesy: Section for Restoration of Musical Instruments of the Germanisches Nationalmuseum, Nürnberg.)

The Earth Apple of 1492–1494

[16.14] Interesting CT images have been obtained for the world's first globe, made by Martin Behaim in Nuremberg 1492–94 and which was then colloquially referred to as the Erdapfel (the Earth Apple) before Christopher Columbus embarked on his voyage resulting in the discovery of America. Behaim overestimated the size of the Mediterranean Sea and the continents of Africa, Asia and Europe were enlarged in proportion to the surface of the globe. As a result, not enough space was left for the missing continents of North and South America. The position where Japan (Cipangu) is shown on the globe is approximately the location of the east coast of North America.

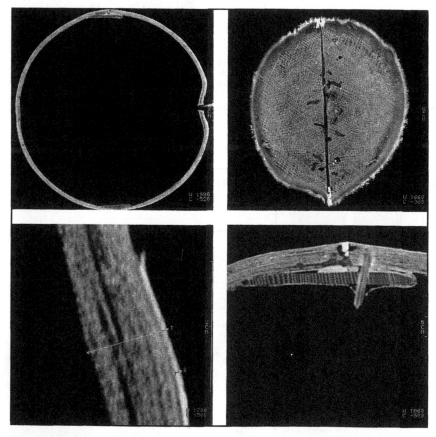

The CT scans revealed new findings concerning the technique used to manufacture the globe. A slice near the axis of the globe (upper left) shows the deformation of the South Pole caused by the weight of the globe and its methods of support while the tissue structure, type of binding and yarn filling are clearly seen (upper right). The interconnection of the two hemispheres at the equator used a wooden ring. The lower scans reveal the technological design and enable measurement of the individual layers of material: five layers of canvas; parchment skin; paper; paint layer; equatorial wooden ring; fixation with wooden nails.

These illustrations and those of [16.11] are also to be found in the 1993 calendar issued jointly by Siemens and the Nationalmuseum, which give for each month of the year, examples of CT images of museum artefacts. They were made using the Siemens SOMATON PLUS-S and HiQ-S CT scanners. (Courtesy: Siemens Med. Erlangen and Germanisches Nationalmuseum, Nürnberg. I am also most grateful to Dr Heinrich Seegenschmiedt for bring the radiological story of the Earth Apple to my notice, as well as those relating to the grand piano of 1749 and to Queen Nefretiti.)

Chapter 17

Industrial Applications

One of the starting points in industrial radiography was the X-ray picture taken by Röntgen of his hunting rifle and this is seen in [17.1], which is taken from the frontispiece of the third edition (1901) of *Practical Radiography* by Isenthal and Ward[1]. Newspapers reported on this new application and the following is from the 1897 *Archives of the Roentgen Ray*, quoting the *Globe*. 'M. Radiguet [10.21] has brought before the French Academy the results of some interesting experiments made by him in the examination of metals, the detection of internal flaws, etc. Blisters could be seen in an aluminium bar one and a half inch thick. The *Globe* for October 22nd mentions some experiments with potatoes, carried out by MM. Bussard and Condon, at the National Agronomic Laboratories. The density of a potato increases as it is richer in fecula, and consequently allows the rays to pass with more difficulty'.

In the 1898 edition of *Practical Radiography*[2] is shown a radiogram of a steel joint taken by John Hall-Edwards of Birmingham, in which it was commented that the radiogram could show the extent of the brazing [17.2]. The First World War, 1914–1918, gave an impetus to industrial radiography and [17.3] shows a radiograph of the fuse of a 75 mm shell[3]. Other items which were radiographed during this period included aeroplane hollow box struts, welds, pistons and other aircraft fittings, and they were sometimes termed radiometallographs[4].

Since the 1920s the industrial use of X-rays has greatly expanded and forms part of standard non-destructive testing procedures for a wide range of materials. By 1934[5] this could be divided into industrial fluoroscopy, radiography of large castings and forgings, radiography of welded vessels and structures, and radiography of small objects. One example[5] of what was classified as a small object is the die-casting in [17.4].

Industrial radiographs of unique historical objects, such as the Liberty Bell [17.5], are of particular interest, as is the fact that in 1933 the fusion welds in the Hoover Dam were examined radiographically[6] using a General Electric 300 kV oil-immersed shock-proof X-ray apparatus—the largest radiographic assignment in the world. The fusion welds totalled more than 75 miles in length with each weld being up to 3 inches thick in penstock sections ranging from 8.5 to 30 feet in diameter. More than 159,000 separate X-ray exposures were required involving the use of more than 24 million square inches of X-ray film. A 400 kV industrial X-ray unit in Russia[7] is shown in [17.6].

Industrial gamma radiography apparatus using radium was not developed until 1930[8] but, with radium replaced by artificially produced radionuclides such as cobalt-60, gamma radiography is a useful non-destructive testing tool for a variety of applications, such as, for example, in situations where a mobile unit has to be taken on site for inspection of oil pipelines.

Another early example of industrial radiography is the use of X-rays for security purposes when a suspect parcel is found. In 1896 the *Strand Magazine* report on 'The new photography'[9] showed a photograph of an explosive book and a radiograph of its contents [17.8]. Shortly afterwards in an 1897 issue of the *Archives of the Roentgen Ray* the following report from the *Globe* newspaper of July 16th was reprinted under the title 'Roentgen rays and French customs'.

'The Roentgen rays, which have been employed to examine the interior of bombs with success by M. Girard, are now enlisted in the service of the French customs by M. Pallain. We are afraid some of the newspapers have exaggerated the dread importance of this latest application. No doubt some contraband articles may be detected in this way, but certainly others cannot. Might not a smuggler of lace, for instance, be able to stow as much as he likes of it in his bag and without fearing detection by the rays? We hear of cigarettes being discovered by this new Ithuriel, but probably the metal box containing them was the cause of it. Cigars and cigarettes, as well as lace, are vegetable matter, more or less transparent to the rays, and one could easily select and pack them so as to escape the vigilance of the "douanier". Perhaps we shall presently see tobaccos and other articles offered for sale "warranted X-ray proof". Professional smugglers may also subscribe to the Röntgen Society.' The *Archives* did not publish the details of the French apparatus, but this is shown in [17.9] from an 1898 French textbook[10]. Figures [17.11] and [17.12] are examples of successful detection of smuggling and Figure [17.10] is a more modern version of the French apparatus in Figure [17.9] which was used to detect what were sometimes called[2] 'infernal machines'.

Another early use of X-rays, which persisted for over 60 years, was to demonstrate the correct fitting of shoes. As early as September 11th 1896, the *British Journal of Photography* described the enterprising display of X-ray photographs by a London bootmaker [3.1] to illustrate the harmful deformation of feet by badly fitting shoes. This pictorial use of X-rays was soon followed by the introduction of shoe-fitting fluoroscopes [17.13] which enabled customers, in shoe shops, to radiograph their own feet to assess the goodness of fit of a pair of shoes. Despite their popularity with customers, shoe-fitting fluoroscopes were eventually withdrawn because of the unacceptable radiation exposure of both employees and customers[12,13].

Röntgen's rifle: 1896

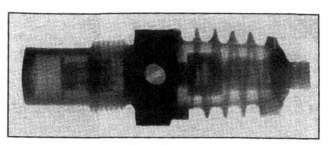

[17.3]   Radiograph of a fuse of a 75 mm shell: 1914–1918 World War[3].

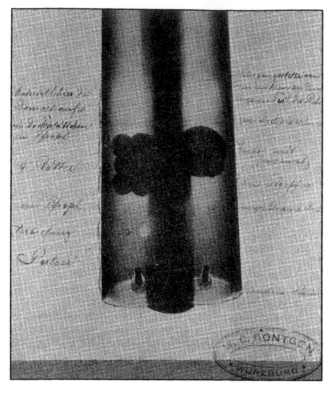

[17.1]   Röntgen is reputed to have enjoyed hunting and to have been an excellent shot. This X-ray picture of his double-barrelled shotgun was made by Röntgen in the summer of 1896 and sent with marginal notes to Franz Exner in Vienna. A copy was given to A. W. Isenthal during his meeting with Röntgen in 1898 (Chapter 1) and subsequently used as the frontispiece of the third edition of *Practical Radiography* by Isenthal and Snowden Ward[1].

Non-destructive testing: 1896, 1916 and 1934

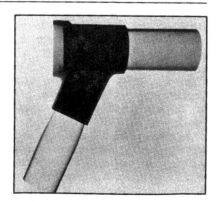

[17.2]   Early in 1896 the German and Austrian Ministries of War called attention to the importance of using Röntgen's methods of finding defects in guns and armour plates, and in the United States, A. W. Wright of Yale University, in January 1896, radiographed a piece of welded metal and showed the welding seam which could not be seen with the naked eye. In February 1896 the Carnegie Steel Works in Pittsburgh used the same method for tests on steel reporting this in the *Electrical Engineer* (New York)[11]. The radiograph[2] shown is an early example taken in 1898 in Birmingham, England.

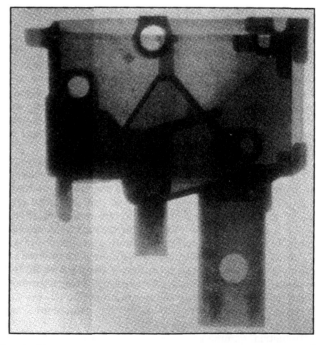

[17.4]   Photograph (top) and radiograph (bottom) of a die cast[6] taken in the USA in 1934.

The Liberty Bell

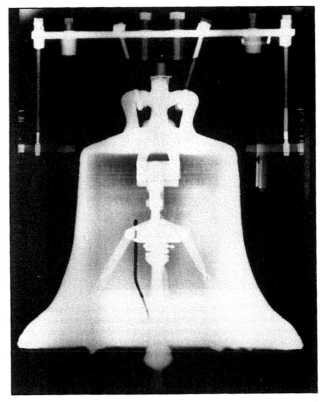

[17.5] Photograph and radiograph of the Liberty Bell. The radiograph was taken in 1975 using a film of size 2.1 metres × 1.2 metres, when it was moved from Independence Hall to a new building in Philadelphia in time for the American Bicentennial celebrations in 1976. (Courtesy: Deutsches Röntgen-Museum, Renscheid-Lennep.)

Industrial apparatus: UK, 1944 and Russia, 1960

[17.6] Industrial radiography in Russia in 1960[7], using a 400 kV machine.

[17.7] Industrial radiography in the United Kingdom in 1944[8]. A 1000 kV X-ray machine is being used to examine a large steel casting.

Customs and smuggling: 1896–1989

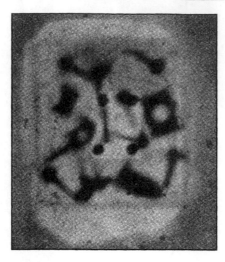

[17.8] An 1896 radiograph[9] of the contents of an 'explosive book' which was described as being constructed on the 'bon-bon' principle. 'One end of a cracker is attached to the book cover and the other end to a box placed in a hollow inside the glued-up book pages. When the book is opened the cracker goes off and ignites the contents of the iron vessel. If this is filled with fulminate of mercury and scraps of iron, the result can be better imagined than described.'

[17.9] The French customs radiographing what was sometimes called[3] 'possible infernal machines' in 1898.[10]

[17.10] One of the first airport X-ray security systems using fluoroscopy was manufactured by Pantak in the United Kingdom in the late 1950s. The test radiograph shown is of a case filled with various items such as shoes and handguns. During a lecture when I showed this radiograph to a group of nurses, a voice from the room said 'Oooh, look he keeps his socks in his shoes'. If that nurse had been a security officer she would have been a failure as the 'socks' were a grenade!

[17.11] Radiograph described as 'Boots, 21/5/20'. Taken by Russell Reynolds, it is a 1920 radiograph of the boots of a Polish soldier who attempted to smuggle gold coins into England.

[17.12] The abdominal radiograph (far right) of a traveller from Nigeria to the United Kingdom was taken in 1989 at Manchester airport. The 28 wrapped packages, taped to his abdomen, when recovered, yielded 153 grams of 39% pure heroin with an estimated street value of £15,000. (Courtesy: Dr A. W. Horrocks.)

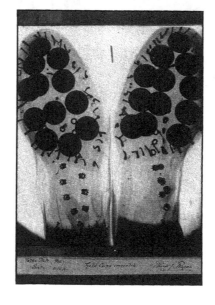

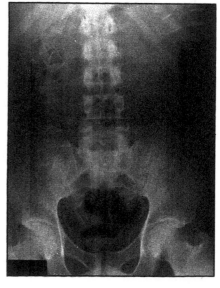

Shoe fitting: the Pedoskop

[17.13] The Pedoskop, a 1950s shoe-fitting fluoroscope designed by Ernst Gross, now displayed in the Deutsches Röntgen-Museum Remscheid-Lennep. Such fluoroscopes were very popular with children but were eventually withdrawn because of the radiation injuries which could result from the uncontrolled use of low voltage X-ray machines. For instance, in the 1949 *New England Journal of Medicine* three types of possible injury were quoted as: interference with normal foot development in children who are repeatedly fluoroscoped; acute radiation burns on the feet resulting in late permanent skin damage; and chronic radiation injury of the blood-forming tissue of shoe store employees who work with inadequately shielded X-ray machines[12]. Radiation measurements in the USA made with a 50 kV X-ray tube operating at 3–8 mA through a 1 mm aluminium filter in a lead- or steel-lined cabinet showed[13] that dose rates to feet were in the range 0.5–5.8 röntgen per second (typical exposures were in the range 5–45 seconds) with a cabinet wall leakage in the range 3–60 milliröntgen per hour and that scattered radiation amounted to more than 100 mr per hour at distances of up to 3 metres from the unit.

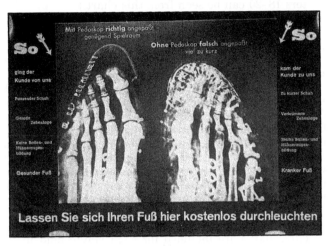

The radiographic illustration on the front of the Pedoskop encouraged customers to 'Have your foot X-rayed for free' with 'before and after' examples on the right and left respectively:

Correct fitting with Pedoskop
How the customer left
Shoe that fits
Toes straightened
No callouses or corns
Healthy feet

Wrong fitting without Pedoskop
How the customer arrived
Shoe too small
Toes bent sideways
Thick callouses and corns
Deformed feet

(Courtesy: I am grateful to Mr Ulrich Hennig, Director, Deutsches Röntgen-Museum for the radiographic illustration and the above translation.)

Mercedes car

[17.14] Radiograph of a Mercedes car taken in 1965 by Agfa. Five films were used and the total exposure was 50 hours. (Courtesy: Deutsches Röntgen-Museum, Remscheid-Lennep.)

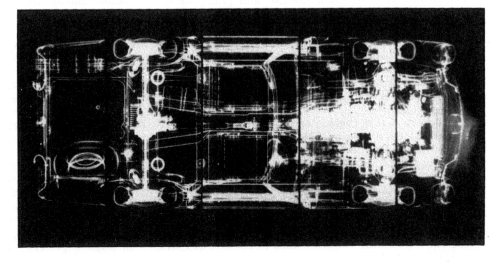

Chapter 18

External Beam Radiotherapy: I

According to the report[1] by Haagensen on an exhibit for the graduate fortnight on 'Tumours' at the New York Academy of Medicine, October 1932, the first proven cure of a cancer patient by X-ray treatment was in June 1899 by Tage Sjögren in Sweden. There was, however, a second successfully treated case at the same time, by Tor Stenbeck, and both were reported at the Swedish Society of Medicine meeting on December 19th 1899[2]. Elis Berven[3] presented these two cases at the 1961 annual meeting of the Radiological Society of North America: Figure [18.1A] shows the patient of Stenbeck with a basal cell carcinoma of the nose in 1899 before treatment and at 30 years after treatment. Figure [18.1B] shows Sjögren's patient with a squamous cell carcinoma of the cheek in 1899, which was cured by X-rays and a minor operation for removal of a small remnant, and at two and a half years after treatment. The apparatus of Stenbeck is shown in [18.2].

The medical literature at the start of the 20th century contained many similar case histories of successful treatments of superficial tumours, such as that in [3.10] which is the first such reported case in the *Archives of the Roentgen Ray* and was treated by J. H. Sequeira of the London Hospital, who stated that he treated his first such patient in June 1899. Figure [18.4] shows the Radiotherapy Department of this hospital in 1905[4].

Because of the lack of penetration of the X-ray beams from the early therapy apparatus, the initial years of X-ray therapy although recording successes with superficial tumours were not satisfactory for the treatment of deep-seated tumours. This had to wait until 'deep' (also termed 'orthovoltage') X-ray therapy apparatus became available and this technological advance did not occur until the early 1920s. However, this did not prevent early attempts at total body irradiation (TBI) for skin disease and [18.8] shows the 'Filterhouse' of Friedrich Dessauer in 1908[5]. This is a very different technique to modern TBI which has been used successfully in conjunction with bone marrow transplants to treat leukaemia. The patient's bone marrow is substituted by healthy marrow from a donor whose tissue is closely matched to that of the recipient and TBI is used to reduce the number of leukaemic cells.

The first X-ray therapy treatments consisted of single-field techniques (a 'field' is defined by the shape of the X-ray beam on the patient's skin surface) with virtually no radiation dosimetry, when compared with what is standard practice in the 1990s: computer treatment planning in three dimensions with radiation dose patterns superimposed on CT scans and special algorithms employed to optimise therapy treatments. Figure [18.4] is an 'ancient and modern' summary chart of the radiation sources used in the first 50 years

of radiotherapy. (The derivation of radiotherapy Chapters 18–20 can be seen from this chart.)

There were, however, many stages of planning technique and dosimetry development between the techniques of [18.3] and [18.6] but two of the most significant were the use of multiple X-ray fields angled towards the tumour, rather than a single direct field, and the proposal that the radiation dose pattern within the body could be shown schematically by the use of isodose curves. These curves, or 'isodosen', were first proposed by Otto Glasser when working in Freiburg in 1920.

Figure [18.11] is an X-ray teletherapy treatment plan of 1919 which just precedes the use of isodose curves. It was already realised that the dose distribution within the body should be available but only a specification of doses at defined points could then be given. Figure [18.12] dating from 1925 is an improvement in that the central axis doses, expressed in percentages (70%–100%) are included for each pelvic field in this treatment plan. The first book of central axis X-ray dosage tables was published by Friedrich Voltz[6] in 1922 and the modern equivalents are published at regular intervals[7] as *Supplements* to the *British Journal of Radiology*.

Isodose curves were explained by Mayneord[8] in terms of an analogy. 'The results of measurements at, say, 1 cm intervals throughout the whole beam of X-rays, may most conveniently be represented by a series of curves where the points are joined which have the same dose per second; i.e. we have drawn "isodose" lines just as in *The Times* of London newspaper weather charts one may join the points having the same barometric pressure or temperature, and hence obtain isobars or isotherms.'

Some of the earliest isodose curves were drawn by Dessauer and by Holfelder [18.13] and the central axis percentage depth doses differed only slightly, but there was a marked difference in the shape of the curves towards the edge of the collimated X-ray beam. The region outside the geometrical edge of an X-ray beam is called the penumbra, and the reason why there is not a sharp cut-off of radiation at the geometrical edge is because of scattered radiation. This is considerable in terms of side-scatter for X-ray beam energies to 250 kV, in which energy range was the great majority of therapy machines until the advent of van der Graaff therapy machines in the late 1950s. Figure [18.13] reveals an interesting detective story as to who was correct, Dessauer or Holfelder? A combination of two Dessauer curves should have involved a skin burn in the region of the overlap. However, this did not occur and the correct dose pattern is that achieved using the two adjacent Holfelder curves.

In Figure [18.11] a radiation unit of 'milliamp-minutes' is used to describe the treatment 'dose' but this was just one of almost 100 radiation units which were proposed for use with X-rays, radium and radon, before the Système International (SI) units were introduced in the late 1970s. Many of these suggestions are given in Chapter 22.

Early protective measures [22.7], when they were considered, usually consisted of a cumbersome lead-lined box, but these were superseded in the 1920s by huge protective cylinders for deep X-ray therapy machines and it was with these machines that the name X-ray cannon[9–11] was used, [18.16–18.19], the most famous of which were the Holfelder cannons. However, they were not ideal for positioning of patients for treatment and smaller deep X-ray therapy machines were designed which were easier to use. One such example from the 1960s is shown in [18.20].

There have been since 1950 and the introduction of cobalt teletherapy (which preceded the use of linear accelerators), three distinct classifications in tele-therapy, determined for X-rays by the generating voltage; see Table 18.1. However, before the introduction of linear accelerators into radiotherapy there were a few X-ray installations operating in the range 300 kV to 1 MV. These included a 750 kV machine installed at Memorial Hospital, New York, which between October 1931 and January 1934 treated 150 patients; and at the California Institute of Technology at Pasadena, two 750 kV transformers were connected in series to a 30-foot-long X-ray tube, which by 1933 had been used to treat 285 patients at 600 kV.

Table 18.1. Teletherapy treatment classifications in terms of X-ray generating voltage.

Classification	Definition
Superficial therapy	10–150 kV
Deep (or orthovoltage) therapy	200–300 kV
Megavoltage (or supervoltage) therapy	Above 1 MV

Since gamma ray energies of cobalt-60 are 1.17 and 1.33 MeV, cobalt teletherapy machines are classified as megavoltage treatment units

The earliest supervoltage (above 1 MV) installations included the 1 MV unit at St. Bartholomew's Hospital, London[12], which was operational in 1944, in which the X-ray tube was 30 feet long and weighed 10 tons. Direct beam protection required 18 inches of barium concrete but on one of the other walls, no additional protection was required: this was because it was the 3-foot-thick wall of the 18th century Great Hall of Barts! Another million volt X-ray unit was described by General Electric, Schenectady, in 1939[13] at the 25th annual meeting of the Radiological Society of North America (RSNA). However, by far the most extensive review of the development of supervoltage radiotherapy (mainly telecobalt) by the mid-1950s is the proceedings *Roentgens, rads and riddles* of a Symposium on Supervoltage Radiation Therapy, held at Oak Ridge Institute of Nuclear Studies in 1956[14].

Prior to the use of linear accelerators, a few radiotherapy centres installed Van de Graaff generators which were built to generate X-rays at either 1 MV or 2 MV. Figure [18.25] shows a 2 MV machine in clinical use at the Royal Marsden Hospital, London, in 1961. This particular installation had the patient's treatment couch on a turntable so that rotational therapy could

be used. The plumb line seen in the centre of the photograph was part of the treatment set-up for such a technique.

Betatrons were also used in some radiotherapy departments for the production of electron beams (as distinct from X-ray beams). This development followed the initial work of Kerst[15] at the University of Illinois who built the first betatron, operating at energies as high as 20 MeV. For clinical applications betatrons have been manufactured with electron beam energies in the range from a few MeV to 45 MeV; with the majority of the units in Europe and only a few in North America. This was partly due to cost since betatrons are expensive in comparison with linear accelerators. It was also realised that the clinically advantageous sharp cut-off of percentage depth dose achieved with low energy beams of 10–15 MeV is lost at very high electron energies. Consequently there is no real clinical advantage in using beams of energies higher than 20 MeV. This condition is also true with electron beams generated using linear accelerators.

The precursors of radiotherapy linear accelerators (linaccs) were those developed for research purposes, of which the first was that of Wideröe[16] in 1928. This was based on a proposal of Ising[17] for a radio-frequency (RF) powered linear accelerator as described by Karzmark and Pering in their review[18] of electron linaccs for radiotherapy.

In the Wideröe linacc a series of spaced, collinear metal tubes were placed in a large evacuated glass cylinder and alternate tubes connected to opposite terminals of an oscillating RF voltage. Successive tubes were of increasing length so that the accelerating particle bunch arrived at the end of each drift tube in phase with the maximum accelerating voltage occurring between adjacent tubes. By selecting an appropriate frequency and voltage, a variety of heavy ions could be accelerated across the space between adjacent tubes and simultaneously bunched. The Wideröe design is of the standing wave type in which the accelerating electric field maxima and nodes remain fixed in space. Sloan and Lawrence[19] in 1931 using such an array of 30 tubes excited to 42 kV at 10 MHz produced a 10^{-7} A beam of 1.26 MeV singly charged mercury ions[18].

Karzmark and Pering[18] also refer to the 1946 work of Alvarez[20] on standing wave linaccs. However, all these early linaccs were for the acceleration of heavy ions or protons and were not suitable for accelerating electrons because such light particles would necessitate very long drift tubes when RF is used as a power supply. A megavoltage electron will traverse a 10 cm drift tube in one-third of a nanosecond. This corresponds to a frequency of 3000 MHz.

It was the developments during and shortly after World War II (1939–1945) on high power, high frequency microwave sources for radar applications that eventually enabled the electron linacc to become available on a production basis for use in radiotherapy. The pulsed diode sources, the magnetron (used up to 20 MeV, 5 MW power) and the klystron (for energies above 20 MeV, capable of 5 MW–30 MW peak RF power) are capable of establishing intense electromagnetic fields in microwave cavities, with intensities high enough to accelerate electrons to megavoltage energies of several MeV. The early development work for acceleration of electrons was undertaken in the late 1940s at Harwell in the United Kingdom[21] and at Stanford University[22].

The basic principle involved is reproduced from

Meredith and Massey's textbook[23] of 1971, using the analogy of the surf bather. 'RF waves like all other electromagnetic radiations, are alternating electric and magnetic fields travelling through space. Since an electric field applies a force to a charged particle placed in it, it follows that, if an electron is injected into a beam of RF waves at an appropriate place and time, it will be acted upon by the force and tend to be carried along by the waves. In broad terms what happens is very similar to what happens to a surfer when he is carried along by a wave. If he launches himself into the wave at the correct moment, he will rush along at the speed of the wave, always provided, of course, that the wave is big enough and powerful enough to carry him. Similarly RF waves must have enough power to be able to "carry" the electrons along. In practice it is not possible to provide this power continuously: it can only be generated in short bursts, or pulses.' (For further reviews on linacc design see Karzmark and Pering[18], Greene[24] and Klevenhagen[25]. These cover standing wave and travelling wave linaccs, magnetrons and klystrons, and include bibliographies.)

First successful treatments of cancer: Stockholm, 1899

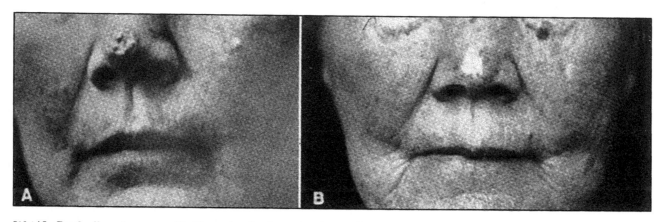

[18.1A]   Basal cell carcinoma cured by X-rays showing the patient in 1899 (A) and 30 years later (B). She was a 49-year old woman whose treatment was given by Tor Stenbeck, commencing on July 4th 1899 for a total of 99 treatments over a period of several months[1,2]. Stenbeck reported that he had treated two cases but gave only the case history of the patient in these photographs. His commentary on the results were that he had 'never in my life seen an ulcus rodens treated in all its extension so nicely and with such a small induration' and the referring surgeon concluded 'that such a lesion can be healed with roentgen irradiation is quite beautiful and it is good to know, particularly in such cases where the patient refuses a surgical treatment, which treatment should come first, particularly when there is reason to expect a greater malignancy'[1].

[18.1B]   Tage Sjögren, like Tor Stenbeck, was a general practitioner in Stockholm, and the photographs show his patient before treatment (A) and at 30 months later (B). After a first series of 50 treatments the epithelioma had healed except for the borders. Further treatment was given but this time a very heavy reaction occurred, although it later subsided. Sjögren's comments on radiotherapy dosage are the first ever made: 'I want only to mention that the dosage of the roentgen rays is at present very difficult. One has namely no safe method to measure their intensity or more precisely to measure the intensity of their active agent. One has therefore more to work on feeling, if I may say so, only on the basis of the experience one has been able to collect'[1].

Apparatus: 1900–1909

[18.2] Tor Stenbeck's Roentgen Institute, Stockholm, about 1900. Stenbeck is standing on the left and the assistant beside the head of the patient is Gösta Forssell. In 1905 with respect to equipping the Roentgen Department at Uppsala it is interesting to note the following: 'It is an advantage to have two roentgen units. The new, stronger apparatus is to be used for diagnostic purposes, while the older and weaker machine is fully adequate and suitable for the treatment of patients. Treatment requires a long exposure time. This places a strain on the apparatus, and it is therefore of great value not to need to use the best unit'[3].

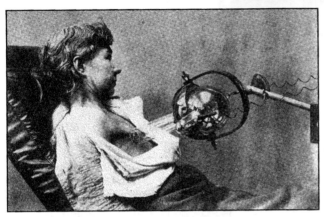

[18.3] Radiotherapy of breast cancer in 1903. The glass cylinder is probably an attempt to limit the X-ray beam to the area prescribed for treatment and is therefore one of the earliest attempts at collimation. However, unless it was made of lead glass, which is not known, it would have been unsuccessful.

[18.4] The London Hospital's Radiotherapy Department in 1905[4]. Lack of radiation protection is obvious and if three patients were treated simultaneously the patient in the central treatment chair would also receive scattered radiation from the adjacent X-ray tubes! Electrical safety also does not look too good!

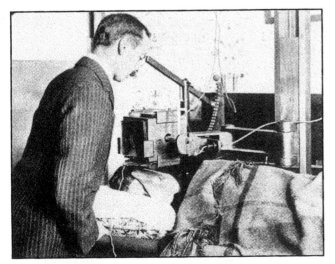

[18.5] X-ray beam shaping and collimation technique of Beck[26] of New York in 1904. 'The sheet of tin is placed over the face to protect the healthy parts from the X-rays and has been pushed forward a little to show the opening in it. The aluminium screen has been slipped in and grounded, but the physician is represented as if he were looking through the box to ascertain whether or not the target is in the proper position with regard to the diseased area.'

[18.6] Beam shaping and collimation: Albers-Schönberg, 1909.

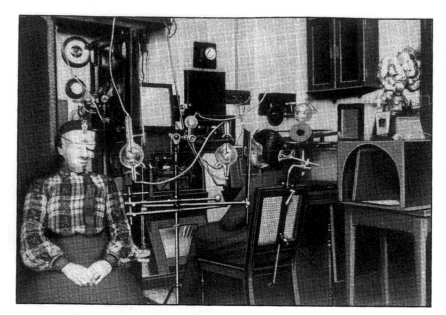

[18.7] In the first X-ray therapy departments (the early radium departments, often within a gynaecological clinic, were separate from the X-ray therapy departments and it was only later that the term radiotherapy department was used) often consisted of a single room in which more than one patient was treated at the same time [18.4]. This X-ray therapy room dates from 1900–1904 and is the department of Hermann Rieder of Munich. Tin shielding masks are used here as in [18.5]. However, in 1909, Albers-Schönberg of Hamburg, writing in the *Archives of the Roentgen Ray* on 'Technique of roentgen irradiatiation in gynaecology' and the use of his compression cylinder [12.5] stated that 'the face and breast should be protected by a separate lead-lined screen, which may have a lead glass window to enable the patient to watch the focus tube'. From around this time, tin was completely discarded in preference to lead as a shielding material—the energy of the X-ray beam being higher in 1909 than in earlier years.

[18.8] Total body irradiation in 1909. Three X-ray tubes are used in this technique of Dessauer[5] but an alternative method by Wetterer[5] used only one X-ray tube and in addition, protected the face with a lead mask. Modern TBI techniques which are used in conjunction with bone marrow transplants employ linear accelerators at focus-to-skin distances in excess of 250 cm with X-ray beam dimensions on the patient's skin surface of some 150 cm. Lung tolerance is considered to be the limiting factor and other organs at risk (which require some form of shielding) include the eyes, kidney, liver, spleen, intestine and gonads[17].

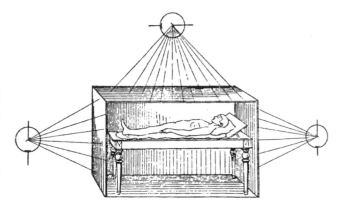

X-ray therapy and radium therapy: 1896–1970s

[18.9] From X-ray therapy and radium therapy of the first 50 years to 'modern' radiotherapy. The terms teletherapy and brachytherapy (from the Greek 'tele' meaning 'distant' and 'brachy' meaning 'near to') now subdivide radiotherapy into its two major modalities. Teletherapy, sometimes called external beam radiotherapy, is the treatment of cancer using apparatus such as a deep therapy X-ray machine, linear accelerator or cobalt-60 machine. The patient is treated at a distance from the machine and either lies on a specially designed treatment couch or is seated in a chair (although the use of a chair is now only for low energy X-ray treatment techniques). When X-rays are used the term is teletherapy, but when a radioactive isotope such as radium or cobalt-60 was used the term was modified to be teleradium or telecobalt. The three major types of brachytherapy, where the radioactive source is close to the tumour, were, for many years, interstitial (sources embedded within the tumour), intracavitary (sources placed within a natural body cavity) and surface or mould brachytherapy (sources placed on a layer of material at a few millimetres away from the skin surface). In the 1980s, however, intraluminal brachytherapy (the sources being placed within a natural body lumen, e.g. the bronchus) came into routine use.

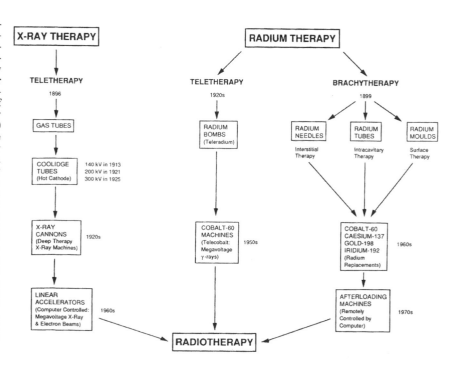

Treatment planning: 1914

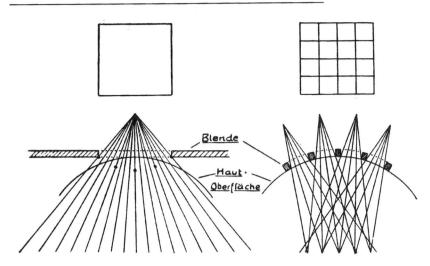

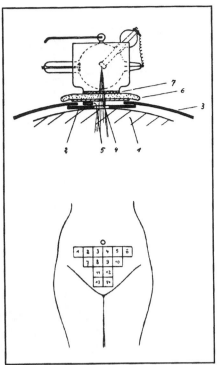

[18.10] These illustrations of multiple beam techniques, from the 1914 catalogue of Reiniger, Gebbert & Schall, pre-date the isodose curve distributions and distributions of dose points as seen in [18.11]. The two treatment field arrangements above both cover an area of 16 cm × 16 cm, on the left with a single field and on the right with 16 fields of size 4 cm × 4 cm. The prescribed dose for both is 10 X (Kienböck) units. The collimation (blende) and skin surface (haut oberfläche) are indicated. An abdominal treatment using 14 fields is shown far right indicating: body of patient (1); lead shield (2); lead rubber (3); filter paper (4); Kienböck's strip (5), see [22.7]; sponge (6); and 3 mm aluminium filter (7). The energy of the X-ray beams available in 1914 was relatively low and had a short depth dose, such that single beam treatments were considered to be inadequate. The non-coplanar multiple beam arrangements had, however, become obsolete by the 1920s.

Treatment plans and isodose curves: 1919–1925 and 1980

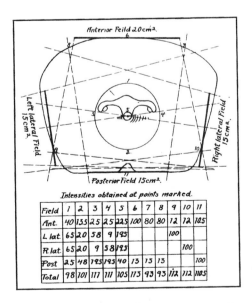

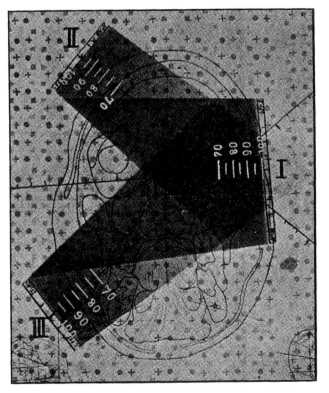

[18.11] Treatment plan[28] using four X-ray beams, Chicago 1919. The full Epilation Skin Dose was stated to be 1050 milliamp-minutes for each X-ray beam. The pelvic tumour volume is the central circle with intensity dose points 1, 2, 3 and 4 on its peripheray and dose point 5 at its centre. The objective was to achieve a homogeneous radiation intensity throughout this volume (98%–111%), but it can also be seen that regions outside this volume received a similar radiation dose.

[18.12] Treatment plan[9] of 1925 with central axis percentage depth doses marked (70%–100%). By convention the maximum-valued central axis dose was always 100% and this would be related to the output of the X-ray therapy machine in whatever units of radiation exposure/dose were being used; there were very many such units until the adoption of the roentgen unit as an international unit of radiation exposure in 1937, see Chapter 22.

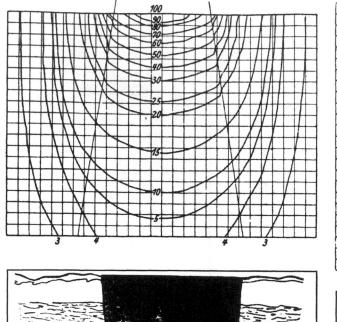

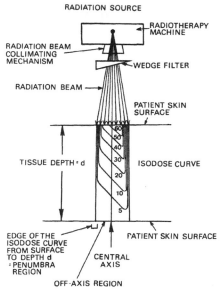

[18.13] Single-beam Dessauer curves (top left), single-beam Holfelder curves (top right), two adjacent Dessauer curves (bottom left) and two adjacent Holfelder curves (bottom right). The skin overdose predicted by the Dessauer curves never occurred, thus proving that Holfelder curves were correct[9].

RADIATION SOURCE

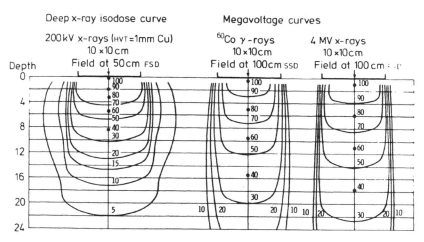

[18.14] Examples of isodose curves[27] from the 1980s. This illustrates how there is significant side scatter with deep therapy X-ray machines but with higher energy photon (X-rays or cobalt-60 gamma rays) beams this feature diminishes and that with a 4 MV linear acceleratot X-ray field the penumbra is much narrower. Side scatter is an undesirable element in a field since it creates a problem when trying to irradiate a tumour volume to a prescribed level whilst minimising the radiation dose elsewhere to unaffected tissues and organs. The reduced central axis percentage depth dose, the maximum dose (100%) at the skin surface (which prevents skin sparing) and the side scatter are major disadvantages of deep X-ray therapy which have been overcome since megavoltage photon beams became a reality in radiotherapy.

[18.15] Schematic diagram[27] of a supervoltage wedged field in which a wedge of material such as copper is inserted in the X-ray or gamma-ray beam to alter the shape of the isodose curves from an open (unwedged) field with a shape such as those in the centre and right of Figure [18.14].

X-ray cannons: 1920–1938

[18.16] A German X-ray cannon. This was the deep X-ray therapy apparatus of Wintz of Erlangen[9] who with Seitz founded the first 'school' of radiotherapy with a technique which called for the entire X-ray dose to be delivered in one continuous session, which, according to the number of fields and the depth dose, extended over a period of hours. Their concept was that a massive X-ray dose would kill all cancer cells at once. It was considered that anything less than a lethal dose would stimulate rather than harm cancer cells. However, this single-day massive therapy protocol was gradually modified to small fractional doses.

[18.17] An American X-ray cannon[9] manufactured by the Wappler Electric Company of New York. The cylinder was lined with one-quarter of an inch of lead and all openings were lead flanged. The treatment table was furnished with a blower which kept a constant circulation of air in the tube chamber, which was carried out from the table by an exit pipe.

[18.18] An example of patient positioning, published in Holfelder's 1938 textbook[11]. (Courtesy: Dr H. Jacobs, Saarbrücken.)

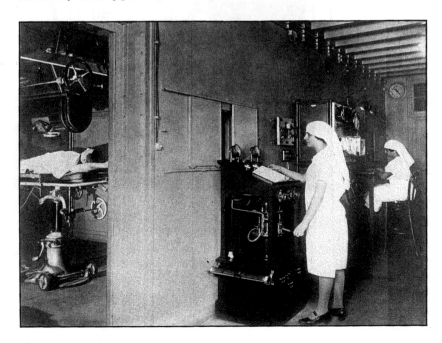

[18.19] A French X-ray cannon of 1920 in the Radiotherapy Department of what is now the Curie-Institut, Paris.

Deep X-ray therapy: 1923–1960

[18.20] A deep X-ray machine of the 1960s, much smaller than the X-ray cannons of [18.16–18.19] This Siemens 250 kV unit at the Royal Marsden Hospital, London, is being used to treat cancer of the breast, using a 50 cm FSD closed ended applicator and bags filled with Lincolnshire bolus to ensure a uniform dose distribution by filling in the air gaps with this tissue-equivalent material. The radiation output was some 30 roentgen/min and the half value layer (HVL) was 3.4 mm of copper.

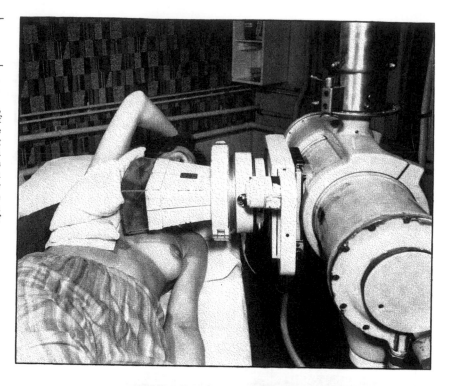

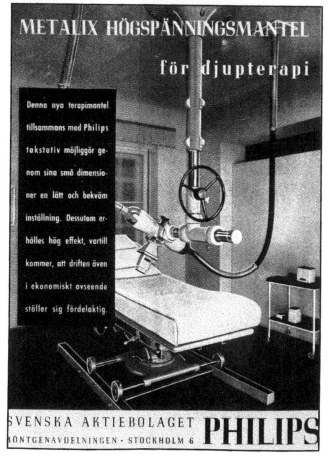
[18.22] Philips Metalix tube for deep X-ray therapy as advertised in 1937 in *Acta Radiologica*.

[18.21] In the early years, some X-ray installations were considered to be useful for both therapy and diagnostic applications and this example from 1923, as advertised in the *Journal of the Röntgen Society*, is one of the last such installations for which this claim was made.

Van de Graaff generators: 1929–1985

[18.23] Robert van der Graaff built his first model of a direct current electrostatic generator in 1929, based on the theory of Lord Kelvin's device in which drops of water carrying small electric charges fell into cups, where the charges accumulated as the drops continued to fall. Van de Graaff, instead of using water, used moving silk belts, carrying charges vertically to be stored on a spherical metal terminal supported by an insulating column. The charges were sprayed on the belt at the base of the generator. The belt carried them upward and into the sphere where they were removed. This first generator which was made from tin cans and a silk ribbon developed 80 kV. The first public demonstration[29] was on November 10th 1931 at a dinner of the American Institute of Physics in New York. It was hailed by the President of Massachusetts Institute of Technology, Karl Compton, as 'the most important development that has ever taken place in the field of extremely high voltages'. By 1933 a huge Van de Graaff machine was constructed at Round Hill, South Dartmouth, Massachusetts and was so enormous that it had to be housed in an airship hanger to accommodate the two generators which developed a voltage greater than 5 MV. The terminal spheres were 15 feet in diameter on 24-foot-high textolite columns and the two generators were mounted on rails 14 feet apart so that the distance between the terminals could be adjusted. Charges were deposited on the spheres by three papers belts each 3 feet wide running at a speed of almost 1 mile per minute.

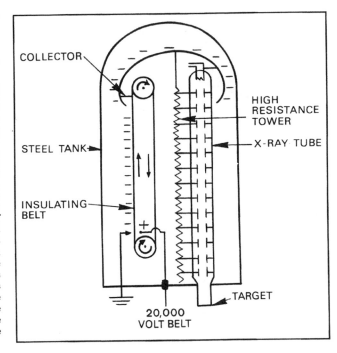

[18.24] The schematic diagram is of a 2 MV generator. The auxillary voltage source of +20 kV is used to spray negative charges onto the moving belt which is driven by a motorised pulley in the direction shown by the arrows. Negative charges from ground potential are attracted onto the belt by the positive potential. Near the top pulley, the negative charges are removed by a collector connected to the spherical cap of the generator. The X-ray tube is placed parallel to the belt, and consists of a filament at the top, a series of accelerating electrodes, and a water-cooled target at the bottom. The upper spherical electrode at a high negative potential is joined to ground by a series of high resistances (the high resistance tower). The high energy electrons impinge on the target and a clinically useful beam of 2 MV X-rays is produced. The generator is enclosed in a steel tank, see Figure [18.25], and is filled with a mixture of nitrogen and carbon dioxide at a pressure of 20 atmospheres[30].

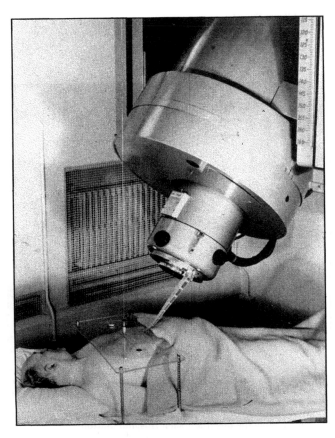

[18.25] A 2 MV Van de Graaff radiotherapy treatment machine in 1961. The first clinical Van de Graaff was installed in 1937 at the Huntington Memorial Hospital in Boston. It operated at over 1 MV and required a room only 25 foot square. The second clinical machine was at the Massachusetts General Hospital, It was insulated by compressed gas, was 50 feet high and operated at 1.25 MV. For a hospital previously only used to deep X-ray units, the increase in energy from some 250 kV to 2 MV made a very significant difference in the ability to treat deep-seated tumours and also tumours of the head and neck. Wedge filters [18.15] could also be used for the first time (since 250 kV machine outputs were too low and the depth dose too small for any practical use of wedges) and, for example, one could plan the treatment of a cervical oesophageal tumour using 2 MV beams wedged in two directions to take into account not only the varying body outline from chin to sternum, but also the slope of the oesophagus from anterior to posterior. This was true innovation in 1961.

Betatrons: 1980–1993

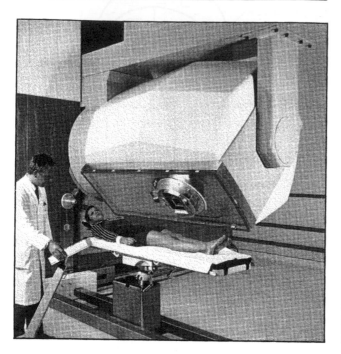

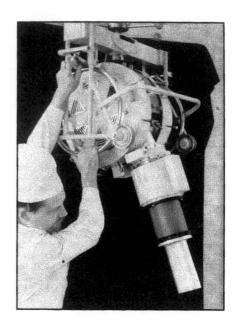

[18.27]   Small betatron developed in the Oncological Research Institute of Tomsk Scientific Centre of the USSR Academy of Medical Sciences, Tomsk, Siberia, for use in intraoperative radiotherapy; for example, irradiation of the tumour bed following surgery. This betatron was described at the 3rd Congress of Oncologists of Byelorussia, Minsk, December 1991. (Courtesy: Prof. B. N. Zyryanov.)

[18.26]   A Brown-Boveri radiotherapy betatron in the early 1980s: the Asklepitron 45. The betatron was used in two modes, X-ray beam or electron beam. Applicators are used for the electron mode. This illustration shows a patient being set up for treatment in the X-ray mode.

   The betatron is a machine in which electrons from a heated filament are accelerated within an evacuated annulus, usually called a doughnut. A changing magnetic field is provided through the doughnut which is placed between the poles of an electromagnet. These poles are shaped so that the electrons are kept in an orbit of fixed radius whilst they are accelerated to high energies. At a selected electron energy a subsidiary electrode is energised so that the electrons deviate from the equilibrium orbit and are extracted through a window in the doughnut. If X-rays are required from a betatron it is arranged that the electron beam impinges on a target.

Linear accelerators: 1953–1992

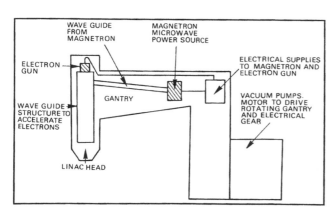

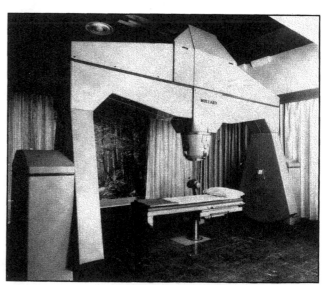

[18.28]   Schematic diagram of the main features of an isocentrically mounted single-gantry 4 MV linear accelerator manufactured by Metropolitan Vickers (AEI Ltd) and first introduced into clinical practice at the Christie Hospital, Manchester, in 1953. The travelling wave was 1 metre length, rotation was $\pm 120°$, the magnetron was 2 MW and there was a 'straight ahead electron beam' with no bending magnet.

[18.29]   Mullard[22] (Philips Medical Systems) 4 MV linear accelerator (the only double-gantry model), with isocentric mounting, rotation of $\pm 105°$, 1 metre length travelling wave, 2 MW magnetron and 'straight ahead electron beam' with no bending magnet. The first installation was in 1953 in Newcastle General Hospital. The photograph is from the Royal Marsden Hospital, Sutton.

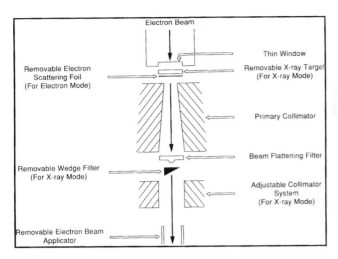

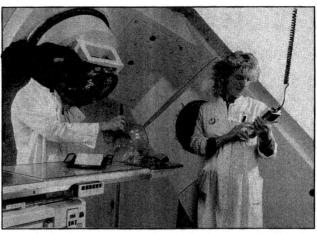

[18.30] Schematic diagram of the main clinical features of a linear accelerator which can be used in an X-ray or an electron mode. (The beam radiation monitor and the optical system for beam definition is not shown in this diagram). Innovations in linacc design as recorded by Philips Medical Systems in 1989 were:

1953: The world's first isocentric medical linacc (installed at Newcastle General Hospital, United Kingdom).
1954: First dual modality unit (X-rays and electrons) (installed at St. Bartholomew's Hospital, London).
1961: Introduction of the concept of a drum gantry.
1973: The world's first dual modality dual X-ray energy linacc (installed at the Antoni van Leeuwenkoek Hospital, Amsterdam).
1979: Microprocessor control incorporated in linaccs.
1981: Concept of motorised wedge giving 0°–60° wedge fields.
1985: The world's first totally computer controlled linacc was launched: (the new SL series).

With a modern linear accelerator[17] movements (either for setting up the patient for fixed field therapy or for using arc rotational therapy) are made about a point called the isocentre. Such isocentric mounted machines (telecobalt and linear accelerators) have a great advantage in terms of accuracy and reproducibility when setting up patients for their treatment position, compared with deep X-ray therapy machines for which there was no fixed geometrical relationship between X-ray tube focus and treatment couch. The focus-to-skin distance (FSD), sometimes termed source–skin distance (SSD) was usually 50 cm for deep X-ray therapy, and for telecobalt machines was often 70 cm, 80 cm or 100 cm, whereas for linear accelerators is virtually always 100 cm. These distances relate to fixed field techniques in which the patient set-up is not isocentric. For isocentric fields the relevant distance is the skin–axis distance (SAD) which is measured from the source to the isocentre.

[18.32] A Siemens KD2 linear accelerator showing a tray fixed to the head of the machine, a treatment cast and a radiographer operating the hand control unit which is used to rotate the accelerator isocentrically around the treatment couch and to raise and lower the couch. The tray is used to fix shielding blocks so that irregular shaped fields may be used and vital organs such as eyes and spine can be protected. The transparent plastic treatment cast is used to fix the patient in position for a head or neck cancer treatment so that they cannot move during treatment. Such cast technology was developed in the late 1950s by C. E. Dickens at the Royal Marsden Hospital, London, and involves first making a plaster of Paris mould of the head and neck and then using a vacuum forming machine to produce the plastic (originally made of Cabulite) cast. For the treatment of tumours in the head and neck (including the cervical oesophagus) where so many radiosensitive organs have to be avoided during treatment, this type of cast fixation provides a real advantage in ensuring that the prescribed treament is delivered accurately. (Courtesy: Dr P. Levendag, Dr Daniel den Hoed Cancer Center, Rotterdam.)

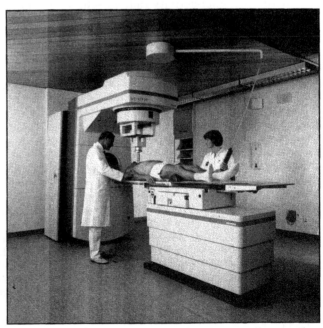

[18.33] A small-field electron beam treatment being set up on a Simens linear accelerator. Applicators are only used with accelerators in the electron beam mode.

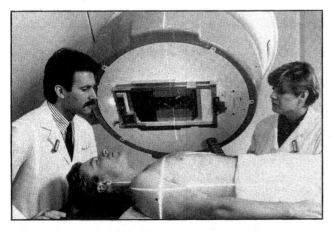

[18.31] Close-up view of the head of a Varian linear accelerator showing the multi-leaf collimator. The white crossing lines in the centre of the photograph are laser lines to assist in setting up the patient in the correct treatment position.

Chapter 19

External Beam Radiotherapy: II

The term 'radium bomb' which was widely used for radium teletherapy for many years [18.9] was initially devised not for teletherapy but for an intracavitary brachytherapy device to treat gynaecological cancer at Memorial Hospital, New York during the years 1917–1919[1,2]. It was termed a 'bomb' because it resembled a small hand grenade. It consisted of a 3.5 cm lead globe containing 1,000 curies of radon surrounded by a 6 mm lead protective cover and supported on a rod [19.1].

The development of radium bomb teletherapy machines in England began in 1919 at the Middlesex Hospital using a 2.5 gm radium source provided by the Ministry of Munitions. At about the same time, similar developments were also made in New York, Paris and Stockholm[3–7]. Some of the earliest radium bombs provided only one radiation beam size of large dimensions and multi-field treatments were impossible. Later designs incorporated higher activity radium sources of up to 10 gm and applicators to collimate the gamma radiation, which could be attached to the radium unit and provide a variety of beam sizes.

With the higher-activity radium sources, the protection problems increased and these were in part overcome by housing the radium source when not in use in a separate lead safe some 3 metres distance from the radium teletherapy unit. The source travelled from safe to unit by pneumatic transfer when the patient was in the treatment position; and was returned by the same method on completion of treatment.

This type of design, which always had a short source-to-skin distance (SSD) of usually not greater than 10 cm, was manufactured by Bryant Symons[8] and for several years was a standard piece of apparatus for treating head and neck cancers [19.9, 19.10]. Problems did, however, sometimes arise with these pneumatic transfer units: notably when the radium source refused to return to the safe. It was not unknown to this author for a long broom handle to come to the assistance of the person who had to free the source 'bobbin' of some 6 cm length from the treatment head or from the transfer tube! These were not everyday problems, but the sound of the radium bobbin thumping hard against the interior of the unit after transit instinctively made one wonder whether there might not one day be the disaster of a shattered source. Fortunately, these Bryant Symons bombs were replaced by cobalt bombs of a more sophisticated design when cobalt-60 sources became available and the potential radiation hazards were considerably reduced.

The radioactive isotope cobalt-60 was first produced in 1941 by J. J. Livengood and Glenn T. Seaborg using a cyclotron, and the first proposal to use this radionuclide as a replacement for radium (see Table 19.1) in

Table 19.1. Advantages of cobalt-60 over radium as seen in 1948[7].

1	Much softer beta radiation: easily filtered out.
2	Homogeneous gamma radiation: 1.1 MeV and 1.3 MeV.
3	Will not leak: no gaseous daughter products.
4	Breakage almost impossible.
5	Suitable alloys of cobalt are chemically inert.
6	Will not localise in bone: quickly eliminated.
7	Magnetic: makes handling easier and safer.
8	Strength determinable before irradiation in the atomic pile.
9	Residual activity can be 'warmed up' in a nuclear reactor.
10	Suitable alloys containing cobalt are inexpensive.
11	Cobalt-60 can be generated in any desired quantities.
12	It is available at moderate cost.

teletherapy was by J. S. Mitchell of Addenbrooke's Hospital, Cambridge in 1946. However, the first cobalt-60 machine, or telecobalt machine as it was termed (the description 'bomb' fell into disuse with cobalt), was loaded with its source in August 1951 in the Saskatoon Cancer Clinic in Canada and the first patient treated in November of that year[9]. (For the most extensive review of the early telecobalt developments see the proceedings of a Symposium on Supervoltage Radiation Therapy held at Oak Ridge Institute in 1956[10].)

The Saskatoon telecobalt machine was built to the design of Harold Johns by the Acme Machine Electric Company. The second telecobalt unit, also Canadian, installed in London, Ontario, was built by the Eldorado Mining & Refining Company (which was a supplier of radium sources) and it was the first unit ever to treat a patient—in October 1951. This company later became Atomic Energy of Canada Limited (AECL) which in 1952 installed telecobalt units in New York and in Chicago, and, from then on, expansion was rapid. By 1959 there were 46 models of telecobalt machine available from 18 different companies manufacturing in nine countries. By 1955 there were 150 telecobalt machines installed worldwide and by 1961 this figure had risen to 1120. However, by 1993 there are only two companies remaining who manufacture telecobalt machines (one of the problems is the supply of suitable kilocurie cobalt-60 sources) and in many centres they have now been replaced by linear accelerator installations.

Other radioactive isotopes were proposed as alternatives to radium for teletherapy units, one example being iridium-192 in 1950[12], but few teleiridium machines were built, mainly due to the short half-life of 74 days which made frequent source replacement essential.

Another alternative was caesium-137 which has a half-life of 33 years and a gamma ray energy of 0.66 MeV compared with the half-life and gamma ray energies of cobalt-60 of 5.3 years and 1.17 MeV and 1.33 MeV. However, using cobalt-60 a much higher specific activity could be obtained than with caesium-137 with the result that telecaesium machines, [19.12], had a relatively low radiation output and because of the source size a significantly larger penumbra than that of telecobalt. These two disadvantages ensured that only a few telecaesium machines were ever built.

The two most recent occasions when gamma ray teletherapy machines have hit the world headlines have been related to radiation accidents, in Mexico[13] and in Brazil[14], for respectively telecobalt and telecaesium.

In Juarez, Mexico, the equipment involved was a Picker Corporation telecobalt machine [19.16] manufactured sometime before 1963 with the most recent supply of cobalt-60 sources in 1969 when the activity was some 3000 curies. The accident occurred in December 1983. Unlike more modern telecobalt machines which have all the cobalt-60 concentrated into a single small disc or rod with maximum dimensions of some 2 cm, the Picker source consisted of about 7000 tiny pellets, each of 1 mm diameter. The machine had originally been sold to the Methodist Hospital in Lubbock, Texas, which, when it was no longer required, sold it to an X-ray equipment company in Fort Worth. They in turn, in 1977, shipped it to the Centro Medico, Juarez in 1977. It was never installed and remained in a warehouse till November 1983. Someone then decided to dismantle it and it was stolen, eventually arriving in a pickup truck at the Junke Fenix, a Juarez scrap metal yard in December 1983. This sequence of events *only became known by chance* in January 1984. A truck loaded with steel rods from scrap took a wrong turning at the Los Alamos National Laboratory in New Mexico, USA, and happened to pass over a radiation sensor in the road outside the laboratory, and set off an alarm. In the month after the break-up of the machine, two steel foundries in Mexico and one in the USA handled radioactive steel, some of which was used to manufacture steel for the legs of restaurant tables. By the time the transport of the steel had been stopped, some 5,000 tonnes of reinforcing rods and some 18,000 table legs

had left Mexico. Also, at least 12 children had played on the highly contaminated pickup truck before it was removed to safety, still containing some of the cobalt-60 pellets.

About the end of 1985 a private radiotherapy institute in Goiania, Brazil, moved to new premises and left in place a telecaesium machine, without notifying the appropriate authorities. The former premises were partly dismantled and two men who thought the machine might have scrap value tried to dismantle it and ruptured the source capsule in the process. The source was in the form of caesium chloride salt which is highly soluble and readily dispersible. This resulted in environmental contamination, and external and internal contamination of several people occurred which eventually led to four fatalities: a 6-year-old girl who had tasted the radioactive powder, a 38-year-old woman and two men aged 22 and 18. It was, without doubt, the worst ever radiological accident involving radiotherapy apparatus.

Radium bombs: 1917–1965

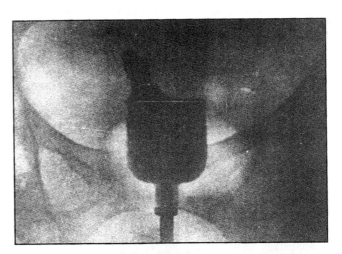

[19.1] Radiograph of a cancer of the cervix patient treated using the Memorial Hospital vaginal bomb containing 1,000 millicuries of radon[1,2].

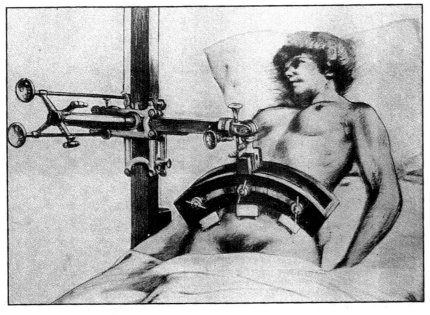

[19.2] The Mallet–Coliez radium bomb[5] designed in 1924. It incorporates three moveable lead applicators which define three radium gamma-ray beams whose central axes converge at a defined point. The applicators were boxes with lead walls of internal dimensions 75 mm length × 37 mm width × 100 mm height. The lead thickness was 20 mm and was stated to guarantee a 60% radiation absorption. Radium tubes were placed in a rigid canal in each applicator and the usual radium loading was 100–200 milligram radium per applicator. A later version was designed using 18 applicators.

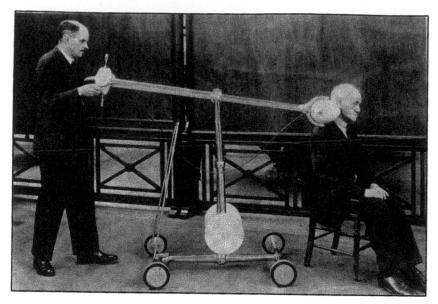

[19.3] A 1 gram radium bomb designed by Flint[15,16] in 1934 for four London hospitals: Royal Marsden, Middlesex, University College and Westminster. In November 1929, the Royal Commission (and subsequently The King's Fund) placed at the disposal of Westminster Hospital 4 gram of radium which had to be used as a single unit under certain stated conditions[17]. In 1931 after 15 months of use the apparatus was abandoned because of the results obtained. This 4 gram of radium was then divided into four 1 gram sources for the radium bomb design shown in this figure. From 1936, this design was superseded by that of the Bryant Symons bomb.

[19.4] In America the terminology for short distance external gamma-ray therapy machines was 'radium pack' rather than 'radium bomb'. A typical example is shown[6] where the radium tubes are placed on blocks of a low atomic weight substance (e.g. wood, cork, rubber) of 3–12 cm thickness. The total radium content could be as much as 2 g.

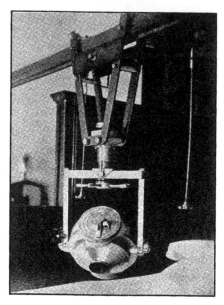

[19.5] In America, Failla[18] was probably the physicist most involved in the design of radium bombs, including those at Roswell Park Memorial Institute in Buffalo and at Memorial Hospital, New York. The Failla radium bomb shown contained 40 radium tubes totalling 4 gram, and had two applicators, positioned such that when one patient was receiving treatment a second patient could be positioned. The unit was suspended from a ceiling rail which ran across the two treatment rooms. The adjoining wall protection was 40 mm lead. Later, in the 1930s, Failla designed a 10 gram radium bomb for the Chicago Tumour Institute and then a 50 gram unit for the Roosevelt Hospital, New York. This latter solved the problem of too low a dose rate, which for the Buffalo 4 gram bomb in 1930 was only 3.5 röntgen/minute.

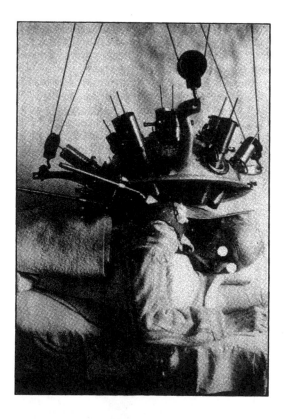

[19.6] A Belgian-designed radium bomb of 1925 from Sluys & Kessler[6,19,20]. The content was originally 1.3 gram radium and the 13 radium foci were enclosed in 13 applicators fixed on a hemispherical cupola. The appplicators were hollowed-out lead cylinders enclosed in brass tubing with the lead 2 cm thick. The radium was enclosed in a brass block, which moved like a piston in the hollow centre. A screw mechanism allowed the radium to be fixed at any position within the applicator. It was used to treat cancers of the larynx, pharynx, oesophagus, bladder and brain but it was stated that the treatment techniques involved 'much delicate adjustment'. Daily treatments were usually of 4–6 hours duration for a total of 150–200 hours. The radium content per foci could be increased from 100 milligram radium to 400 milligram for a total of 5.2 gram.

[19.7] A treatment technique using the Sluys–Kessler radium bomb[20] and five of the 13 applicators.

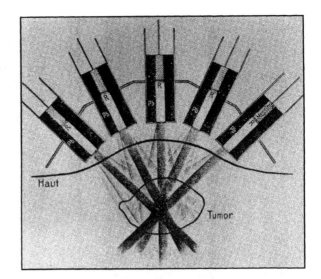

[19.8] Physics dosimetry for the Sluys–Kessler apparatus. A spherical aluminium ionisation chamber of size 2 cm was used, termed in 1929 a 'small chamber'. The instrument with which the measurements were made used a filament electroscope of the type shown in [22.25], although in this instance the two ionisation chambers were replaced by a single chamber in which the pressure could be varied. Protection for this working physicist was a lead wall of 15 cm placed between the electroscope and the ionisation chamber[6].

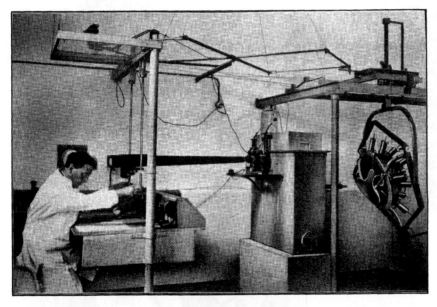

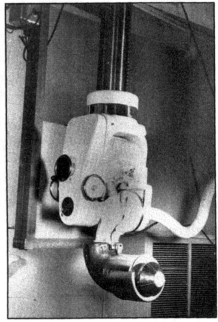

[19.9] A Bryant Symons radium bomb at the Royal Marsden Hospital, London, in the 1960s. The pneumatic transfer tube can be seen on the right of the photograph. A few of these units were later modified to contain cobalt-60 sources.

[19.10] A Bryant Symons radium bomb treatment at Westminster Hospital in the 1940s. The patient is seen wearing a plaster of Paris cast to ensure no movement during treatment.

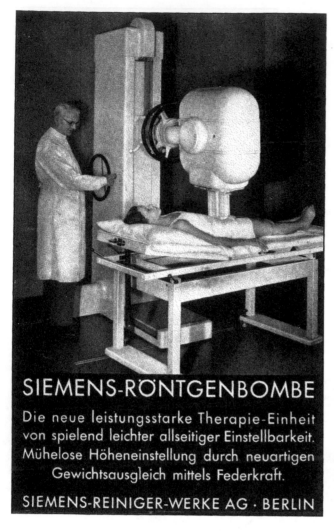

SIEMENS-RÖNTGENBOMBE

Die neue leistungsstarke Therapie-Einheit von spielend leichter allseitiger Einstellbarkeit. Mühelose Höheneinstellung durch neuartigen Gewichtsausgleich mittels Federkraft.

SIEMENS-REINIGER-WERKE AG · BERLIN

[19.11] Although external beam radium machines were called bombs for many years there is only one example (such as in this advertisement in a 1939 issue of the journal *Röntgenpraxis*) of an external beam X-ray machine being called a 'bomb'.

Telecobalt machines: 1955–1980

[19.13] The Cancer Hospital of the Shanghai Medical University was the first hospital in China to use radium sources for the treatment of cancer (1929). The teletherapy machine shown is the first cobalt machine used in Shanghai. It was made in the Soviet Union in the early 1950s but its design was such that it could not be used for rotational therapy. This deficiency was remedied by using the 'swing bed' shown in the photograph. (Courtesy: Prof. Liu Tai Fu.)

Telecaesium machine: 1965

[19.12] The world's first telecaesium machine[17,18] went into service in 1957 at the Royal Marsden Hospital, London, using a 1500 Ci caesium-137 source obtained from Windscale (now Sellafield) and continued in routine clinical use for over 25 years. The Bryant Symons telecaesium machine illustrated above, installed in the 1960s, also at the Royal Marsden was unusual in that it had two treatment 'ends': one for tumours of the head using an SSD of 30 cm and one for tumours of the neck using an SSD of 20 cm. In addition, this machine was also used to treat breast cancer. It was a caesium-137 source from a telecaesium machine that was the cause of the radiation accident in Goiania, Brazil, in 1987.

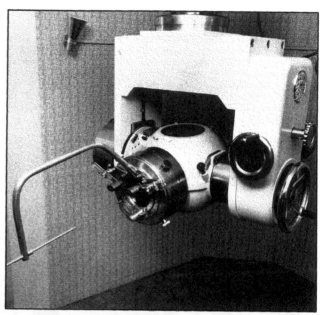

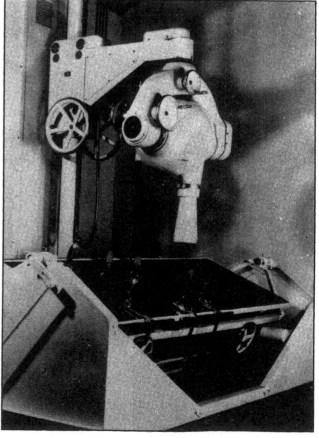

[19.14]   An AECL Theraton telecobalt machine, 1957[21].

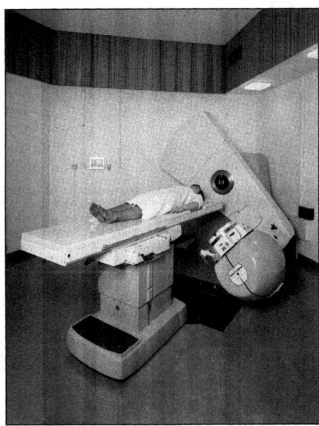

[19.15]   A TEM Mobaltron F100 telecobalt machine. The pit in the floor beneath the treatment couch enables the machine to have a larger arc of rotation around the patient than would otherwise be possible. Two metal crash bars can be seen protruding on either side of the head of the unit. These are a safety mechanism which if they touch the treatment table or patient, immediately cause the machine to stop its motion. The large dimensions of the shielded head of the machine are necessary for radiation protection purposes, since when the cobalt-60 source is withdrawn from the treatment position into the head of the machine it must be fully shielded.

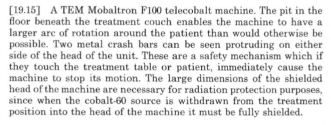

[19.17]   Dose distribution curves in a plane through the nasopharynx. This 1960s treatment involved the use of seven cobalt-60 external beams and included protection of vital organs such as the eyes, mouth, larynx and spine. This distribution was calculated before the availability of dedicated treatment planning computer systems. Modern computer planning systems routinely produce three-dimensional dose distributions linked to CT scans. In addition, treatment simulators are now routinely available for external beam verification, of correct size and position of a beam, prior to delivery of the actual treatment[22]. All treatment planning dimensions are in cm.

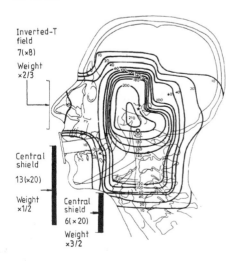

Inverted-T field
7(×8)
Weight ×2/3

Central shield
13(×20)
Weight ×1/2

Central shield
6(×20)
Weight ×3/2

[19.16]   The head of a Picker telecobalt machine of the 1960s (left). This would be of a similar type to that involved in the radiation accident in Juarez, Mexico, in the early 1980s. The Canadian postage stamp commemorates the design by Harold Johns of the world's first cobalt-60 external beam radiotherapy machine. The collimator system of shutters which defines the gamma-ray beam irradiating the patient is included and the diagram is similar to that in [18.15] although in this instance there is an 'open' (that is, 'unwedged') radiation beam. The decay scheme of the radioactive isotope cobalt-60 is included, specifying the dual gamma-ray energies of 1.17 MeV and 1.33 MeV.

Chapter 20

Brachytherapy

The first successful brachytherapy treatment for cancer was in 1903 in St. Petersburg by Goldberg and London[1] who treated two patients with facial basal cell carcinoma. Radium brachytherapy has already been referred to in previous chapters, such as the earliest radium applicators of Wickham and Degrais[2], Paris, 1904, [3.16]; the first international radium standard, 1912, [3.15]; and a chart [18.9] detailing progress in both brachytherapy and teletherapy; and in Chapter 23 some of the early hazards in the use of radium and early radiation protection practices are featured.

Figure [20.1] is typical of the illustrations[2] that appeared before about 1910 in the very early textbooks reporting the results of radium brachytherapy: they were artists' impressions in colour of the patient before and after treatment. Figures [20.2] and [20.3] are photographs of two radium brachytherapy patients treated in the early 1920s at the Institut-Curie, Paris. Figure [20.2] shows a typical design of radium plaque of this period, including its method of fixation. This was in the era before the Manchester System of Paterson and Parker (1930s) when it was not yet realised that to achieve a homogeneous radiation dose distribution on the skin surface (or at 0.5 cm depth) beneath a radium plaque (or mould as they were sometimes called) required a non-uniform distribution of radium content on the plaque. Thus all early plaques had a uniform distribution of radium content.

The series of three photographs in [20.3] is unique in that it shows a baby girl just before treatment (1923), at 28 months after treatment and finally when she was about the age of 10 years. The photographic records do not show what condition she was treated for, but similar case history drawings by Wickham and Degrais (1908)[2] were for angiomata and epitheliomata. Nevertheless, whatever the condition, it was extremely disfiguring and the success of treatment can be judged from the follow-up some 10 years later.

The first brachytherapy applications were surface moulds and plaques such as in [20.2] and they remained in use for many years until superseded in part by superficial X-ray therapy or electron beams from linear accelerators. However, there has recently been a moderate increase in their use, in conjunction with remote afterloading machines, because of the guaranteed radiation protection and because the labour-intensive planning and radioactive loading of such plaques is no longer necessary. Remote afterloading takes the labour out of loading the plaques and modern treatment planning computers remove the labour of former years which was required for radiation dosimetry for the treatment of large irregularly shaped lesions.

There was, however, an intermediate stage between the virtual absence of dosimetry in the very early days and the high technology remote afterloading and computer planning of today. This was the development of standard systems of dosimetry for surface moulds (and for interstitial brachytherapy using radium needles) of which the most famous was the Manchester System[3] [20.10]. This system could also be applied, with varying degrees of success, to radium-replacement radionuclides such as caesium-137 tubes and needles, radon seeds [20.13], gold-198 grains [20.14] and tantalum-192 hairpins [20.15].

In the 1960s, though, a further dosimetric system was devised, known as the Paris System[4,5]. It is based on the principle of equidistant parallel linear sources of uniform linear activity. In this system the dose delivered to the target volume (tumour dose) is specified along an isodose line defined as a percentage of the basal dose (by convention 85%). This isodose is called the reference isodose and must encompass all the tumour volume and the safety margin. Figure [20.18] is a radiograph of an iridium-192 wire implant using the plastic tube technique of Pernot[6] for a tumour of the soft palate for which dosimetry is according to the Paris System. This system is now more popular than the Manchester System, in part because with treatment planning computer software, dose distributions can be computed in three dimensions for any arrangement of sources and there is thus no need anymore for such rigid standardisation of positioning and radioactive loading of sources such as required by Manchester, and in part because the Paris System 'rules' are simpler to use in practice.

However, to return to the earlier years, before the 1930s, when no physics-based dosimetry systems existed, the major treatment site using brachytherapy became (and today, in some centres, still remains) the cervix, because this was a readily accessible tumour site with a relatively high cancer incidence in many countries, exceeded only for women by cancer of the breast (and later by cancer of the lung when female populations increased their habit of cigarette smoking). The various techniques developed differed in the design of the containers for the vaginal radium sources (applicators), the radioactive loading of the vaginal and uterine sources, and the number and duration of the brachytherapy intracavitary insertions in vagina and uterus. Cancer of the corpus uteri (endometrium) was, and is, not so prevalent as cancer of the cervix uteri, and most of the efforts in devising brachytherapy protocols were for cervical carcinoma.

The first centres of note[7] for this treatment were in Paris, where Regaud established his technique at the Fondation Curie in 1919, in Sweden at the Radiumhemmet (Radium Home) in 1910, where the Stockholm technique was developed, in Munich where

Döderlein treated patients from 1908, and in Baltimore where Howard Kelly established his radium clinic in 1908 and treated cancer of the cervix from 1911. The Manchester technique[3] using rubber ellipsoids as applicators was developed in the 1930s from the Paris method[8] and the Fletcher[9] method was in turn in the 1950s developed from the Manchester method. Figures [20.19]–[20.24] show some examples of cervical carcinoma applicators from 1912–1993 and [20.27] shows the classical Heyman technique for the treatment of endometrial cancer.

The earliest brachytherapy treatments were surface applications of radium or intracavitary treatments for cancers of the uterus and vagina, but interstitial brachytherapy with radium inserted into the tumour soon followed. It was probably the Munich physician, H. Strebel, who first devised and used an interstitial technique [20.9] as reported in 1903[11] although the suggestion is often attributed to Alexander Graham Bell, the inventor of the telephone. Bell corresponded in the same year, via a letters column of *American Medicine*, with a Dr Z. T. Sowers, as reproduced below.

Dear Mr Sowers,

I understand from you that Röntgen rays, and the rays emitted by radium, have been found to have a marked effect upon external cancers, but that the effects upon deep-seated cancers have not thus far proved satisfactory.

It has occurred to me that one reason for the unsatisfactory nature of these latter experiments arises from the fact that the rays have been applied externally, thus having to pass through healthy tissues of various depths in order to reach the cancerous matter.

The Crookes' tube, from which the Röntgen rays are emitted is of course too bulky to be admitted into the middle of a mass of cancer, but there is no reason why a tiny fragment of radium sealed in a fine glass tube should not be inserted into the very heart of the cancer, thus acting directly upon the diseased material. Would it not be worth while making experiments along this line?

[Signed]   Alexander Graham Bell

We now know how successful these 'experiments' were and this chapter contains illustrations of interstitial brachytherapy for cancers of the breast, prostate, lung, oesophagus and brain in [20.32]–[20.44].

The term 'afterloading' has already been used several times and the introduction of modern afterloading techniques both manual as in [20.16]–[20.18] and remote controlled as in [20.31], [20.34] primarily evolved because of the radiation safety advantages of this principle. Source guides/applicators could be safely positioned without any radiation hazard because the radioactive source would be afterloaded. However, as the techniques developed, other advantages were realised, not least the increased degree of control (particularly with high dose rate (HDR) techniques) of the brachytherapy dose distribution with remote afterloading and computer treatment planning software incorporating optimisation techniques. The principle, though, is not new, and it was first propounded in 1903 by Strebel[11], followed in 1906 by the New York surgeon Robert Abbe[12]. Abbe's afterloading applicator tubes were made of celluloid or of rubber and were used because of the fragile nature of the then radium sources in the hope that such afterloading would prevent source breakage[13].

There was though, 10 years later in the USA, still some argument about whether or not the use of radium should be encouraged and in the minutes[14] of the 1914 'Radium hearing before the Committee on Mines and Mining United States Senate' one medical expert declared 'The field of usefulness for the substance has been quite limited and there is at present no basis for the belief that recent developments so glaringly displayed in the newspapers will bring any immediate advances of a startling character.' This, though, was dismissed by Howard Kelly of Baltimore who had just returned from a visit to Paris. 'I have spent four days in Paris and have seen all the radium men. I say "men" advisably, for Madame Curie is sick. Do not be afraid that radium will play out. It is an established thing and is doing wonderful things in curing bad cases, many of which have stood the test of years. The testimony in London is the same as here.'

Such doubts had completely vanished within the next five years and it was reported in the New York City *Mid-Week Pictorial* of June 2nd 1921 that President Harding presented to Marie Curie, in the East Room of the White House, one gram of radium worth $100,000 for medical use in Europe, the money having been raised by the women of America (Chapter 2 and [2.19]).

Since the first two remote controlled brachytherapy afterloading machines were devised by Nalstam[50] and by Henschke[15] for gynaecological applications sources [20.50], several designs of remote afterloader have been manufactured but many have now become obsolete as technological improvements have been achieved in the field of miniature radioactive source production, computer technology and brachytherapy treatment planning software. The first remote afterloaders operated at low dose rates (LDR) but with the advent of the Cathetron[16] [20.51] high dose rate (HDR) brachytherapy became practical and opened up an entire new field of brachytherapy advantages as detailed in [20.57] after Joslin[5] who pioneered the use of HDR brachytherapy using the Cathetron.

Most cancer body sites were treated in the early days of brachytherapy with varying degrees of success, although there was usually at least one anecdotal case history showing excellent results, whatever the technique. However, it became obvious that some body sites were not really amenable to brachytherapy and by the 1940s the modality was largely limited to gynaecological cancers and those in the head and neck and breast. The availability of reliable high technology HDR afterloaders and associated treatment planning software has brought about a renaissance in brachytherapy, not least because of the short treatment times and dose distribution optimisation software.

Thus we are now increasingly seeing intraoperative brachytherapy techniques for sites such as the pancreas and brain, and excellent palliative results for cancers of the lung. This is important progress because of the high incidence and poor prognosis of this tumour means that the alleviation of lung cancer symptoms is a real gain in terms of quality of life if, as has been shown[18], heamoptysis, breathlessness, cough and lung collapse can be effectively treated. Brachytherapy techniques for lung cancer for the years 1929 and 1989 are compared in [20.37] and [20.38].

Radium patient treatments: Paris, 1908–1923

[20.1] Patient with an epithelioma of the parotid region: before and after treatment in 1908. The tumour was described by Wickham and Degrais[2] as 'hard as plaster and extending transversely over 9 cm from external ear to inner quarter of cheek and vertically over 12 cm from temple to edge of lower maxilla. At the most prominent part it was nearly 5 cm high.' The radium treatment was given using applicators number 1, 2 and 4 in [3.16]. The dramatic effect on the tumour is obvious and it is no wonder that case histories such as this caused such great enthusiasm amongst the early brachytherapists, because without radium, such a patient would have been untreatable as the tumour was inoperable. These dramatic responses to radium brachytherapy were often hailed as cures, which even if they were not, were certainly treatment successes leading to improved quality of life.

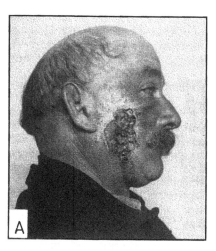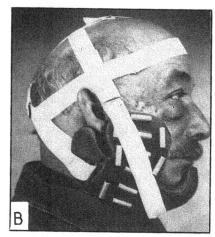

[20.2] [A] Patient before radium treatment: March 1922. [B] Patient with radium plaque in position: February 1923. [C] Patient cured: 1923. (Courtesy: Institut-Curie, Paris.)

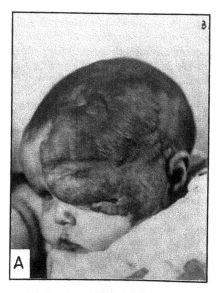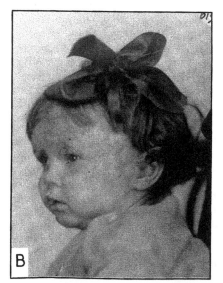

[20.3] [A] Baby girl before radium treatment: December 1923. [B] Follow-up: April 1926. [C] Follow-up: circa 1933. (Courtesy: Institut-Curie, Paris.)

Radium surface moulds: 1905–1929

[20.4] Radium plaques being applied in the Skin Department, St. Vincent's Hospital, Melbourne, Australia in 1905[19].

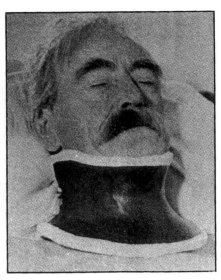

[20.6] Radium 'collars' were a form of surface mould and that shown, after Cade[21] in 1929, is a Columbia paste collar for the treatment of the cervical area. It is 15 mm thick and is in contact with the skin in the submaxillary and carotid areas, but fits loosely over the clavicles. The radium needles (not shown) are applied on the surface and the whole area is covered with adhesive plaster.

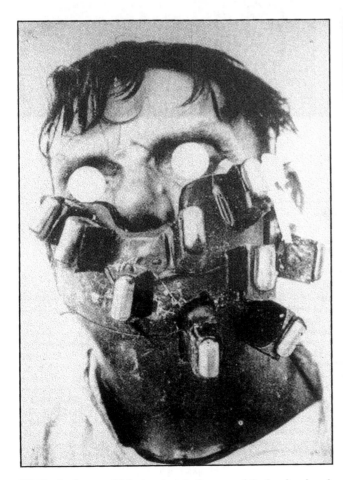

[20.5] Surface mould for treatment of cancer of the head and neck in the 1920s. Most of these techniques were in this anatomical region and that shown in this photograph is reproduced from the textbook[20] issued in 1929 by the Union Minière du Haut Katanga, which was the major supplier of radium until the early 1930s.

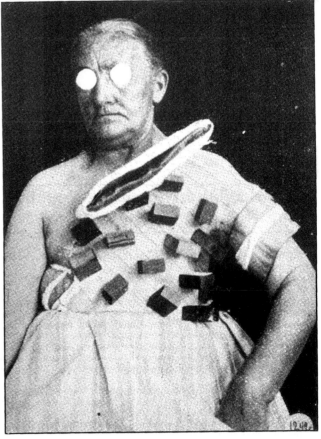

[20.7] Surface mould applicator used to treat breast cancer in 1929[20]. Such moulds were still in use in the mid-1940s in some centres[22].

Interstitial brachytherapy: Dublin, 1914

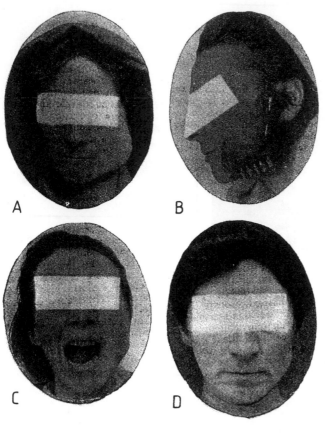

A   B

C   D

[20.8] Although the use of interstitial brachytherapy had been suggested in 1903[10,11], it was not until 1914 that any illustrated case histories were reported. These earliest reports were from Dublin, Ireland, by Stevenson and Joly[23] who used radon in glass capillary tubes placed within 'ordinary steel serum needles as supplied by any medical instrument maker: external diameter 1.4 mm, thickness 0.3 mm, diameter of bore 0.8 mm'. Stevenson's first illustrated case (left) was the treatment of an inoperable parotid sarcoma. (A) Patient before treatment. (B) Case showing position of needles. (C) Showing amount patient could open her mouth after one month treatment. Before treatment she could not separate her teeth. (D) 52 days after first treatment.

Stevenson's second illustrated case (below) shows the position of needles in a fibrous scar. This patient was treated for a burned and lacerated wrist which had prevented her working for five years because pieces of tendon were cut away when first treated and the fibrous scar was a source of constant pain. The radon treatment 'made the scar soft', the pain vanished and several weeks later she restarted work.

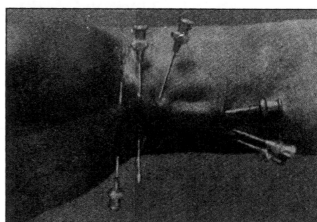

Interstitial afterloading: Munich, 1903 and New York, 1906

[20.9] The first proposals for afterloading have been referred to earlier in the introduction to this chapter, quoting Strebel[11] from Munich and Abbe[12] from New York. It is also interesting to note in Strebel's paper on 'Proposals concerning radium therapy' (I am most grateful to Prof. Karsten Rotte for his translation from the original German of this 'Vorschlage zur Radiumtherapie') that he treated at least one patient with interstitial afterloading brachytherapy (although no photographic record was published), thus preceding Stevenson and Jolly[23] and probably predating Alexander Graham Bell's proposal[10]. Strebel appears to vanish from the literature after this 1903 communication and has possibly been overlooked for this reason.

Strebel claimed that he 'was probably the first to whom it occurred to actually use radium for therapy and that he placed a quantity of radium enclosed in a capsule made from cardboard and paraffined paper on an area of lupus and left it in place for hours. A moderate reaction was obtained with subsequent fading of the lupus node. With further radiation an ulcer was formed which healed only poorly, but the lupus was not healed'. He reasoned that the unsatisfactory results were due to 'low radiation capacity of the radium available'. He then noted that Danlos and Bloch of Paris announced their successful treatment of lupus vulgaris after his own unsuccessful attempts: a claim which makes Strebel one of the earliest, if not the earliest, of radium brachytherapists.

He noted that the radium sources available in 1903 with a strength of 1.5 million uranium units were capable of producing dermatitis lasting 8–14 days after only 10 minutes exposure. It was to avoid this problem that Strebel instituted his interstitial afterloading technique.

'I am now in a position to increase the effectiveness of radium for deeper seated pathological condition quite significantly, without causing undesirable effects on the skin itself. Instead of surface applications, intratumoural application is carried out by inserting the radium, which is enclosed in the drilled tip of a small aluminium rod, directly into the centre of the tumour, with the help of a previously inserted trepan. This increases the size of the irradiation beam in that the radium irradiates evenly in all directions'. Strebel then continues to advocate the trepan (afterloading guide) needle being inserted in different directions, always using the first insertion opening and states that this method is suitable for cancers of the uterus, stomach and liver; and 'can be used successfully as a substitute for X-rays when it is a question of the treatment taking place within small spaces such as the nose, larynx and bladder.'

Paterson and Parker Manchester System: 1934–1938

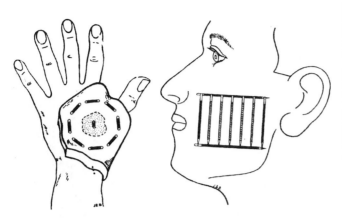

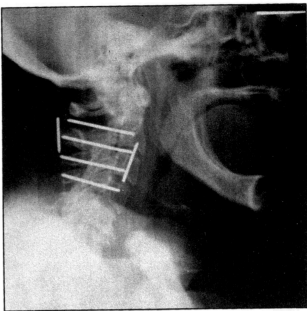

[20.10] Radium surface applicator arrangements in the Manchester system[3] devised by Paterson and Parker in the 1930s. The arrangements for the cheek surface mould were equally applicable to interstitial brachytherapy treatments with needles implanted in a single plane, although in this case it was sometimes impossible to implant both crossing needles (e.g. for a tongue cancer implant where no base needle near the mandible could be implanted).

In this system the distribution of radium depended for circular moulds on D/h, the ratio of the diameter (D) of the mould to the treating distance (h) which is the distance from the plane of the radium source to the plane within the tumour on which the dose is to be prescribed. Thus for D/h < 3 a single circle is sufficient; for D/h equal to or greater than 3 but less than 6, 5% of the radium should be placed at the center and 95% round the periphery. For larger values of D/h there should be a central 'spot' of 3% of the total radium and for D/h values of 6, 7.5 and 10 respectively, the percentage of the total radium in the outer circle should be 80%, 75% and 70% with the remainder being distributed round a circle of half-diameter. Rules were also devised for square, rectangular and irregular shaped planar implants, for double-plane implants, and for intracavitary radium treatments for cancer of the cervix. The aim for the surface and interstitial treatments was to deliver a prescribed 'uniform' dose to a given plane, where 'uniform' was defined as ±10% and a 'look-up' table was provided for the dose to be computed. In practice, the dose rate was often 1000 roentgen per day. For the intracavitary Manchester system the dosimetry was referred to specified points A and B defined in relation to vaginal and uterine anatomy. Point A is often still used today[24] as a dose specification point, although the pre-loaded radium Manchester system itself has now vanished into history.

The Manchester system was not the only radium dosage system to be developed in the 1930s, but it was by far the most comprehensive and the most used of all such systems. Other systems[25], with different rules, were developed by J. Murdoch and E. Stahel in 1931 of Brussels, by the London Hospital surgeon H. S. Souttar in 1934, by G. C. Laurence in 1936 who was a member of the research staff of the Canadian National Research Laboratories, by W. V. Mayneord in 1932 of the Royal Cancer Hospital in London, and by Edith Quimby[26] of New York. This latter system was extensively used in North America for many years.

The basic calculations underlying these systems were dose distributions around ideal geometry radium sources (line, disc, annulus, sphere, cylinder) of uniform radioactive density, all of which commenced with assuming the source consisted of an infinite number of small point sources from which a composite dose distribution could be calculated. The Sievert integral[27] was a turning point in this field of theoretical dosimetry.

[20.11] Lateral radiograph of an interstitial radium needle implant to the neck, 1965, with a separation of 1 cm between the four parallel needles.

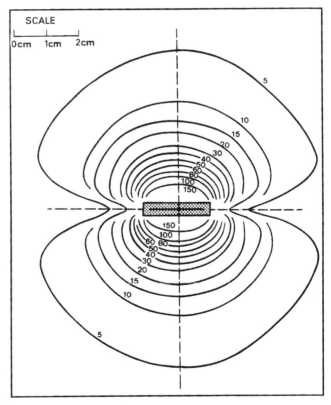

[20.12] Isodose distribution around a 15 mg radium Amersham G-tube: all G-tubes were of 2 cm geometrical length. The curves are in rads/hour and were calculated (1966) on the basis of Sievert integrals[27]. The end-effect due to the inactive ends of the tube is clearly seen. Similar end-effects occur with radium needles due to the point end and the eyelet end. Radium G-tubes, and later caesium-137 G-tubes, were often used for interuterine linear sources and for vaginal sources in the treatment of cancer of the cervix. See also [3.17, 20.56].

Radon seeds, gold grains and tantalum hairpins: 1965

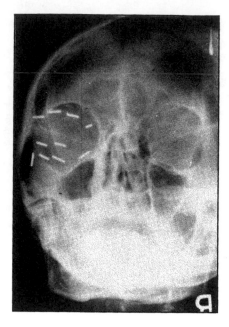

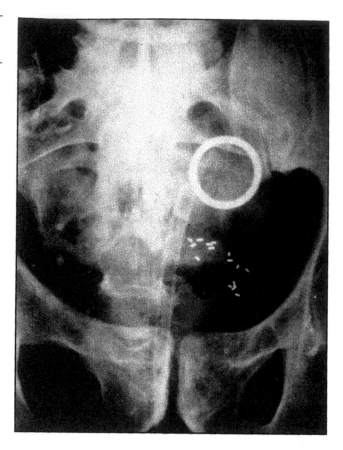

[20.13] Radon seed implant to the orbit, 1965. (Courtesy: Dr M. Lederman, Royal Marsden Hospital.) The smallest radium needle or tube was 2 cm in length and therefore for some small area surface moulds or implants, the use of radium sources was impractical as overdosage would occur. Small radon seeds were used for such implants since they were available in lengths of 5 mm, 7.5 mm, 10 mm and 12.5 mm (from Amersham as unthreaded or threaded seeds) and the Manchester system rules could be approximated. A radon production plant at the Institut-Curie in Paris is seen in [23.17]. In America, the first radon plant was designed and installed at Memorial Hospital, New York, by G. Failla in 1917. The vault for storage of the radium was skillfully connected with a pumping apparatus, so that the daily withdrawal of the emanation (radon) for therapeutic use was rarely interrupted, according to the *Annals of Memorial Hospital*. For a review of radon techniques, see the 1948 textbook[28] of Jennings and Russ.

[20.14] To avoid the potential hazards of the use of radon, which is a gaseous radionuclide, gold-198 grains (seeds) were developed since this radioactive isotope has a half-life similar to that of radon, 2.7 days compared to 3.8 days. Implantation was achieved using a specially designed 'gold grain gun[29]' (which was incorrectly implicated by Scotland Yard as a weapon of assassination in the Georgi Markov murder by a small metallic grain containing ricin, on Lambeth Bridge, London, 1977, in what is now called the Poisoned Umbrella Murder). The lateral radiograph shown is of a gold-198 grain implant to the bladder. Two problems were inherent in this technique of the mid-1960s. One was the difficulty, if not impossibility, of obtaining a Manchester system distribution of activity (treatment planning computers were not then available for individual calculations) and the other was the probability of an enthusiastic nurse in the operating theatre who as soon as the grains were implanted, and the blood welled up in the bladder, would use the sucker not only to remove the blood but also to remove the grains! This would only be detected on a check radiograph and by this time the grains could be stuck in the U-bend of the operating theatre sink.

[20.15] Tantalum-182 wire hairpin implant for bladder cancer, H. J. G. Bloom, Royal Marsden Hospital, London, 1965[30]. Tantalum-182 wire was the forerunner of iridium-192 wire.

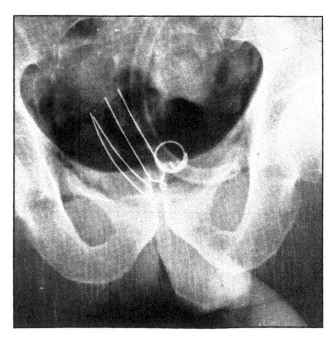

Manual afterloading gold and iridium techniques: 1953–1989

[20.16] The first patient treated with post-operative afterloading[31] was in 1953 by Ulrich Henschke at the Ohio State University Hospital. A patient previously treated in 1951 by surgery for cancer of the lower gum recurred in 1953 with large nodes in the left neck and a radical neck dissection proved impossible. Henschke suggested suturing stainless steel wires through the neck nodes and then three days later these were used to pull nylon ribbons loaded with gold-198 seeds through the nodes. The radiograph of the implant is shown and 7500 rad were given and the neck nodes regressed completely.

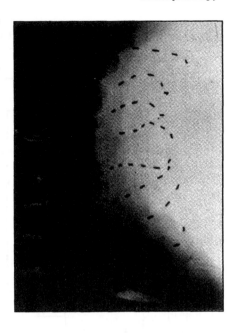

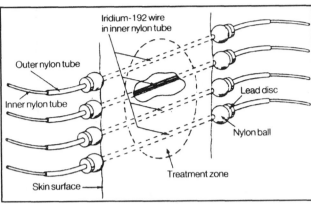
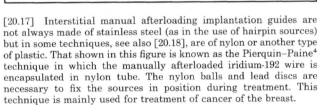

[20.17] Interstitial manual afterloading implantation guides are not always made of stainless steel (as in the use of hairpin sources) but in some techniques, see also [20.18], are of nylon or another type of plastic. That shown in this figure is known as the Pierquin–Paine[4] technique in which the manually afterloaded iridium-192 wire is encapsulated in nylon tube. The nylon balls and lead discs are necessary to fix the sources in position during treatment. This technique is mainly used for treatment of cancer of the breast.

[20.18] Iridium-192 wire implant for cancer of the soft palate, involving the pillars on both sides[6]; Pernot's plastic tube afterloading techniques.

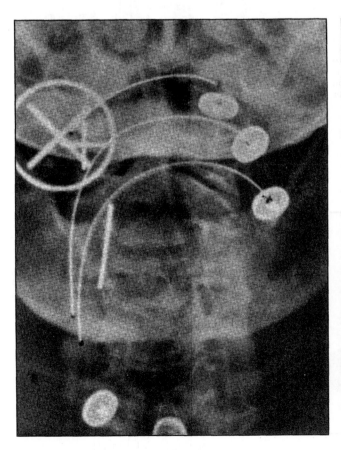

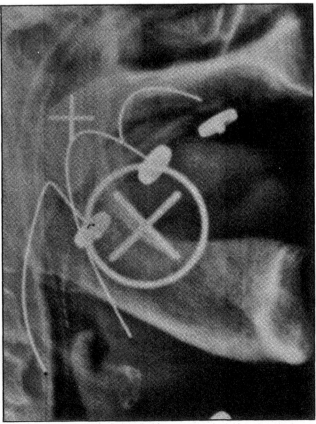

Gynaecological intracavitary applicators: 1905–1992

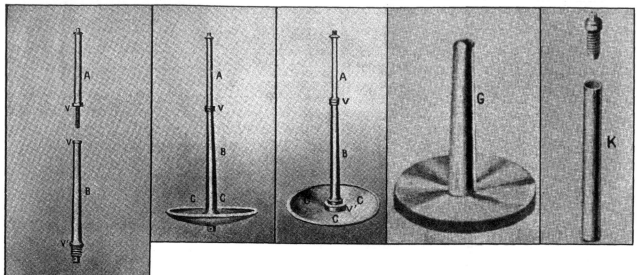

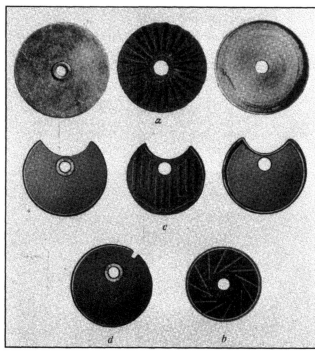

[20.19] First published photographs of gynaecological applicators: they were used by Wickham and Degrais[2] of St. Louis Hospital, Paris, from the beginning of 1905. 'Three parts form the apparatus. The base is screwed on. The depressions on the rods A and B are intended to be filled with radiferous varnish. The varnish deposited in depression C is independent of the centre of the cup. The end which passes through cup C is perforated for the insertion of a band. The stem is inserted in the uterus: as a whole if it is desired to treat the body, or without part A for the cervix alone. The cup is applied to the cervix.' Top right: lead or silver screen which fits on the stem and cup, and screen-tube serving for the whole system, or one of its parts.

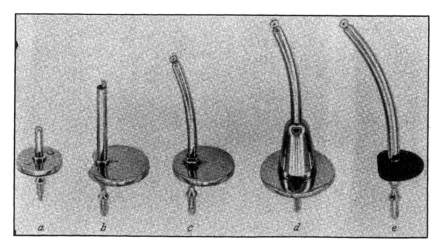

[20.20] Applicators designed in 1912 in Freiburg by Carl Josef Gauss and later, 1928–29, modified in Würzburg by Gauss and Carl Theodor Neef[32]. They were called 'intercervical tubes (left) connected with portio plates (above)' and remained widely used in the German-speaking world for many years. Within the portio plates were grooves to accommodate the radium sources.

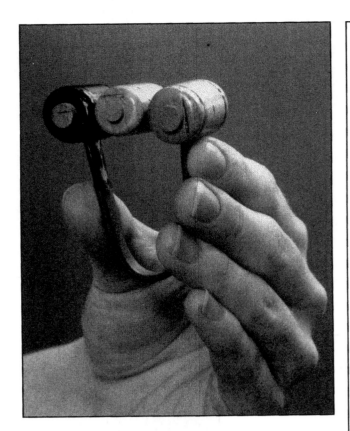

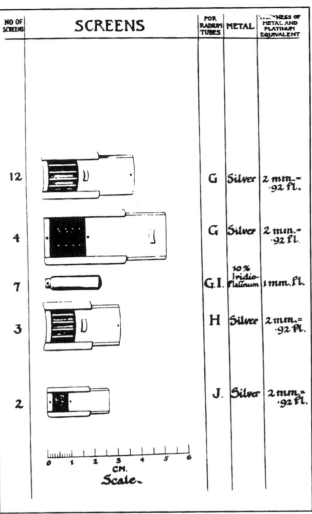

[20.21]   The vaginal radium in the Paris method was loaded into cylindrical cork applicators, two of which were joined together by a flexible metal spring. A third cork could also be used for large vaginal vaults. The method was one continuous treatment for 120 hours. The uterine tube contained 33.3 mgm radium and the two vaginal applicators contained 13.3 mgm radium each.

[20.22]   Typical Stockholm vaginal boxes used at the Marie Curie Hospital, London[7]. There were several variations of the Stockholm method but basically the vaginal boxes, made of silver with grooves for the sources, contained 60–80 mgm radium and the uterine tube 50 mgm radium. Three treatment fractions were given, each of 22 hours, separated by 1–2 week intervals.

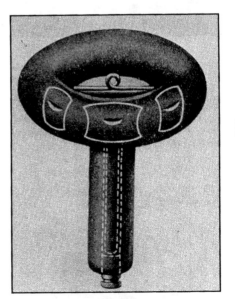

[20.23]   Ring applicator described in 1929 by von Seuffert of Munich[33].

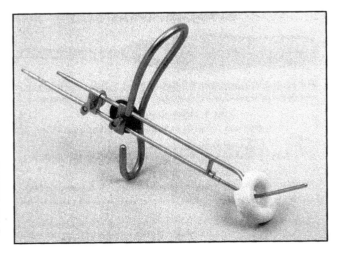

[20.24]   Modern design of Ring applicator for use with the remote afterloading Selectron-LDR/MDR which uses caesium-137 pellet trains. The perineal bar helps keep the applicator in position during the low dose rate treatment. A Ring applicator can also be used for high dose rate treatment, such as with the microSelectron-HDR, but in this case, with the much shorter treatment times, a perineal bar is not required, but a rectal retractor is useful to reduce the dose to the rectum.

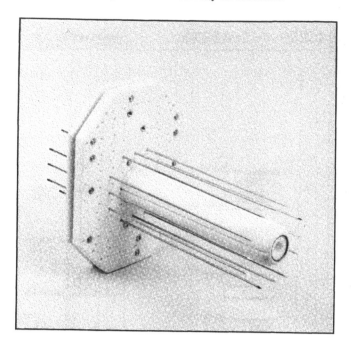

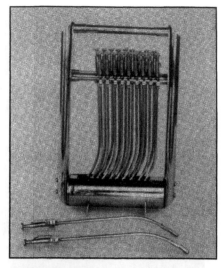

[20.25] In modern interstitial brachytherapy special templates are sometimes used when there are a large number of needles involved in the technique such as the MUPIT (Martinez Universal Perineal Interstitial Template) configured here for use with vaginal cancer and a microSelectron-HDR remote afterloader. The MUPIT can also be used for treatment of cancer of the prostate or cancer of the rectum and anus. Several template designs are in existence and another example was used for the patient in [20.36].

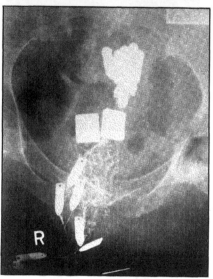

[20.27] Heyman technique[35] for treatment of cancer of the uterus. The apparatus (top) consisted of a series of curved steel rods containing a radium capsule at one end which was fixed to the rod by a taut steel wire. At the other end of the rod was a release mechanism so that after the radium capsule was inserted into position the rod could be withdrawn. The steel wire, no longer taut, was used at the end of treatment to withdrawn the radium capsule. The radiograph (bottom) shows Heyman capsules in the uterus, two Stockholm vaginal boxes [20.22], and the metal tags at the ends of the steel wires.

[20.28] Schematic diagram showing a 1990s remote afterloading modification of the original Heyman technique, developed in Stockholm at the Radiumhemmet, for the treatment of cancer of the uterus[35]. As many radium capsules as possible are packed into the uterine cavity and vaginal boxes can also be used in conjunction with the capsules [20.27]. This Heyman technique was also earlier modified by Norman Simon of the Mount Sinai Medical Centre in New York for manual afterloading[36].

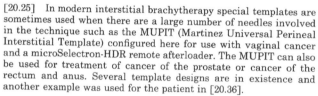
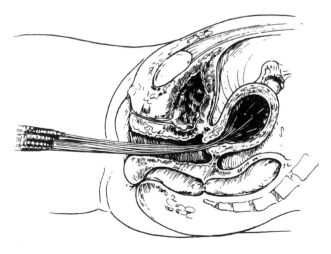

[20.26] A 1914 design of radium and mesothorium applicator that could be used to treat cancer of the vagina and cancer of the rectum[34].

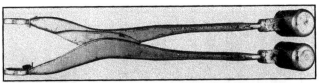

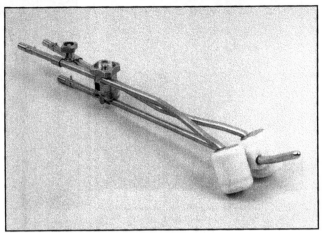

[20.29] Manchester System applicators[3]. The vaginal applicators (termed colposats in North America) were given the special name 'ovoid'. Three pairs of ovoids are shown (small, medium and large) and a washer (bottom left) and a spacer (bottom centre) and three uterine tubes (long, medium and short). In this method, two treatment fractions each of 72 hours duration were prescribed over a period of 10 days. The choice of uterine tube (lengths 6 cm, 4 cm and 2 cm) and vaginal ovoids depended on the anatomy. The uterine sources were loaded 15 + 10 + 10 mgm, 15 + 10 mgm and 20 mgm radium. Each ovoid loading was 22.5 mgm, 20 mgm and 17.5 mgm radium.

[20.31] The earliest design (top) of Fletcher applicator colpostats (ovoids), developed in the late 1950s by Gilbert Fletcher of the M. D. Anderson Hospital, Houston, incorporated tungsten shielding to reduce the rectal dose. Its design was also revolutionary for this period, not only because of the rectal shielding, but also because it was the first design in which fixed geometry could be obtained during treatment and ovoid and intrauterine tube positioning did not depend solely on the expertise of the physician or surgeon's use of gauze packing in the vagina. Later designs were developed for manual afterloading using radium or caesium sources, initially by Herman Suit and then by Luis Delclos. Modern design (bottom) of a Fletcher–Suit–Delclos type of applicator, in this case, for use with the Selectron-LDR/MDR remote afterloading machine.

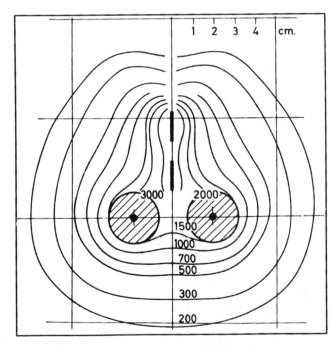

[20.30] Typical 'pear-shaped' isodose distribution for a Manchester insertion[37]. The uterine source is 15 + 10 mgm radium and the vaginal radium is 2 × 20 mgm. The dose levels correspond to 1000 rad to the Manchester point A.

Breast cancer: 1929 and 1992

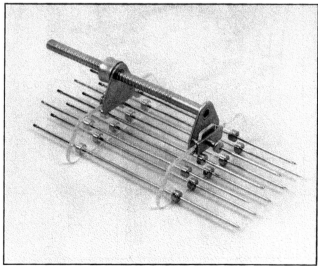

[20.32] This is a breast template such as used for the patient in [20.34] who was treated using a microSelectron-LDR/MDR remote afterloader.

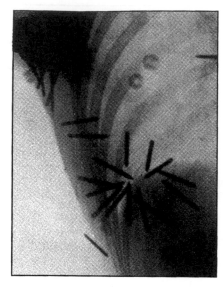

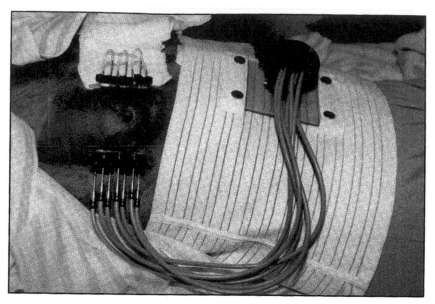

[20.33] Interstitial implant for breast cancer[21] in 1929. The radiographs shows the distribution of radium needles. The small circular shadows are beads attached to the distal end of the threads for identification purposes. In this Westminster Hospital technique, 40–50 needles containing a total of 75–100 mgm radium are inserted. The average total dose prescribed was 16,000–21,000 milligram-hours.

[20.34] Remote afterloading interstitial brachytherapy in the 1990s. This Academic Hospital Utrecht technique[38] is used for breast conservation therapy and ensures that it is not necessary for the patient to have mutilating breast surgery, such as often occurred in the first half of this century. The remote afterloading machine is the microSelectron-LDR which uses caesium-137 ribbon source assembles. A template is used (left) to assist in positioning the needles accurately, the source transfer tubes (centre) and the afterloading coupling (right) are also shown.

Prostate cancer: 1922 and 1992

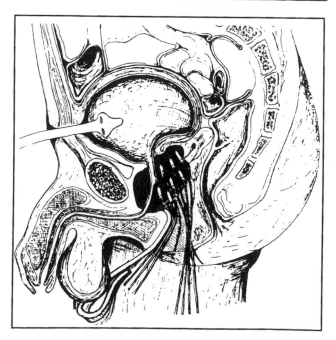

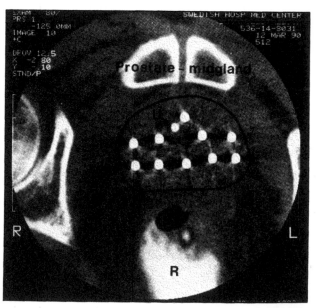

[20.35] Treatment of cancer of the prostate by an open perineal method first described in 1922[21]. The radium needles were left in the position 18–21 days and the average dose was stated to be 7,000 milligram-hours of radium.

[20.36] Templates are also used in interstitial brachytherapy for body sites such as the prostate and the current sophisticated techniques, which often also make use of CT scans, bear no resemblance to the early 'straightforward' implantation of radium needles for a prescribed time to deliver a given number of milligram-hours dose [20.35]. The CT scan[39] is a post-implantation image to verify the needle positions (indicated in the centre of the CT scan by white dots) before the iridium-192 stepping source is afterloaded in this microSelectron-HDR technique at the Swedish Hospital Tumor Institute in Seattle. U and R denote urethra and rectum respectively.

Lung cancer: 1929 and 1992

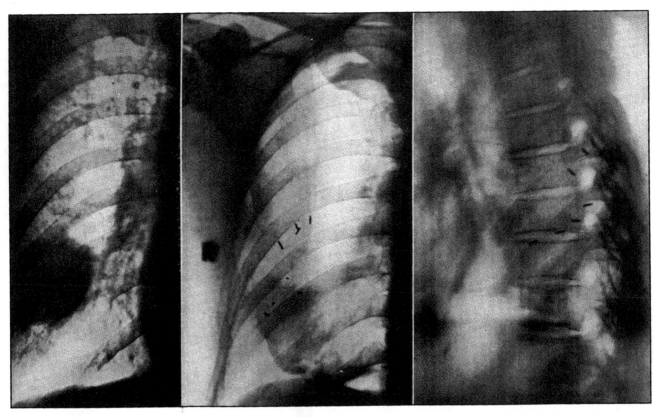

[20.37] Radiographs before and after radon seed implantation of the lung[42] in 1929 at the London Hospital. This technique was first used in 1921 by Yankauer[44] in New York and Henry Pancoast[45] after whom Pancoast's tumour was named, stated that 'The implantation of radon in the lower respiratory tract for bronchial carcinoma is an advance in the treatment of what has been a hopeless condition. Even this, however, is probably not the last word in the treatment of bronchial carcinoma.' Pancoast was correct as seen from [20.38].

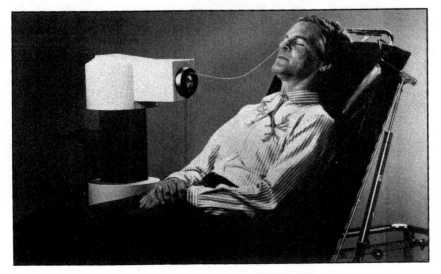

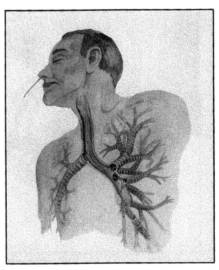

[20.38] Artist impression of the treatment of endobronchial cancer using the micro-Selectron-HDR remote afterloader with its miniature iridium-192 source. On the right is shown the catheter passed through the nose and down into the bronchial tree beyond the tumour mass.

Brain and pituitary tumours: 1929–1992

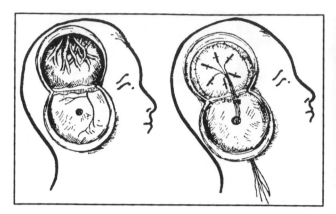

[20.39] Interstitial brachytherapy with radium needles at the London Hospital[42] in 1929 for treatment of an angioma of the meninges. This is an early example of intraoperative brachytherapy. Six needles each of 2 mgm radium were implanted and brought out after being implanted for five days, through the central hole in the bone flap.

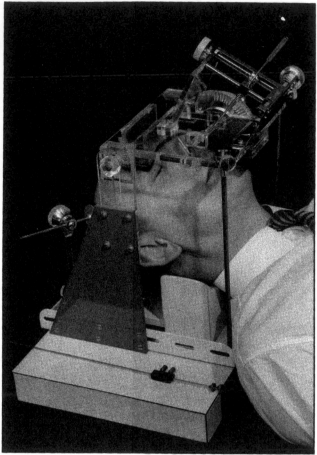

[20.40] Apparatus for a radon seed implant to the pituitary using the nasal cavity route, Westminster Hospital, London, 1956. An improvement on this technique in the 1960s was the use at the Royal Marsden Hospital of small yttrium-90 rods: one technique using the inner canthus route to the pituitary. Yttrium-90 is a pure beta emitting radionuclide with a maximum energy of 2.27 MeV and a half-life of 64.2 hours.

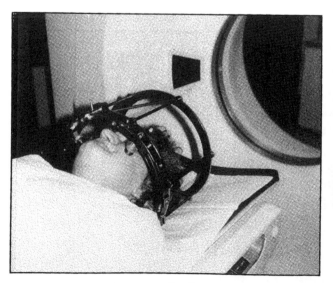

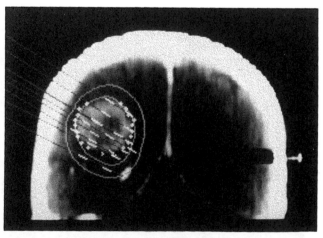

[20.42] Iodine-125 seed interstitial brachytherapy for a brain tumour: coronal view of a planned implant showing catheters, seeds and isodose lines for 30 cGy/hr and 50 cGy/hr, University of California, San Francisco[49].

[20.41] Interstitial brachytherapy for brain tumours using micro-Selectron-HDR afterloading flexible catheters at the University of Würzburg Strahlenklinik[43]. The photograph shows the patient with a stereotactic ring positioned during CT scanning.

Oesophagus cancer: 1904 and 1913

[20.43] An applicator[40] designed in 1904 by Hartigan, a London surgeon. It was described as 'permitting the application of radium to hitherto inaccessible situations, eg. oesophagus, larynx and bladder'.

[20.44] St. Bartholomew's Hospital, London, 1913 technique[41] for intralumenal bachytherapy of cancer of the oesophagus. The apparatus consists of a silver rod 61 cm in length down which was inserted a small radium tube source. It was necessary to bend the silver rod at the pharyngeal angle so that the patient could move their head into a reasonably comfortable position. Its use also required the silver rod to be heated to redness and then cooled so that it became more flexible.

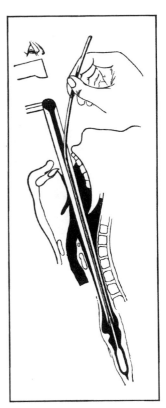

Head and neck cancer: 1915–1933

[20.45] Radium tube arrangement[46] for the treatment of floor of mouth cancer at the Royal Marsden Hospital, London, 1915. The equipment was obtained from Siemens and the notation is as follows. N: applicator in protective material with handle. O: radium tube. P: rubber filter. Q: protective material. S: cross-section. T: flat applicator arranged in a similar manner.

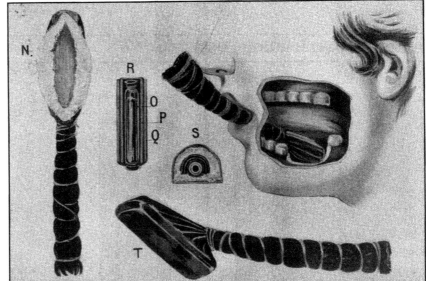

[20.46] Radon seed technique for treating a base of tongue cancer: Memorial Hospital, New York, 1933[47].

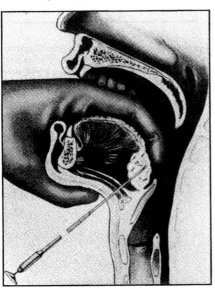

[20.47] Fenestration operation for intraoperative placement of radium sources for treatment of cancer of the larynx, Westminster Hospital, London, 1929[21].

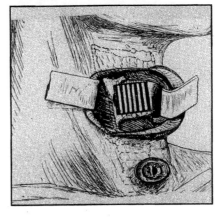

Bile duct cancer: 1992

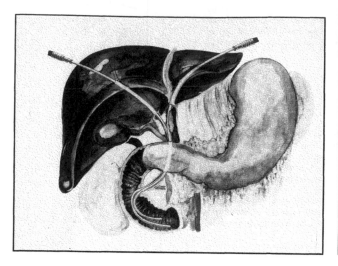

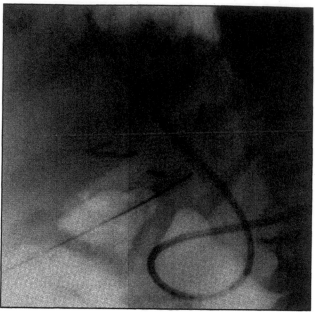

[20.48]   One of the few body sites which could never be treated using brachytherapy when only pre-loaded radium sources were available: the bile duct. However, with the miniature iridium-192 source of the microSelectron-HDR and a long catheter, this is now practical, as shown at the Beth Israel Medical Center in New York[48]. A schematic diagram of the catheter pathway is shown (left) together with a radiograph (right) of a 10 French nasobiliary tube in the bile duct.

Remote afterloading: 1962–1993

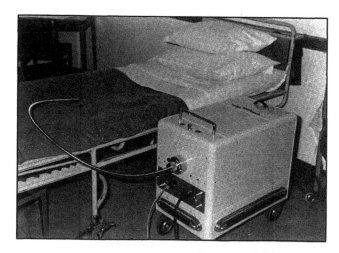

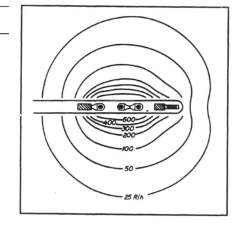

[20.49]   The original experimental apparatus (left) from the Radium-hemmet in Stockholm[50] which was the first ever remote afterloading brachytherapy machine. The trolley (into which is built a protective lead container and a transport cable mechanism) is seen and the 140 cm flexible tube terminated with a stainless steel tube for application in the patient. Only some 10 patients were treated with this remote afterloader. The apparatus contained a three-linked source system of three 50 mg and one 9 mg radium tubes. These are seen in the isodose distribution (above). Ulrich Henschke[15] was the first to design a remote afterloading system with an oscillating source, but the Rune Walstam design[50] was the first to be reported (1962) in the literature.

[20.50]   The Henschke remote afterloading machine[15].

Brachytherapy

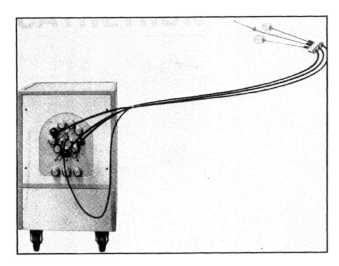

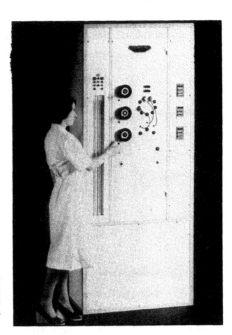

[20.51] The Cathetron HDR afterloader. The size of the control panel (right) can be visualised because of the radiographer standing next to the panel and this can be compared with a later HDR remote afterloader, the microSelectron-HDR in [20.38]. The source safe is also shown (left) with source transfer tubes connected to an applicator arrangement containing two vaginal ovoids and a uterine tube. The Cathetron[16] was developed in 1963.

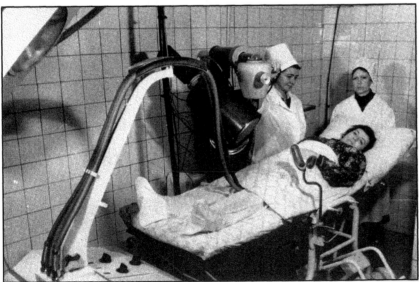

[20.52] The AGAT-V remote afterloading machine[51] at the Research Institute of Oncology and Medical Radiology, Minsk, Byelorussia. (Courtesy: Dr N. Okeanova.)

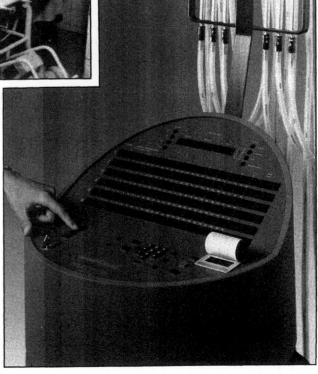

[20.53] Selectron-LDR/MDR afterloading machine. On the right are six transfer tubes for the caesium-137 pellet trains. The six channels can be programmed with any combination of active and inactive pellet sources and the channel displays (centre) show the selected source positions and times. The cylindrical body of the machine (only the top of which is shown in this photograph) contains a sorting mechanism for the pellets, a main safe, a distributor mechanism and an intermediate safe. Applicators are connected to the source transfer tubes and examples are seen in [20.24] and [20.31]. Dose rate plays an important role in brachytherapy technique, although all centres do not in practice use the same definitions for low, medium and high dose rates. As an example the ICRU Report No. 38 defines[52] LDR as 0.4–2 Gy/hour, MDR as 2–12 Gy/hour and HDR as exceeding 2 Gy/minute.

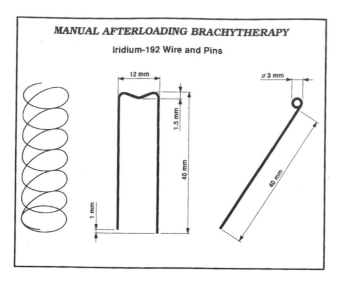

MANUAL AFTERLOADING BRACHYTHERAPY

Iridium-192 Wire and Pins

[20.54] The iridium-192 wire hairpins and single pins were used with a guide gutter manual afterloading technique. If plastic tube techniques were to be used, iridium-192 wire sources were cut to size from a wire coil of typical length 500 mm. The platinum coating of 0.1 mm for these iridium-192 sources acted as a beta-ray filter.

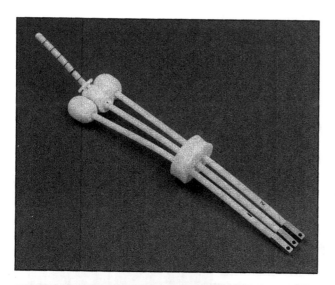

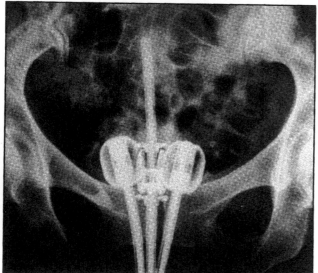

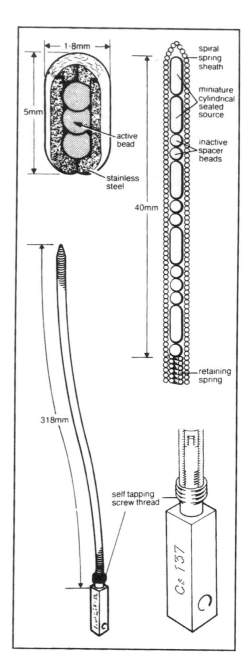

[20.55] The Amersham caesium-137 afterloading system[53] for gynaecological brachytherapy is an example of 1970s manual afterloading technology devised primarily for use in developing countries. The applicators are made of plastic and are disposable. They are based on Manchester ovoid and interuterine tube applicators with similar dose-fractionations to point A. The radiograph shows the applicators in position. The diagrams illustrate the caesium-137 source train assembly. The source train in its entirety is shown bottom left, with the source train handle bottom right. At top left is a cylindrical caesium-137 source which incorporates three active beads. Top right shows an intrauterine source train in which inactive spacer beads can be seen.

NON-AFTERLOADING BRACHYTHERAPY

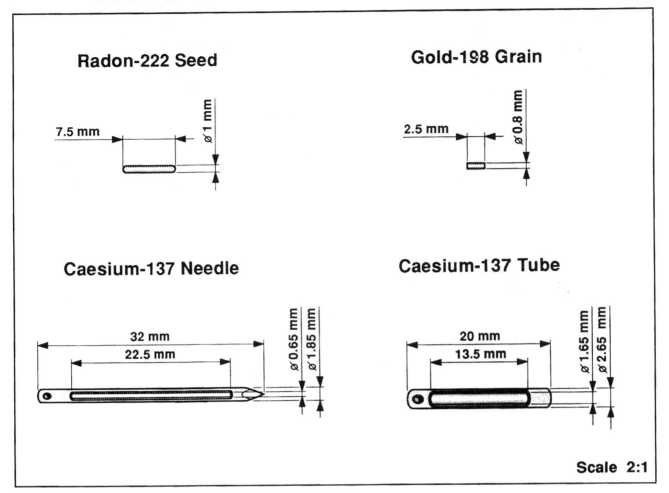

Radon-222 Seed

7.5 mm · ∅ 1 mm

Gold-198 Grain

2.5 mm · ∅ 0.8 mm

Caesium-137 Needle

32 mm · 22.5 mm · ∅ 0.65 mm · ∅ 1.85 mm

Caesium-137 Tube

20 mm · 13.5 mm · ∅ 1.65 mm · ∅ 2.65 mm

Scale 2:1

[20.56] Comparison of different brachytherapy source designs. Gold-198 grains replaced radon seeds in the 1960s, and later caesium-137 tubes and needles replaced radium tubes and needles (see also [3.17, 20.12]. The Amersham caesium G-tubes which replaced the radium G-tubes were available with caesium-137 activities of 15, 30, 45, 60 and 75 mCi. The Amersham caesium-137 needle is shown here with an external length of 32 mm and 22.5 mCi activity. Three other lengths were available: 24.5, 40 and 55 mm with single-strength activities of respectively 1.5, 3 or 4.5 mCi. Double-strength activity needes were also available but had similar external and active length. In the 1970s and 1980s these were typical source ranges for pre-loaded brachytherapy techniques, although the caesium G-tubes were also used for manual afterloading gynaecological techniques.

Advantages of HDR brachytherapy over radium for the treatment of cancer of the cervix and uterus:

- It practically eliminates radiation hazards.
- The position(s) of the isotope source(s) and can be readily checked and maintained constant during treatment.
- It reduces treatment times and inconvenience to patients.

Patient advantages:
(1) Not being confined to bed for hours or days during irradiation with the consequent indignity of bed pans, etc.
(2) No indwelling catheters or vaginal packing.
(3) Not being labelled 'radiation risk zone' to relatives, visitors and staff.
(4) Day case treatments are possible.

Clinical advantages:
(1) Anatomical relationships not altering during treatment.
(2) The position of the source applicators is easily maintained during treatment.
(3) Patient preparation for treatment is simple.
(4) No specialised nursing.
(5) A relatively high throughput of patients on each machine.

Physical advantages:
(1) Short treatment times and minimal irradiation protection problems.
(2) Similar irradiation procedures to external beam irradiation.
(3) Optimisation of dose distribution being relatively straightforward.

[20.57] High dose rate brachytherapy advantages, after Joslin[17].

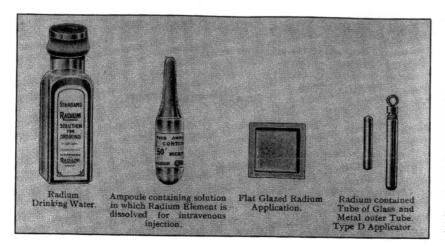

Radium
Drinking Water.

Ampoule containing solution
in which Radium Element is
dissolved for intravenous
injection.

Flat Glazed Radium
Application.

Radium contained
Tube of Glass and
Metal outer Tube.
Type D Applicator.

[20.58] Pre-dating Amersham as a supplier of radium sources was the Radium Chemical Company which produced 13.6 gram of radium element in 1918, and whose representative in the United Kingdom was Watson & Sons (Electro-Medical) Ltd of London from whose 1920 catalogue these items are reproduced. The radium tubes (far right), 40 mm long and 5 mm in diameter, were constructed on the principle proposed by Dominici in Paris in the first decade of this century. The flat applicators were 2 cm × 2 cm and when 'full strength' contained 5 mgm radium per cm². They were recommended for use in the range one-quarter full strength to 5 × full strength and it was emphasised that curved surface applicators 'were out of the question'. Standard radium solution for drinking and radium solution for intravenous injection were also advertised by Watson & Sons.

REMOTE AFTERLOADING BRACHYTHERAPY

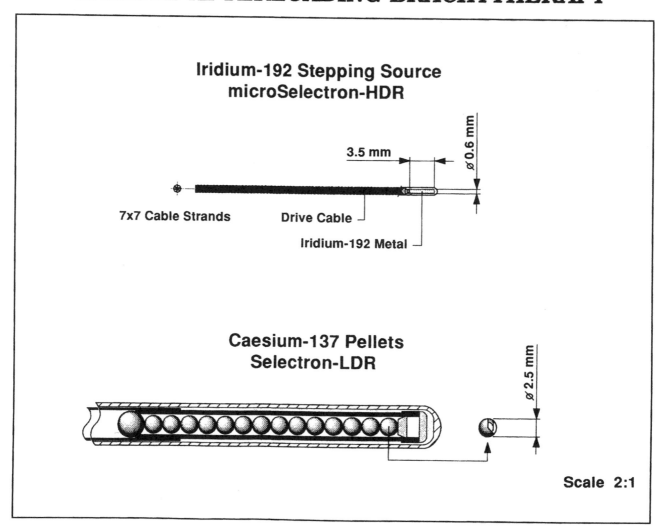

Iridium-192 Stepping Source
microSelectron-HDR

3.5 mm

⌀ 0.6 mm

7x7 Cable Strands

Drive Cable

Iridium-192 Metal

Caesium-137 Pellets
Selectron-LDR

⌀ 2.5 mm

Scale 2:1

[20.59] Two types of remote afterloading sources are shown. The low dose rate caesium-137 pellet train (top) is used with the Selectron-LDR/MDR [20.53] and the miniature iridium-192 high dose rate source (bottom) is used with the microSelectron-HDR [20.38, 20.60]. This source is a stepping source which can be programmed to move through any of 18 channels. This series of illustrations from [20.56] to [20.60] shows how far brachytherapy equipment innovation and radionuclide source technology has advanced over the last 20 years.

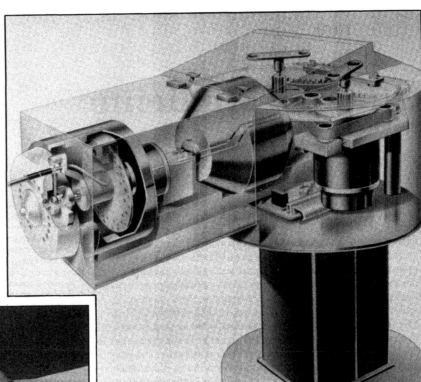

[20.60] The microSelectron-HDR remote afterloader as used in [20.38] for the treatment of endobronchial cancer showing (right) cut-away view of the head and (above) the front face containing the indexer with 18 channels clearly identifiable. The photograph shows an applicator transfer tube being placed in a channel prior to treatment.

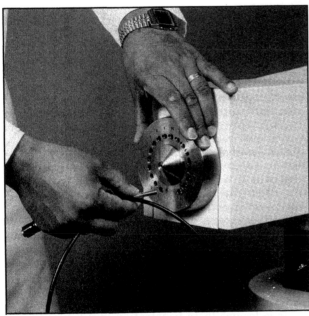

Patient case history: Paris, 1986–1992

[20.61] A unique case history, totally different to those of Kaplan [23.25], is that of this young lady who was treated for vaginal clear cell adenocarcinoma in 1986 at the Institut Gustave Roussy in Paris[54]. The treatment consisted firstly of an ovarian transposition as far away as possible from the area to be later irradiated and a pelvic lymphadanectomy. Secondly the primary adenocarcinoma was treated using a low dose rate caesium-137 brachytherapy technique with a moulded applicator and a prescribed dose of 60 Gy. Three years later the ovaries were positioned back into their proper anatomical site and the patient eventually became pregnant and on November 27th 1992 gave birth to a perfectly normal healthy girl. This successful outcome was widely reported in the French press[55]. (Courtesy: Dr Alain Gerbaulet, with the patient's permission.)

Chapter 21

Nuclear Medicine

The specialty of nuclear medicine is relatively recent compared with the use of X-rays in diagnosis and therapy and of radium in therapy, in that it effectively dates from the Manhattan Project in World War II, which led in peacetime to the production of artificially produced radioactive isotopes in sufficient quantity for medical applications[1]. Previously, attempts to use radioactivity for diagnostic purposes had proved a failure as early as 1904[2] as seen in Figures [3.7] and [3.8] when only radium and radon were available for imaging purposes, although in 1927 Blumgart and Weiss[3] successfully injected aqueous solutions of radon intravenously and monitored the velocity of blood flow between one arm and the other with a cloud chamber. The description 'nuclear medicine' was introduced only gradually and departments of medical physics were originally divided into radiotherapy and isotope sections with the latter including both therapy and diagnosis using radionuclides such as iodine-131. This was largely due to the early medical uses being the responsibility of radiotherapy physicians.

The phrase 'effectively dates from the Manhattan Project' is arguable in that several vital discoveries occurred prior to this date, without which clinical nuclear medicine could never have been established. These include the discovery in 1932 by James Chadwick of the neutron, the discovery in 1934 by Frédéric and Irène Joliot-Curie[4] of the transmutation of atoms that produced radioactive daughters (artificially produced radioactive isotopes), the discovery in 1938 of nuclear fission by Otto Hahn and Fritz Strassmann (separation of barium as a product after neutron bombardment of uranium) and the work of George de Hevesy[5] who in 1923 used tracer principles for the first time in a biological problem. This was the use of naturally occurring thorium-B (lead-212) to study the absorption and translocation of lead nitrate in bean plants. In 1934 he was also the first to use stable isotopes in clinical isotope dilution studies: the use of deuterium oxide to measure body water and its turnover. de Hevesy also, in 1935, initiated the use of artificially produced radioactive isotopes (phosphorus-32) in biological problems and in 1936 described the principles of neutron activation analysis and then published the results of this technique in 1938. In 1942 he described the first in vitro radioisotopic labelling of erythrocytes with its clinical use for measuring blood volume.

Routine radioactive tracer work, which sometimes became the province of the chemical pathologist, preceded radioactive imaging techniques and it was only when the latter came to the fore with the use of rectilinear scanners and gamma cameras that diagnostic radiologists took over this work from radiotherapy physicians, apart from the therapeutic use of iodine-131

Table 21.1. List of contents of *Journal of Nuclear Medicine*, Volume 1, Number 1, January 1960.

Therapy of carcinoma of the prostate metastatic to bone with ^{32}P labelled condensed phosphate.
Some problems regarding permissible doses with radioactive isotopes.
Radiosulfur (^{35}S) labelled Congo red dye.
The use of radioactive fat in the study of fat metabolism.
^{131}I Diodrast for the determination of renal clearance and renal tubular mass.
Fallout: one of several sources of radiation exposure to the total population.
Depth determination of radioactivity with biodirectional collimation.

to treat thyroid cancer. Physician specialists in nuclear medicine, as distinct from radiology, also began to appear in hospitals during the late 1960s and the early 1970s, but the structural organisation of this specialty varied widely in different countries.

The *Journal of Nuclear Medicine* published its first issue in January 1960, with papers on iodine-131, phosphorus-32 and sulphur-35, as listed in Table 21.1. Ten years earlier one of the first extensive review of medical radioactive isotopes had been published for the British Institute of Radiology by Mayneord[6] and iodine-131 and phosphorus-32 were among the major five isotopes considered for detailed review: the others were iodine-131, iron-55, iron-59 and sodium-24. Iodine-131 is now still in as widespread use as ever and the use of phosphorus-32 for the treatment of polycythemia vera is also common, but its use for the treatment of surface lesions (beta penetration is 5–6 mm) has been abandoned. One of its practical problems was lack of a good decontamination process. It was extremely difficult to remove and the cutting of small pieces of phosphorus-32 sheet for therapy use or the setting up of P-32 lipiodol lymphangiograms were fraught with radiation contamination hazards.

By the early 1980s the impact of nuclear medicine diagnostics in health care was very significant and the International Atomic Energy Agency[7] was able to list the following types of studies in which diagnostic procedures are undertaken with the use of radiopharmaceuticals: thyroid studies, hepatic studies, renal studies, mineral metabolism, regional blood flow, cardiac studies, pulmonary studies and gastrointestinal studies and bone studies. They further demonstrated the wide scope of the applications by listing the types of tests and treatments in daily use for just thyroid studies. These were: thyroid radioiodine uptake, turnover rate of iodine, turnover of organofied iodine in plasma, treatment ·response of thyrotoxic patients, response to antithyroid drug therapy and treatment response after radioiodine therapy or sub-total thyroidectomy.

At this time, Pochin[8] in his monograph on *Nuclear Radiation: Risks and Benefits* concisely reviewed the medical applications of artificially produced radionuclides and their dependence on essentially two of their properties:

(A) that the nuclides are chemically identical with the corresponding stable forms, so that the radioactive isotope will be metabolised in the body in the same way as the normal stable form;

(B) that the presence and amount of the radionuclide can be measured with such ease and sensitivity by simple counting of the frequency of the radioactive emissions, whether it be determined in samples of body fluids or excretions or by gamma radiation transmitted through the body wall.

These two features permit radionuclides to be used in several important ways[8], including the following.

(1) To estimate the speed or efficiency of a body organ function when this is reflected in the rate at which the radioactive form of an element, or any compound containing it, is concentrated in the organ or discharged from it. Similar principles apply to the determination of the volume of blood or other body fluids through which a tracer substance becomes distributed.

(2) To examine the sequence of chemical transformations of an administered substance, and the speed and efficiency with which they are carried out, or the partitioning of such metabolites between different body tissues. A similar principle determines the value of radioactive labelling agents in 'radioimmunoassays' in which the chemical distribution of the labelled compound allows a very low concentrations of biologically important substances to be measured in blood or other samples withdrawn from the body.

(3) To examine the spatial distribution of labelled materials throughout the body or within a body organ, particularly by determining the outline of the organs or the presence of tumours in which they become concentrated, by 'scanning' over the body surface to map out the gamma radiation emitted from the underlying sites of radionuclide retention.

(4) To measure the concentration, or detect the presence of, various chemical elements in human tissues, or blood samples, by rendering these elements radioactive as a result of neutron irradiation.

(5) To use selective concentration of certain radionuclides in cancer or other tissues, to deliver intense radiation locally to these tissues, particularly by beta or other radiations of low penetrations, with good prospects of destroying the abnormal tissue without undue irradiations of any other body tissue.

Radioimmunoassays, see (2) above, and related procedures are very sensitive methods that can be used for the measurements of hormones, enzymes, hepatitis virus and certain serum proteins, some drugs and a number of other substances[7]. Drops of the patient's blood serum are used in vitro to measure the hormones and other substances of interest and the radionuclides are added after the blood specimen have been taken. In 1981 according to the IAEA a total of some 40 million such tests were used per annum in the USA alone[7], and in the decade since this number must have risen dramatically. Examples are assays for thyroxine, digitoxin and human growth hormone.

A wide range of radionuclides and radiopharmaceuticals have been used in nuclear medicine and Table 21.2 details from unpublished working notes some of the investigations undertaken at the Royal Marsden

Table 21.2. Typical nuclear radioactive isotope tests of the early 1960s.

Absorption of vitamin B_{12} using cobalt-58: half-life of 72 days.
Vitamin B_{12} is stored mainly in the liver and if labelled with Co-58 can be detected by scintillation counting. Counting over the liver and lower abdomen (when this falls to a constant low level it indicates clearance from the gut). Used for patients with pernicious anaemia and those with liver malabsorption of vitamin B_{12}.

Absorption of vitamin B_{12} using cobalt-57: half-life of 272 days.
Oral dose of radioactive vitamin B_{12} is given and the fraction appearing in the urine is taken as a measure of the absorption of the vitamin. This is the Schilling test.

Blood volume using chromium-51: half-life of 27.7 days.
Chromium is rapidly taken up by erythrocytes at room temperature, which can be labelled with Cr-51. If the red cells are tagged and reinjected, the red cell volume and the plasma volume can be found.

Extra-cellular fluid volume using bromine-82: half-life of 36 hours.
Extra-cellular volume would most logically be determined by measuring the chloride space since chloride is a predominantly extra-cellular ion. As no suitable chlorine isotope is available bromine-82 is used instead.

Fat absorption using radiotriolein and oleic acid labelled with iodine-131.
These tests are used to detect the presence of impaired fat absorption and to determine its cause. In the presence of malabsorption of triolein, normal absorption of oleic acid suggests that the primary fault lies in the pancreas.

Iron clearance and utilisation using iron-59: half-life of 45.1 days.
Ferric citrate solution injected IV is bound to transferrin and is then cleared from the plasma into the iron pool and the erythroid cells of the blood-forming organs. The rate at which the iron is cleared from the plasma is an indication of the plasma iron turnover and of the erythropoietic activity.

Red cell half-life using chromium-51: half-life of 27.7 days.
The red cells are tagged in the same way as for blood volume measurements and reinjected into the patient. Since the chromium-51 is firmly tagged to the red cells it is only released when these are destroyed. Blood samples are taken each day and the rate of red cell destruction is determined.

Total exchangeable sodium and potassium using sodium-24 and potassium-42: half-lives respectively 15 hours and 12.45 hours.
Doses are given orally as early in the day as possible and urine collections are made at known times. Counting is made using a Geiger–Muller counter and scaler. Sodium and potassium can be measured simultaneously by counting with and without a brass shield. This shield cuts out all the potassium-42 counts but only about half of the sodium-24 counts.

Thyroid function, four-hour neck uptake using iodine-132: half-life of 2.2 hours.
The method depends on the generation of iodine-132 from the decay of tellurium-132 which has a half-life of 3.25 days: a 10 mCi consignment therefore decays to 0.1 mCi in three weeks. (This is an example of one of the earliest 'columns' provided by Amersham and preceded by several years the technetium-99m column which was to be very widely used for organ imaging [scanning] worldwide). The dose is given orally and four hours later the thyroid uptake is measured using a scintillation counter.

Total body water using tritium: half-life of 12.3 years.
Measurements are made on a sample of serum or urine at least three hours after a known oral dose of tritiated water has been given. It is assumed that the tritiated water (H_3O) is distributed in the body in the same way as H_2O.

PVP (polyvinylpyrrolidone) excretion using iodine-131: half-life of 8 days.
PVP has a molecular size about the same as that of plasma albumin. The amount excreted into the gut over a given period may be taken as a measure of protein loss in such conditions as hypoproteinaemic due to protein losing enteropathy. The compound may be iodinated with iodine-131 (or iodine-125).

Hospital, London, in the early 1960s. This gives an overview of tests and radioisotopes used at a leading nuclear medicine centre of 30 years ago.

Counting equipment is essential in nuclear medicine

for both in vitro and in vivo measurements and [21.1–21.7] illustrate some of the early counters, Geiger and scintillation, which were used. These include equipment for measurement of thyroid uptake [21.1, 21.6], of renal dynamic function [21.7], and for radioimmunoassay [21.4, 21.5].

Another class of counting equipment is that designed for whole body counting. The first design employed liquid scintillators and was built at the Los Alamos Scientific Laboratory[9] in the 1950s. It had a liquid volume of 200 litres and a total of 108 two-inch photomultiplier tubes for detection of the scintillation photons and transformation of them into electrical pulses. However, it was not possible to perfectly balance such a large number of photomultiplier tubes and so its spectral resolution was poor, even though it had good measuring geometry and high sensitivity. Because of physical and technical reasons, whole body counters using liquid scintillators can never reach the high resolution of those with sodium iodide crystals.

Whole body counters measure the total radioactivity in the body and in nuclear medicine and physics departments they are used principally to measure directly the variation in the retention of an administered radionuclide with time. This avoids laborious indirect studies involving collection and monitoring of urine and excreta, which was the only available method up to the mid-1960s in many hospitals. There are several designs of these counters, including a shadow-shield type shown in [21.8] which is a 2π walk-in liquid scintillation counter (shown without the shield) in which 15,000 normal persons were studied during the four-year period 1959–1963 to measure their caesium-137 content[10]. That in [21.9] is the Mayo Clinic whole body counter of 1966[11] which shows eight plastic scintillator detectors and two sodium iodine (thallium activated) crystals.

The third whole body counter shown[12] [21.10] is the improvised counter used in Goiania, Brazil, following the September 1987 accident when a strongly active caesium-137 source (50.9 TBq: 1,375 Ci) was removed from its protective housing in a teletherapy machine in an abandoned clinic. Four casualties subsequently died and 28 people suffered radiation burns. The NaI (Tl) crystal was 20 cm in diameter and 10 cm thick and was collimated using a 5 cm thick lead shield. The counter design incorporated a fibreglass leisure chair. Two-minute counting times were used and this enabled a detection level of 9.1 kBq (247 nCi) to be counted with a 95% confidence level.

Shielded-room whole body counters are different from the shadow-shield type in that they are housed in a specially constructed room, usually with steel walls some 10–20 cm thick which are surrounded by concrete. The steel serves as protection against low energy photons and the concrete against high energy photons. Modern steel, from the post-World War II era, usually contains too many radioactive impurities for such a low background installation and therefore earlier produced steel has to be used; sometimes this is from scrap battleships! These shielded whole body counter rooms are usually only to be found in large university hospitals, in national radiological protection centres and in facilities containing nuclear reactors.

The equipment used for the count rate distributions in [21.6] was the forerunner of the automated systems which became known as rectilinear scanners. The technique was initially known as area scanning and in

1951–52[13,14], Mayneord of the Royal Marsden Hospital, London, devised a system in which Geiger counter detectors were moved automatically over a matrix of 100 points. Good pictures were obtained for a patient with thyroid carcinoma following administration of 3 mCi of iodine-131, but because of the use of Geiger counters this system was relatively insensitive. Mayneord later[15] built an automated system using scintillation detectors and test scans were obtained with a thyroid phantom containing 20 µCi of iodine. However, by this time (1954) Cassen[16,17] at the University of California, Los Angeles, had also developed an automatic scintillation scanner which was able to produce good diagnostic images of thyroid glands after a patient had been administered about 150 µCi of iodine-131.

The elements of a scanning system are: the source of radiation, collimation of high energy photons, the radiation detector, the nuclear spectrometer, recording systems, signal processing and display systems, and control equipment[18–20]. Over the useful lifespan of rectilinear scanners [21.11–21.16], which effectively ended when gamma cameras became widely available, a major research field was in the design of collimators with optimal spatial resolution and sensitivity for the various radioactive isotopes used and the various depths within a patient at which scanning was required. For example, different collimators would be required to scan a thyroid and a brain. The pinhole collimator was the simplest. This consisted of a single cylindrical hole to define a small diameter region of view as was suitable for thyroids. However, the focused collimator with many holes was mainly used once scanning became established not only for thyroid cancer but also for tumours of the brain, liver, lung and other deep-seated organs. It was not easy to maximise the geometrical efficiency of a focused collimator, as this depended on many factors including the depth of focus, the diameter of the crystal scintillator and the radius of view at the focal depth.

The large majority of scanners moved the scintillation detector relative to the patient in a simple rectilinear array in what was effectively an automated version of the manual movements which obtained the results of [21.6]. The detector was motor driven along a series of parallel lines with a line spacing which was usually 0.5 cm but could be increased to 1 cm or 2 cm. In order to obtain maximum information content in minimum time, scan speed and line spacing had to be appropriately chosen, as also did the amount of activity administered to the patient. Even so, with the technology available in the mid-1960s a liver scan could take up to one hour.

Output display systems [21.17–21.21] initially consisted of a series of black marks on white paper (two or three carbon copies were also sometimes obtained at the same time) by using a mechanical printer. Once a preset number of counts had been accumulated a single dot was printed: hence the term 'dot factor'. The development of colour displays is generally attributed to Mallard at the end of the 1950s[21]. The mechanical printer now printed coloured dots in a range of colours (black-green-amber-red-mauve-blue was one colour scheme[19]) and the boundary between colours represented an isocount line. Colour also proved to be in practice a useful aid to increase the visual contrast.

An alternative to the colour scan on paper was the photoscan, first developed in 1956[22]. This was a photographic display in which the intensity of a light source or the rate of flashing of a neon tube, is controlled by the count rate and illuminates a film. The region of the film thus illuminated is moved in synchronisation with the linear scanning motion. The photographic density on the film is then related to the detector response to the gamma ray distribution within the organ being imaged.

The gamma camera which eventually made the rectilinear scanner obsolete was developed by Hal Anger[23], hence its sometime name of the Anger scintillation camera, and first displayed at the Fifth Annual Meeting of the Society of Nuclear Medicine in June 1958 in Los Angeles [21.22]. This original camera contained a single NaI (Tl) crystal measuring 4 inches in diameter and 0.25 inch thick, a total of seven photomultiplier tubes each of 1.5 inches diameter and a pinhole collimator. Many improvements have been made in design in the subsequent 40 years, including an increased number of photomultiplier tubes, larger diameter and thickness of sodium iodide crystal scintillators and better persistence monitors to retain the image. The use of computers has also revolutionised image processing and extended the range of dynamic studies (as distinct from static imaging) far beyond the scope of what could have been envisaged in the late 1950s, not least in cardiac investigations.

Figure [21.23] shows a gamma camera of the late 1960s/early 1970s, manufactured by Nuclear Chicago Ltd. The basic system consists of two parts, the detector stand containing the camera head and the control console which houses the electronics of the system and the CRT monitors, one of which was used in conjunction with a Polaroid camera to obtain hardcopy output. Nine photomultiplier tubes were used in this camera. Polaroid prints were the earliest of the hardcopy outputs but these were later overtaken by the use of film, and cine cameras could be also attached to the system. The computer hardware in the 1970s included magnetic tape storage and teletypes, but as computer technology developed, tape storage became obsolete and disc storage the standard, with line printers rather than teletypes.

Figure [21.24] is of a gamma camera of the mid-1980s and three examples of its use are shown in [21.25, 21.28, 21.29]. A recent whole body gamma camera scan of the skeleton [21.27] is included for comparison with early whole body X-ray images in previous chapters [3.5, 4.1]. Brain and liver gamma scans of the late 1970s[24] [21.26, 21.30] once, with bone imaging, provided the major workload of a nuclear medicine imaging facility and for this reason anatomically realistic liver and brain phantoms were developed for testing gamma camera performance [21.34–21.38]. However, gamma camera brain scanning has now been replaced by CT and MR images, and liver studies by ultrasonic scans, leaving bone, lung, renal and cardiac imaging the major workload of modern nuclear medicine imaging.

Iodine-131 was the first radionuclide used for imaging an organ, the thyroid, but because of its high principal gamma energy of 360 keV and its half-life of 8 days, it was not ideal for imaging purposes. There followed a wide range of different radionuclides and also different pharmaceuticals which were tagged with the radionuclides. By 1967[25] for brain scanning these included mercury-197 neohydrin, mercury-203 neohydrin, iodine-131 labelled albumin (RISA), gallium-68 EDTA, arsenic-74, copper-64 and technetium-99m pertechnetate. For liver scanning there was iodine-131, labelled Rose Bengal, and gold-198 colloid and for the pancreas, selenium-75 methionine. Scanning of the pancreas was also the earliest routine example of subtraction scanning with selenium-75 (gamma energies 66–401 keV, half-life of 121 days) locating in the liver and pancreas and technetium-99m locating in the liver but not the pancreas and therefore a subtraction scan enabled the effect of the liver image to be removed and a better assessment of pancreatic disease to be made.

However, the use of technetium-99m described in 1964[26], with its monochromatic 140 keV gamma ray energy and 6 hour half-life, soon exceeded that of all other radionuclides in total, and today, technetium-99m remains the most useful radionuclide. Examples of its use have included: for thyroid and brain in the form of pertechnetate; for bones as pyrophosphate or polyphosphate; and for liver as a sulphur colloid. Its preparation in a nuclear medicine laboratory, often termed a 'hot lab', is from a special generator, or 'cow', which is eluted or 'milked' using saline to obtain the technetium-99m pertechnetate. Figure [21.31] shows in diagrammatic form an early design from the Radiochemical Centre, Amersham, in the 1970s.

The generator system is one whereby a longer-lived 'parent' radionuclide (for technetium-99m production the parent in molybdenum-99) decays to the shorter-lived 'daughter' radionuclide. The technetium-99m generator is an alumina column with molybdenum-99 absorbed on to it and is kept in the laboratory with the parent continually decaying to technetium-99m.

Similar designs of Amersham generators were also available for strontium-87m (388 keV, 2.8 hour half-life) and for indium-113m (392 keV, 99.5 minute half-life). Sterile saline was injected at the top of the generator and after a certain period of time technetium-99m pertechnetate was collected in a vial. Generators usually lasted for one to two weeks, depending on the initial technetium-99m yield on first elution, which then reduced for successive elutions. A later design of Amersham generator [21.32] was delivered with vials of sterile saline which only had to be placed on top of a needle at the top of the generator, thus obviating the need for a syringe needle assembly as in [21.31]. The radiopharmaceuticals could then be prepared from a 'kit' in the laboratory, using the radioactive material which had been generated. Such an example was technetium-99m sulphur colloid for liver scanning, which if not prepared properly has been known to end up straight in the stomach rather than the liver!

All radiopharmaceuticals prepared for patient administration have to be measured to ensure that the correct amount of activity is present and the standard type of measuring equipment for the last 30 years has been what is termed either a radionuclide dose calibrator or activity meter. A schematic diagram[27] is shown in [21.33] of what is in essence a well-type ionisation chamber, into the well of which radioactive material is introduced for measurement. The activity of the material is measured in terms of the ionisation current produced by the emitted radiations which interact in the gas. The chamber is sealed, usually under pressure, and has two coaxial cylindrical electrodes maintained at a voltage difference. There is an associated electrometer, the ionisation current is converted to a voltage sisgnal, amplified and processed, and displayed usually in digital form in units of activity, Bq or Ci.

Quality control and quality assurance are essential not only for radiopharmaceuticals (only a part of this QC is activity measurement) but also for nuclear medicine instrumentation[27] and to achieve this, test objects are required for instruments such as rectilinear scanners and gamma cameras. Such test objects are usually termed phantoms[28] and the first of these devices to be widely used was the Picker thyroid phantom which was made of perspex and when filled with a quantity of iodine-131 provided a scan analogous to a thyroid scan [21.34]. With larger organs such as the liver being routinely imaged using rectilinear scanners, the principle of the emission-type Picker thyroid phantom was extended to various designs of 'liver slice' phantom of which [21.35] is a typical image[29].

However, these so-called liver slices in practice gave an image which was nothing like a true liver, for which there are many anatomical shape variants[30]. Figure [21.35] shows the experimental arrangement for an anatomically realistic liver phantom and two gamma camera images of this anthropomorphic phantom[31], the London liver phantom, which was used as a test object for a World Health Organisation interlaboratory comparison study[32]. It has an interesting development story in that the underlying idea was the use of radiotherapy mould room treatment cast technology to produce plastic shells of two halves of a post mortem liver from the Royal College of Surgeons of England which when sealed together could contain technetium-99m and in which solid perspex balls could be inserted to represent tumours. The first 'normal liver' was presented for scan reporting amongst scans from actual patients to the then leading nuclear medicine consultant in the United Kingdom, who duly signed and reported it as 'a normal liver'. The London liver phantom was born and the consultant has periodically requested his original signed report so it can be destroyed!!

The phantoms in [21.34–21.36] are all emission-type phantoms in that they contain radioactive material in liquid form and therefore have the disadvantage of having to be filled with radioactivity each time they are required for use. Many of the later phantoms[28] developed for use with gamma cameras to determine spatial resolution were of the transmission-type and examples included the Anger pie phantom which was constructed from a lead plate by drilling a pattern of holes of different diameters in six sectors [21.37] or a quadrant bar phantom in which each quadrant has different lead bar separations [21.38]. The radioactive 'flood source' for transmission-type phantoms was originally a hollow perspex disc uniformly filled with technetium-99m and placed on the upturned gamma camera head [21.39] but these have often now been replaced by solid-disc flood sources using cobalt-57 (principal gamma energies of 122 and 136 keV, half-life of 272 days) to simulate technetium-99m and barium-133 (principal gamma energy of 356 keV, half-life of 10.8 years) to simulate iodine-131 or indium-113m (392 keV gamma rays and half-life of 99.5 minutes). These flood sources are also used to measure gamma camera field uniformity and have proved more practical than the technetium-99m-filled perspex discs, which apart from the possibility of radioactive contamination of the camera, were very unwieldy due to their weight.

Planar scan images provide a two-dimensional pattern of the three-dimensional distribution of a radionuclide within an organ and suffer from low image contrast due to the radionuclide activity in the organ being surrounded by background activity. Tomographic radionuclide imaging virtually removes the background activity and with a high contrast scan image reflects the true three-dimensional distribution of activity. The acronym ECAT (emission computer assisted tomography) includes what has now become known as SPECT (single photon emission computed tomography) and PET (positron emission tomography). However, the earliest tomographic techniques, which preceded the use of the acronym SPECT, were termed section scanning [21.40, 21.41], where the image plane was either parallel to (longitudinal section) or perpendicular to (transverse section) the detector face. In SPECT the image plane is perpendicular to the detector face, analogous to CT images in diagnostic radiology which, following their introduction in the early 1970s, made the then radionuclide tomography scanning instrumentation virtually only of interest as a research tool in a few centres. However, by the mid-1980s, SPECT experienced a resurgence of interest and most workers in this field performed thallium-201 studies and some abdominal imaging investigations[33,34]. Brain images with SPECT entered general use with the availability of iodine-123-iodoamphetamine and technetium-99m-HMPAO (hexamethyl propylene amine oxime). In instrumentation there have been various technological improvements for SPECT since the 1960s, now resulting in high resolution and high sensitivity devices [21.42, 21.46] which have proved extremely useful in a variety of applications. These include cardiac and skeletal studies, and brain function investigations not only for tumours, but also for non-malignant disorders such as Alzheimer's disease and the dementias and other psychiatric illnesses.

A refinement of the Anger gamma camera is the positron camera[35] which is based on the detection, by means of opposed detectors and coincidence counting techniques, of the two 511 keV photons which are simultaneously emitted in almost opposite directions when a positron emitted by a positron-emitting radionuclide interacts with an electron in the body. Short-lived 'physiological' radionuclides are now the standard in PET scanning: carbon-11 (half-life of 20 minutes), nitrogen-13 (10 minutes), oxygen-15 (2.1 minutes) and fluorine-18 (1.82 hours), which are produced by small dedicated on-site 'medical' cyclotrons. However, in the first years of PET scanning the positron emitters tended to be copper-64 [21.47] or arsenic-74 (half-life of 17.4 days), which was used extensively at the Massachusetts General Hospital[36] to study a variety of brain tumours [21.48].

Table 21.3. Some clinical applications of PET: established and potential (after Ter-Pogossian[38]).

Neurology/Neurosurgery/Psychiatry
  Dementia (Alzheimer's disease)
  Epilepsy
  Stroke
  Huntington's chorea
  Schizophrenia
  Brain receptors
Cardiology
  Myocardium viability
  Stenosis/Restenosis
  Early evaluation of high risk patients
Oncology
  Tumour evaluation
  Tumour staging
  Assessment of tumour extent

Imaging modalities other than PET, such as CT, MR and SPECT, have been more readily accepted in clinical practice and there are in 1993 still fewer than 100 PET scanners worldwide[37]. One of the problems has been the practicality of installing a dedicated cyclotron near the PET facility, a practicality which is more concerned with funding than with cyclotron technology. However, functional PET imaging is growing in interest and Table 21.3 lists some current clinical applications of PET[38].

Figure [21.49] shows a PET scanner of the 1990s and [21.50] PET images of areas of activity in the brain.

Geiger counters and scintillation counters: 1940s–1980s

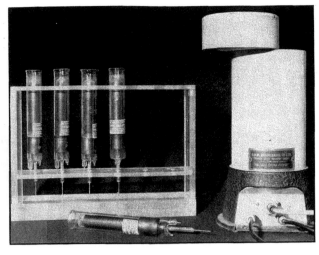

[21.2] Geiger counters for measurement of radioactive liquids, 1970. The lead 'castle' has a moveable lead top so that a Geiger tube may be inserted. Contact is made with mercury electrodes.

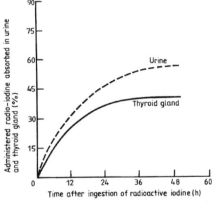

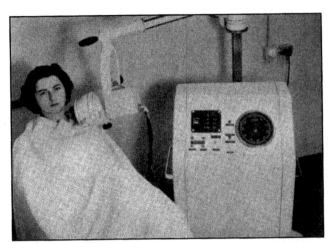

[21.3] Two Geiger counters on an electronic test bed, 1970. Several of the longer-length Geiger tubes could be incorporated into a lead-shielded counter to measure, for example, Winchester bottles containing radioactive urine. However, it was of course essential to ensure that all Geiger tubes were functioning properly and a test bed (the scaler/rate meter is not shown) enabled easy measurement of the Geiger plateau and the threshold voltage by recording counts versus applied voltage. Geiger counters can also be used as radiation survey monitors.

[21.1] Geiger counter measuring system (top) at the Royal Marsden Hospital in 1950. The Geiger tube is carried in a lead housing on an X-ray tube stand and the scaling circuits and electronic gear are housed in a cabinet suitable for clinical conditions[6]. Representative normal values[6] (bottom) for the sequential changes in thyroid iodine uptake and cumulative excretion of radioactive iodine in urine, after oral administration of iodine-131.

[21.4] In the 1960s well counters were designed such that one *in vitro* sample was counted at a time, as in the diagrammatic cross-section through detector and shielding arrangement. However, by the early 1980s well counters were available with 8–20 crystals (usually 1 inch) for measurements of 8–20 samples. The electronic operation had also been considerably simplified such that they were push button operated, and several models were microprocessor controlled with hard copy printouts of results available.

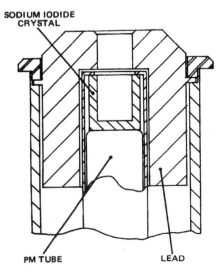

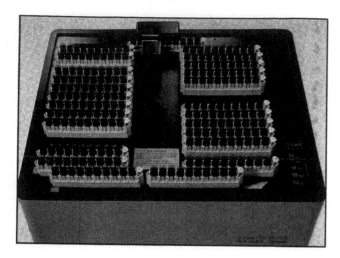

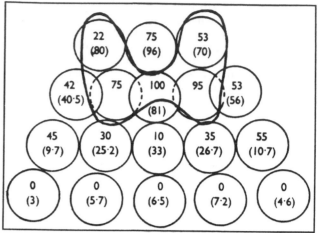

[21.5] Automatic gamma sample-changer, the Berthold LB MAG 312, which can accept up to 310 samples in 31 magazines, early 1980s.

[21.6] Thyroid uptake measurements were first made using Geiger counters, as in [21.1] but Geigers were soon replaced by collimated scintillation counters and this diagram (above) shows from the Royal Marsden Hospital, London, one of the earliest experiments[6] with a 'model' thyroid (such 'models' are now called 'phantoms' of which there is an extensive range for quality assurance testing, including those for representing body organs such as the thyroid and the liver[40]). The upper figures are count rates with a scintillation counter designed in the late 1940s and the figures in brackets are the count rates with a Geiger counter. Both sets of figures are expressed as percentages of the maximum count rate, corrected for background. The thyroid phantom was made of agar into which 20 µCi or iodine-131 had been incorporated, and the agar was embedded in paraffin wax.

In 1948 at St. Bartholomew's Hospital, London[40], a Geiger counter with lead shielding was used to study an intrathoracic goitre by point-to-point counting using 90 µCi of iodine-131. The equipment was similar to that in [21.1]. Even with a maximum count rate of 242 counts/120 seconds and a point source resolution of 4 cm FWHM, the goitre was able to be outlined. The count rate pattern is superimposed (below) on a photograph of the patient's neck.

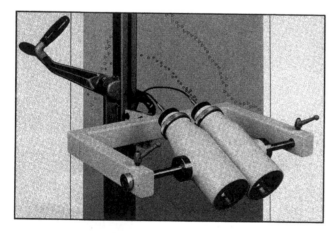

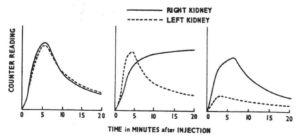

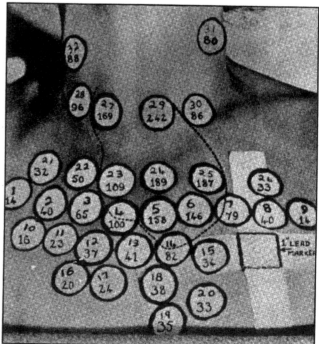

[21.7] The thyroid function tests referred to in [21.1] and [21.6] are not dynamic uptake measurements. However, renal function tests are dynamic and use a pair of scintillation detectors which are placed over each kidney to measure the clearance from the kidneys of iodine-131 labelled Hippuran. (Photograph courtesy: Nuclear Enterprises Ltd, 1980.) This chemical substance, ortho-iodohippurate (Hippuran) is exclusively and quickly removed from the blood by the kidneys and excreted by them. Since the counters are shielded each measures the clearance from a single kidney. The electrical output from each kidney is fed into a pen recorder voltmeter and an ink trace records the measurements, examples are shown in the three graphs (right) and after Meredith and Massey (1971)[41] showing uptake into and excretion from the kidneys of iodine-131 Hippuran. Left plot: both traces are normal. Centre plot: indication of acute obstruction of the right kidney. Right plot: normal right kidney but non-functioning left kidney.

Whole body counters: 1958–1987

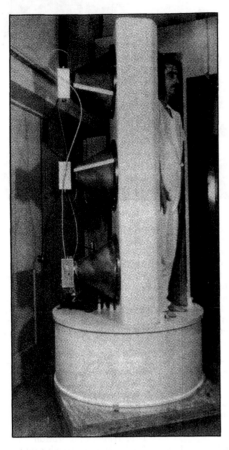

[21.8] A 2π walk-in liquid scintillation counter[10], 1958.

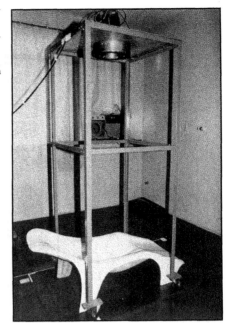

[21.10] An improvised whole body counter used in the Goiania accident, Brasil[12], 1987. Whole body counters also proved to be extremely useful following the Chernobyl accident, April 1986, when it was necessary to investigate intake of the 30-year half-life caesium-137.

Rectilinear scanners: 1950–1966

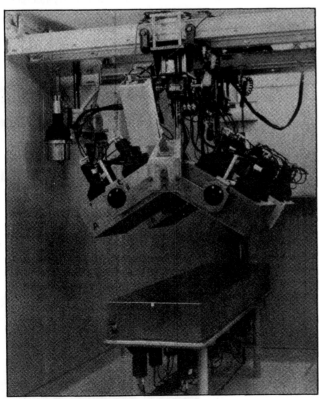

[21.9] The Mayo Clinic whole body counter[11], 1966.

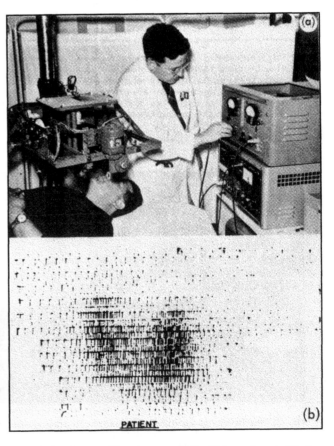

[21.11] The first rectilinear scanner[17,18] which used a scintillation detector and the black dot image of a patient's iodine-131 thyroid scan: 1950.

Positive Scintigraphy

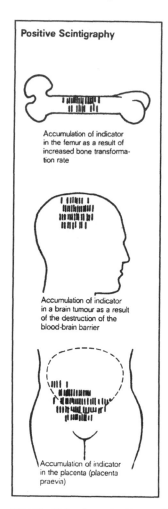

Accumulation of indicator in the femur as a result of increased bone transformation rate

Accumulation of indicator in a brain tumour as a result of the destruction of the blood-brain barrier

Accumulation of indicator in the placenta (placenta praevia)

Negative Scintigraphy

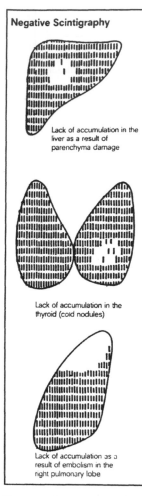

Lack of accumulation in the liver as a result of parenchyma damage

Lack of accumulation in the thyroid (cold nodules)

Lack of accumulation as a result of embolism in the right pulmonary lobe

[21.12] This schematic diagram[42] illustrates the two types of rectilinear scans in clinical practice: here called positive and negative scintigraphy, scintigraphy being an alternative term for scanning and scintigrams an alternative for scans. Colloquially these two types were described in practice as 'looking for hot spots or cold spots', using either colour scans or photoscans.

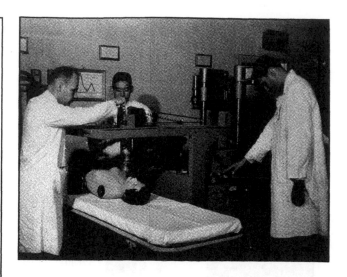

[21.13] Clinical training in 1960[43]. This photograph was described as 'Manikins containing mock-iodine "thyroid glands" with unknown percentage of uptakes and body background activities, originally constructed for a thyroid uptake calibration study, have been repeatedly used as training aids. Skill in external measurements and facility in scanning are acquired in the absence of live patients'. These manikins were some of the earliest nuclear medicine test phantoms: see [21.34–21.38]. The 1960 ORINS (Oak Ridge Institute of Nuclear Studies) Medical Division Course for AEC (US Atomic Energy Commission) Qualification was given over a period of three weeks as shown in the table below.

Preclinical: Weeks 1 and 2.	Clinical: Week 3.
Basic physics	Thyroid function
Mathematics and statistics	Cr-51 red cell label
Instrumentation and	Fe-59 utilisation
measurement	I-131 albumin plasma volume
Radiological safety	I-131 triolein absorption
Tracer methodology	Co-60 vitamin B_{12} absorption

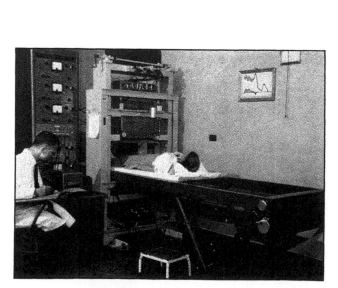

[21.14] The Oak Ridge Institute of Nuclear Studies linear scanner[44] designed in the late 1950s by Marshall Brucer. In its first five months of clinical use a total of 49 patients were studied, of whom 20 had iodine-131 (in the range 0.5–160 mCi) administered for thyroid carcinoma, for either diagnosis or therapy. Nine other radioactive isotopes were studied in the remaining 29 patients.

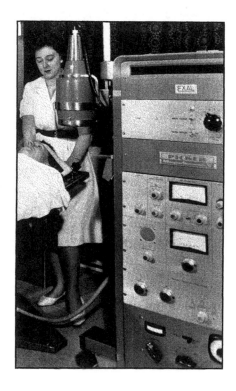

Nuclear Medicine

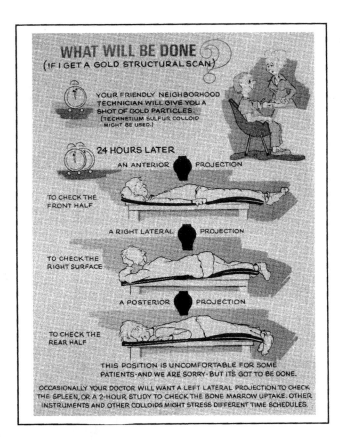

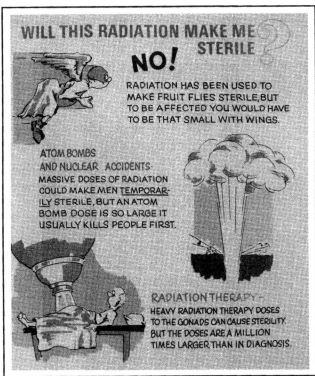

[21.15] Marshall Brucer who designed the ORINS linear scanner in [21.14] was also responsible during 1972–76 for a series of light-hearted informative pamphlets for nuclear medicine patients. These were published by Mallinckrodt Nuclear of St. Louis, were all titled 'What are you can expect from your Throid Uptakes & Scans/Lung Scan/Brain Scan/Liver Scan/Bone Scan' and subtitled 'Answers to the Questions Most Commonly Asked by Patients'. In addition, this series also contained 'The Technician's Battle Against Radiation Contamination', subtitled 'Radiation Protection Procedures in the Nuclear Medicine Laboratory'. Educational pamphlets for patients are now commonplace in nuclear medicine, diagnostic radiology and radiotherapy, but in the mid-1970s the Brucer/ Mallinckrodt series was the first of its kind.

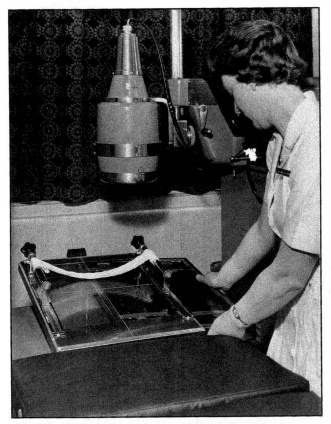

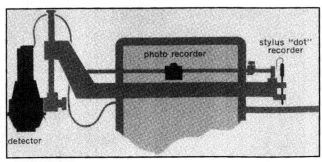

[21.16] A Picker Magnascanner in 1966 showing (above) the scanning head and a photoscan cassette being positioned and (left) the patient being immobilised prior to a brain scan: 3 inch or 5 inch crystals were used.

Picker scanners were the most popular worldwide of all commercially manufactured scanners and dual-head versions were available as well as the single-head version. A selection of collimators was available and each could be screwed into the bottom of the scanning head and locked into position. The controls of the scanner are: pulse height analyser, discriminator, high voltage supply, ratemeter, controls for scan motion, colour dot recording and photoscan recording. The scanning head (schematic) contains a preamplifier, photomultiplier, crystal and collimator. The scanning table on which the paper is placed for the colour dot scan is on the right. Within the electronics for the scanner is the photoscan system.

Rectilinear scans: lung, brain, bone, liver

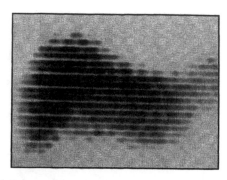

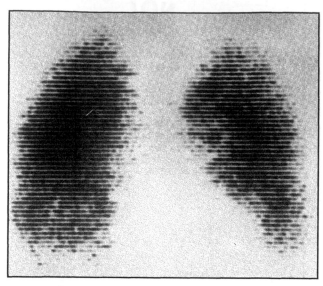

[21.19] Liver scan[22]: Western Reserve University School of Medicine, Cleveland, Ohio, 1960. The patient was a 44-year-old male with a clinical diagnosis of carcinoma of the larynx with probable mild Laennec's cirrhosis. 270 µCi of gold-198 colloid was administered.

[21.17] Normal lung scan[45]: 1973. Lung scanning depends on the mechanical pulmonary micro-embolisation of intravenously administered particles of approximately 20 µm in size: macroaggregated labelled (originally using iodine-131 which was later replaced by technetium-99m) human serum albumin (MAA). These particles cannot pass the capillaries of the pulmonary artery and remain temporarily fixed there until they are cleared away mechanically or by phagocytosis. Scanning is used to reveal perfusion disturbances (lung infarction, embolisms, tumours) in the pulmonary circulation. Normally the scan shows a homogeneous distribution which is practically the same for both lungs.

[21.18] Abnormal brain scan, indicated by the increased uptake in the superior aspect, superimposed on a lateral radiograph: 1965. This is taken from an advertisement for a Picker Magnascanner [21.16]. The scan was made at the Institute of Neurology, London. The company EXAL was the then UK distributor for Picker. Technetium-99m-labelled sodium pertechnetate cannot pass through the blood–brain barrier so that after intravenous injection the scan of a normal brain shows minimal uptake. In cases of brain tumour or haematoma, for example, the blood–brain barrier is insufficient and the abnormality is indicated by an area of increased technetium-99m uptake.

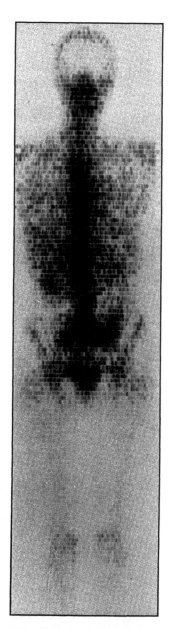

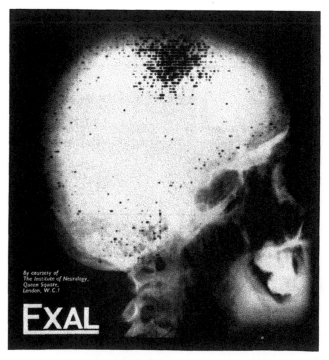

[21.20] Whole body skeletal scan, 1973, with a Siemens Scintimat-2 scanner using strontium-87m which has a half-life of 2.8 hours and a gamma ray energy of 388 keV. It is obtained from an yttrium-87 'cow' and is therefore more readily available than cyclotron produced fluorine-18.

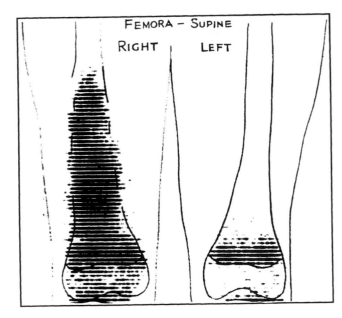

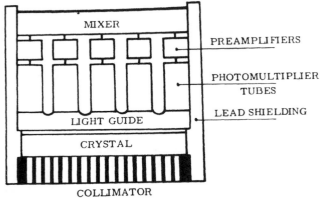

[21.23] The schematic diagram illustrates the head of an Anger gamma camera. Each PM tube receives an amount of light related to its distance from the scintillation in the crystal and generates an electrical impulse which is proportional to the light it receives. The signals from the PM tubes are summed to give the $+Y$, $+X$, $-Y$ and $-X$ deflection signals and these act upon the oscilloscope trace of the display and the trace is deflected to correspond physically with the position of the original scintillation event. The camera head consists of a collimator, sodium iodide crystal of some 27 cm diameter in the late 1960s (but increased in later years for large field of view cameras [LFOV]), light guide, photomultiplier tubes and preamplifiers. With a parallel-hole collimator only those photons travelling in directions essentially perpendicular to the plane of the crystal are detected and are located at crystal positions which bear a 1:1 correspondence with their positions of origin in the organ being imaged, such as a liver or brain. Other photons are stopped by the collimator septa, which are the lead portions of a collimator as distinct from the holes. A choice of collimators is available from a manufacturer and the minimum requirement in the late 1960s was two: one for low energy radionuclides such as technetium-99m and one for high energy radionuclides such as iodine-131. Pinhole collimators were used for thyroid imaging but multihole collimators for brain liver, bone and other organs.

Nuclear Chicago model 6406 9-photomultiplier tube gamma camera of the late 1960s/early 1970s. The principal advantage of the gamma camera over the rectilinear scanner was the time taken to produce a picture; with a scanner this could take from 15 to 90 minutes to complete, (see below)

[21.21] Bone scan[46] of an osteogenic sarcoma following intravenous injection of 2 mCi of fluorine-18: Westminster Hospital, London, 1967. The limits of the tumour involving the lower end of the right femur can be clearly seen. Fluorine-18 (511 keV gamma rays) was one of the first radioactive isotopes used for bone scanning, was cyclotron produced and with a half-life of only 1.9 hours, but not readily available on account of the logistics of delivering the isotope from cyclotron to patient.

Gamma cameras: 1958–1980s

[21.22] Hal Anger and his first gamma camera: from a photograph taken by W. G. Myers at the 1958 annual convention of the American Medical Association in San Francisco[47].

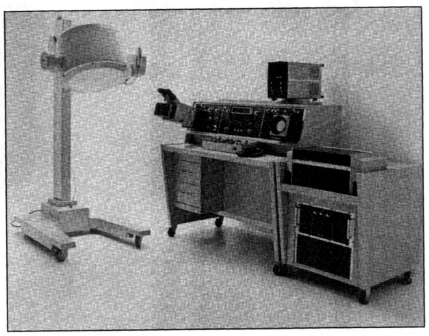

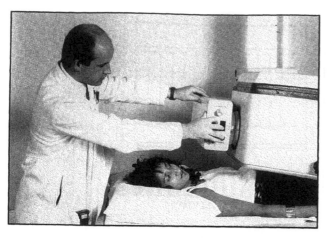

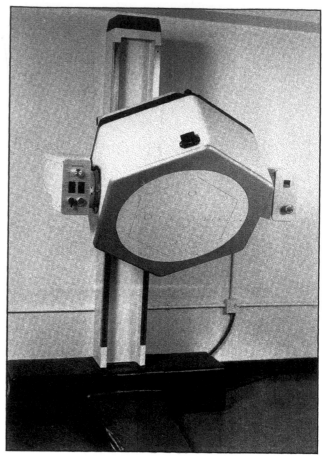

[21.24]   Large field of view (LFOV) gamma camera (right) which is typical of the mid-1980s. This was a Scintronix camera (developed from earlier Nuclear Enterprises designs) and was linked to a Data General computer: Westminster Hospital, London. The system incorporates two visual display units, one of which displays the images in colour whilst the other displays any operator inputs and computer instructions. The camera head can be rotated about two perpendicular axes and can be raised and lowered to the required height, and may be used, either with the patient lying prone or supine (above) on a table, or sitting in a chair; the choice depends on the imaging study.

Gamma camera scans: lung, bone, heart, brain, liver

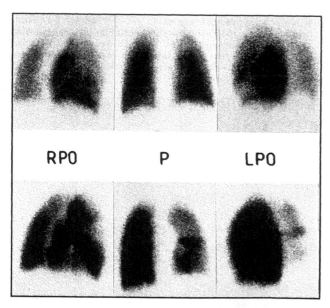

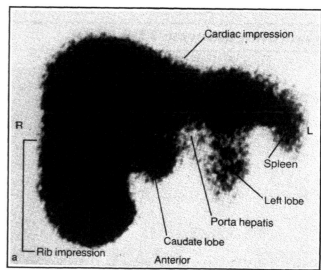

[21.25]   Two types of lung study can be performed: ventilation and perfusion. To produce ventilation images patients are required to breath an aerosol of the imaging agent (e.g. 50 MBq technetium-99m DTPA) for about five minutes prior to the imaging procedure which takes about 20 minutes. Up to 30 minutes later, the perfusion imaging agent (e.g. 100 MBq technetium-99m MAA) is administered intravenously and the same views are taken as for ventilation. The total gamma scan duration is approximately 1 hour. Such studies were impractical with rectilinear scanners. Top: ventilation images for right posterior oblique, posterior and left posterior oblique views. These are normal. Bottom: perfusion images. The appearance of the left lung is normal but that for the right lung shows multiple incomplete segmental defects which are consistent with extensive pulmonary embolism of the right lung. The patient had been admitted with right chest pain.

[21.26]   Normal liver and spleen scan[24]. When tumours larger than 3 cm are present in right or left lobes these appear as circumscribed white 'holes'. Tumour resolution depends on the gamma camera and also on the site of the tumour.

[21.27] Whole body scan with 25 mCi technetium-99m MDP to determine the location of bony metastases in previously irradiated sites: Modesto, California, USA, 1992. The image shows a solitary focus in the right hemi-pelvis which was persistent and painful despite 50 Gy external beam radiotherapy six months previously. (Courtesy: Dr L. L. Doss[48].)

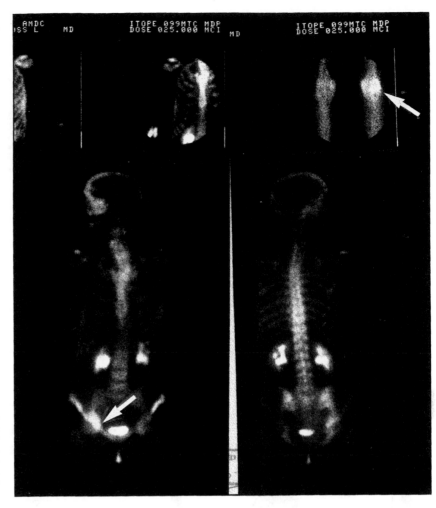

[21.28] Bone scanning performed with 650 MBq technetium-99m HDP. After intravenous injection of the radiopharmaceutical, a delay of at least two hours is necessary to facilitate bone uptake and blood clearance. During this time the patient should eat normally and drink plenty of fluids in order to increase blood clearance and should empty their bladder prior to imaging. Seven views are normally acquired, each taking about 3 minutes. Lesions are seen in the skull, sternum, pelvis, both femurs and in many vertebrae. This was the second scan of a malignant melanoma patient who had been scanned three months previously. The aim was to determine if there had been marked progression of metastatic skeletal disease.

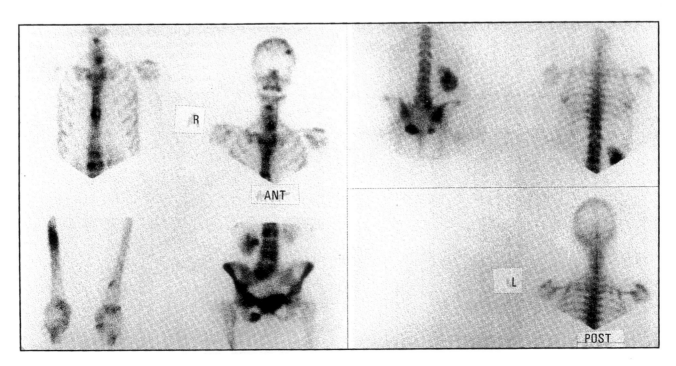

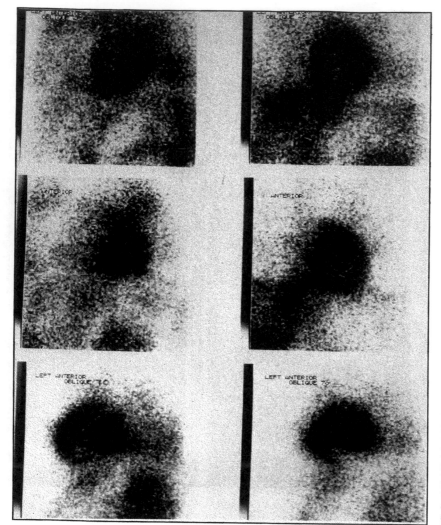

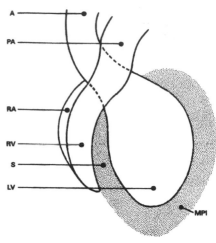

[21.29] Cardiac scanning using 100 MBq thallium-201 chloride. Two sets of images are acquired with the patient under stress and about three hours later when the patient has rested: reperfusion images. The patient is stressed either by exercises or more commonly by a slow IV injection of a vasodilator (Dipyridamole). Immediately following this the radiopharmaceutical is administered IV and the stress images obtained as in the left column, from top to bottom: left anterior oblique 45° view, anterior view, LAO 70° view. The rest images, right column, are of the same views and this step takes about 30 minutes. The stress images show considerable abnormality with an extensive region of uptake in the lateral wall of the left ventricle (LV) on the anterior view, the septum and apex on the LAO 45° and apex and inferior wall on the LAO 70°. The rest images show a relative increase of radioactivity in the lateral wall of the LV on the anterior and to some extent the septum on the LAO 45°. There remained conspicuous anormalities at the apex and in the inferior wall on the LAO 70° view. The defects of perfusion of the left ventricular wall suggest severe damage to the region of the apex and particularly the inferior wall of the LV. In addition there is evidence of transient ischaemia over an extensive area of the lateral wall of the LV. The schematic diagram (right) is a representation of cardiac anatomy as seen in the left anterior oblique view[49]: LV = left ventricle, S = septum, RV = right ventricle, RA = right atrium, PA = pulmonary artery, A = aorta, MPI = myocardial perfusion imaging.

[21.30] Typical image of a primary brain tumour, a glioma, in the left fronto-temporal region: anterior and left lateral views[24].

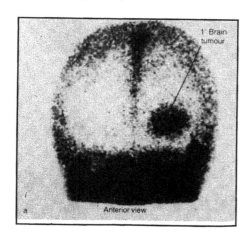

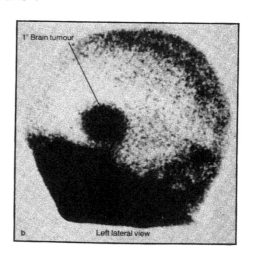

Technetium-99m generators: 1970s–1980s

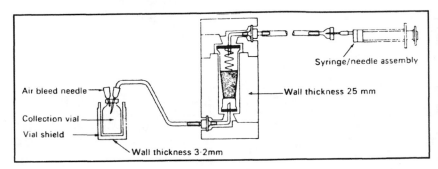

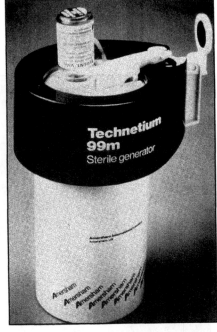

[21.31] Schematic diagram of an Amersham technetium-99m generator of the early 1970s. The wall thicknesses refer to lead shielding. The central column is of alumina with molybdenum-99 absorbed on it. This isotope continually decays to technetium-99m and this daughter product is washed out (eluted) with a saline solution when required. The eluate containing the technetium-99m must always be checked for an excess of the long-lived parent, for aluminium from the column, for sterility and for pyrogens.

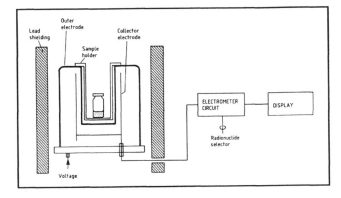

[21.32] An Amersham technetium-99m generator of the early 1980s.

[21.33] Schematic diagram of a radionuclide calibrator[27].

Phantoms: 1960s–1980s

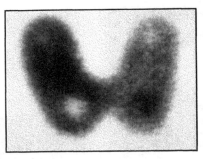

[21.34] Picker thyroid phantom (left) of the 1960s which was designed with two cold spots and two hot spots for use with rectilinear scanners and photoscan (right) using this phantom.

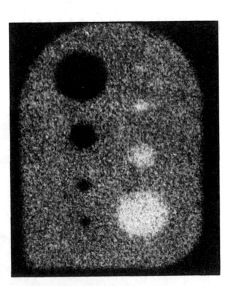

[21.35] Liver slice phantom scan[29]: 1976, developed for use with rectilinear scanners but also used with gamma cameras. The depth of the hot and cold spots are 10 mm and the diameters are 40 mm, 20 mm, 10 mm and 7 mm. This particular design was known as the Williams liver phantom.

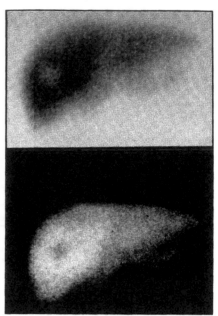

[21.36] The London liver phantom[31] in which the plastic model of the liver is filled with technetium-99m and in which are sited a number of perspex spheres to represent tumours. The surrounding body tissues are represented by water and in this photograph (left) the water tank is placed on a laboratory stool and the head of the gamma camera, which is just visible, is then lowered to the top of the tank. The two gamma camera images (right) show 'tumours' in the centre of the right lobe and in the periphery of the left lobe.

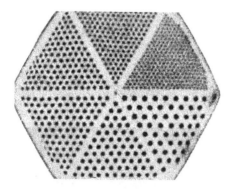

[21.37] Gamma camera image of an Anger pie phantom. Other hole-type phantoms included the Smith orthogonal hole phantom which consisted of an octagonal sheet of lead perforated with an orthogonal array of holes, with the lead pattern mounted between two lucite sheets; and the BRH (Bureau of Radiological Health) test pattern phantom which is a lead plate containing 7200 holes with various hole separations and lead spacings between holes. Transmission-type phantoms were also developed by the College of American Pathologists which were basically liver-slice and brain-slice phantoms, and not anthropomorphic. They were used in conjunction with a flood source and were very useful for interlaboratory comparisons[32] in that they were 'black boxes' with an unknown number of tumours and therefore image assessments could be independently scored for performance: that is, unless some unscrupulous persons radiographed the phantom to determine the tumour locations prior to gamma camera imaging. However, they were usually found out because their results were too good to be true!

[21.38] Quadrant bar phantom (flood source not shown) on gamma camera head.

[21.39] Technetium-99m flood field source in which the radioative liquid was in a disc volume: 1982. The polythene sheeting was essential as a safeguard against contamination of the camera in the case of radionuclide leakage.

Single photon emission computed tomography, SPECT: 1964–1993

[21.40]   The radioisotope section scanner devised by Kuhl[50] in the early 1960s at the University of Pennsylvania. The mechanism includes a yoke, a rotation arm, and a pedestal mounted on tracks by flanged wheels. The patient is supported on a cantilever platform between two opposing 3 inch diameter and 2 inch thick sodium iodide scintillation detectors mounted at the extremities of the scanner yoke. The scanner has three movement modes: rotary, transverse and longitudinal.

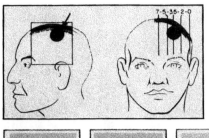

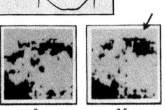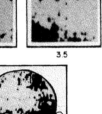

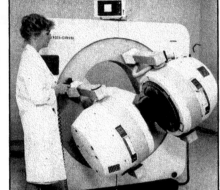

[21.41]   Top: sagittal section brain scans in planes on either side of, and through, a brain tumour in the posterior left frontal lobe: September 1963. The radiopharmaceutical administered was 2.5 mCi of mercury-192 chlormerodrin and the collimator was 127-hole with a 0.5 inch resolution distance at a focal length of 3 inch. The scanning speed was 0.5 cm/s. Bottom: November 1963 rectilinear brain scans confirming a left motor area tumour identified in the earlier scans. Craniotomy revealed the tumour to be a grade IV astrocytoma.

[21.42]   SPECT camera of the mid-1980s[33]: the Siemens ROTA camera with dual detectors with either 37 photomultiplier tubes per detector for high sensitivity or with 75 photomultiplier tubes per detector for high resolution. The major clinical applications of this period were cardiac, brain, bone and liver.

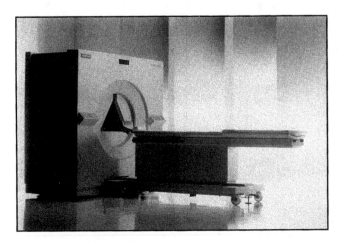

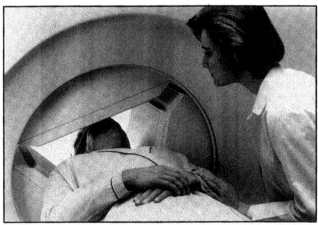

[21.43]   SPECT cameras of the 1990s now have three detector heads, rather than one and in the Siemens MULTISPECT all three heads are housed in the gantry, giving it an appearance similar to a CT scanner. This is a large field of view (LFOV) device: 31 cm × 41 cm, and can therefore be used for pelvis, liver and lung imaging, as well as for brain and cardiac studies.

[21.44] Image taken using a Tomomatic dedicated SPECT brain camera[51] of a patient with Alzheimer's disease: note the asymmetry of reduced flow and frontal lobe involvement. (Courtesy: Medimatic A/S, Hellerup, and Dr G. Waldemar, Rigshospitalet, Copenhagen.)

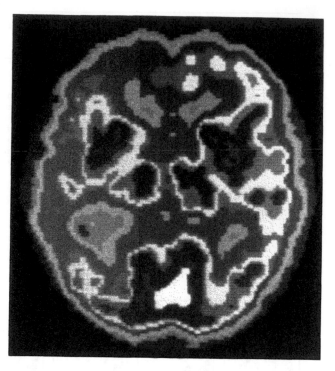

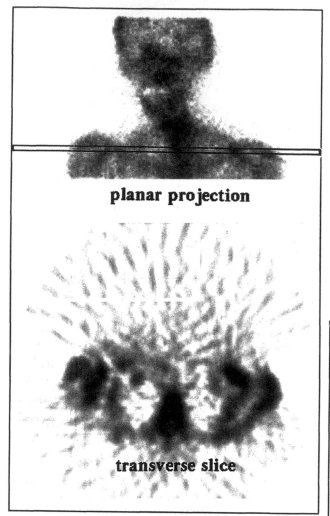

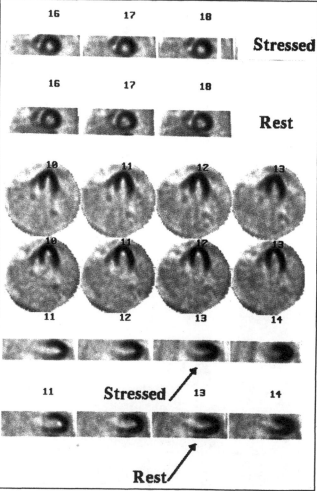

[21.45] SPECT images of a patient with cancer of the thyroid and a bone metastasis. She had received several treatments with high doses of iodine-131 and also external beam radiotherapy. Bone scintigraphy was performed using technetium-99m pyrophosphate. The planar image on the left shows unormal uptake of the tracer. The tomographic slice on the right which was taken at the position indicated by the bars shows much better contrast of the lesion. The whole set of transverse images of that region permits precise localisation of a lesionas well as an accurate estimate of the extent of the lesion, preceding radiological signs by about four weeks. (Courtesy: Prof. H. Bergmann, Vienna University Hospital AKH.)

[21.46] SPECT myocardial scintigrams. This patient presented with chest pain and an inconclusive stress electrocardiogram. SPECT of the heart was performed using a three-headed gamma camera and the tracer used was technetium-99m labelled MIBI. The stressed images demonstrate a small myocardial area inferiorly with reduced perfusion. The rest images show partially reversible distribution of the tracer in this area, confirming reversible ischemia. (Courtesy: Prof. H. Bergmann, Vienna University Hospital AKH.)

Positron emission tomography, PET: 1964–1993

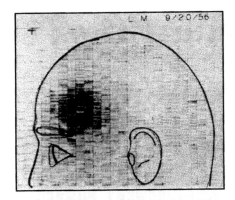

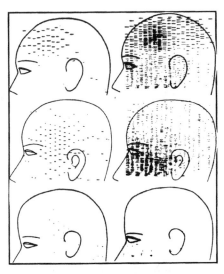

[21.47] Positron scans of the brain: University of Cologne, 1964[52] showing (top) a left-sided precentral meningioma, (centre) a left temporal glioblastoma, and (bottom) a left temporal astrocytoma. Copper-64 was used (half-life of 12.8 hours) and the procedure lasted 15 minutes. From the degree of uptake, the histology of the tumour could be assessed with the highest uptake always being in meningiomas.

[21.49]   PET scanner: Siemens 1993.

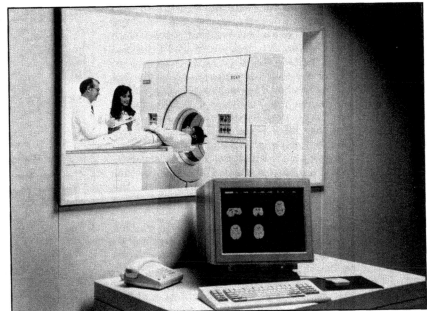

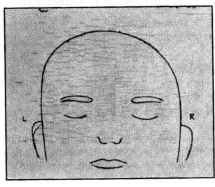

[21.48]   Appearance of an olfactory groove meningioma on a lateral positron scan using arsenic-74: Massachusetts General Hospital, Boston, 1959. The scan time was 40 minutes[36].

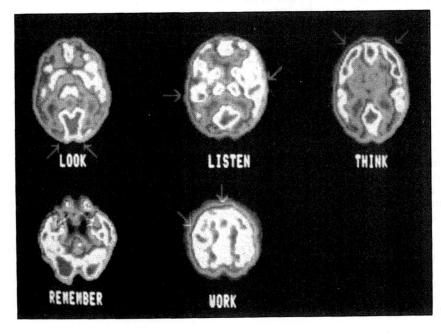

[21.50]   Grey-scale reproduction of PET colour images showing areas of activity in the brain. (Courtesy: University of California, Los Angeles and Siemens.)

Chapter 22

Radiation Units and Quantities, and Radiation Measurement

Radiation measurement techniques of the early years, for both X-rays and radium, are not surprisingly linked with the early proposals for radiation units, although in certain instances the techniques were far from accurate and informative and occasionally bordered on the impractical, such as the suggestion by Pullin and Wiltshire in 1927[1] that 'the depilatory property of X-rays is most remarkably constant in its effect and should be used as a method of measuring dosage.' They did, though, recognise the 'obvious disadvantage of making the necessary observations on a human subject' and suggested using mice instead. There was also a commercial aspect entering into radiation measurement in 1907 when Kassabian of Philadelphia[2] was discussing units based on the ionisation effect. He suggested that 'an absolute unit could be the Becquerel or the Curie, while a commercial unit might be known as one ray.'

The 'radiation measurement' in the title of this chapter mainly refers to 'absolute' measurement of ionising radiation and not to the whole spectrum of radiation measuring devices, several of which have been illustrated in the previous chapter on nuclear medicine. These include the Geiger counter, and the rectilinear scanner and Anger gamma camera which are dependent upon a sodium iodide scintillation detector. This type of instrumentation requires calibration for a measurement of a dose rate or of activity; the Geiger counter, for example, directly recording only counts per unit time.

The subsection titles of this chapter, not all of which are illustrated, are either the names of particular radiation units or are methods of measurement which have formed the basis of either an X-ray unit or a gamma-ray unit. This is a differentiation which is necessary because it was not until the year 1937 that the röntgen, as it was then defined, was accepted as a unit of measurement for both X-rays and gamma-rays. However, all the basic methods of measurement upon which definitions of radiation units depended were known in the early years of the 20th century and therefore the history of radiation units is essentially the history of the first 40 years following the discoveries of X-rays and of radioactivity. The röntgen (R) unit, based on the ionisation effect, has been refined over the years until it passed into history with the advent of SI units, as did the Curie (Ci) as a unit of activity. The current SI specially named radiation units are the becquerel (Bq) for activity and the gray (Gy) for absorbed dose, which is dependent on the thermal effect

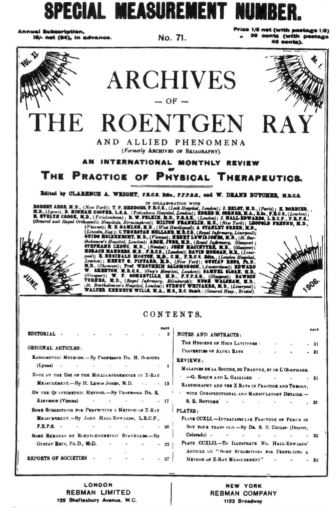

[22.1] The importance attached to X-ray dose measurement is demonstrated by the publication of a special issue of the *Archives of the Roentgen Ray*, devoted to this topic, in 1906. Major contributors to this issue were Kienböck of Vienna, Bordier of Lyons and Hall-Edwards of Birmingham. The *Archives* (see Chapter 4) had by this time ceased being associated with the Röntgen Society which in 1904 started the *Journal of the Röntgen Society*[105]. This was mainly due to the resignation from the Röntgen Society of 15 London physicians (10 of whom are named on this 1906 cover as collaborative editors) who had wanted the Röntgen Society to be exclusively devoted to medical matters. When this was not agreed they formed the British Electrotherapeutic Society. The words Radiotherapy, Phototherapy, Electrotherapy and Thermotherapy, seen on this special issue cover of 1906, were first used in 1903.

Table 22.1. Some radiation unit conversion factors.

Quantity	SI unit name, symbol	Non-SI unit	Conversion factor
Exposure	$C\,kg^{-1}$	röntgen (R)	$1\,R = 2.58 \times 10^{-4}\,C\,kg^{-1}$ $1\,C\,kg^{-1} \approx 3876\,R$
Absorbed dose, Kerma	$J\,kg^{-1}$ gray, Gy	rad (rad)	$1\,rad = 0.01\,Gy$ $1\,Gy = 100\,rad$
Dose equivalent	$J\,kg^{-1}$ sievert, Sv	rem (rem)	$1\,rem = 0.01\,Sv$ $1\,Sv = 100\,rem$
Activity	s^{-1} becquerel, Bq	curie (Ci)	$1\,Ci = 3.7 \times 10^{10}\,Bq$ $1\,Bq \approx 2.7 \times 10^{-11}\,Ci$

rather than the ionisation effect, and the sievert (Sv) for dose equivalent.

The quantities which in the mid-1990s are now in widespread use and are subject to measurement by standards laboratories[3] are for photons: exposure (units of coulomb/kg), kerma (joule/kg or Gy) and absorbed dose (joule/kg or Gy), and for electrons: absorbed dose. The sievert as the unit of dose equivalent is used in the field of radiation protection. Some radiation unit conversion factors are given[3] in Table 22.1. Detailed information can be obtained from the latest publications of the International Commission on Radiation Units and Measurements and on those of the International Commission on Radiological Protection. In addition, for a review of the changeover to SI units in the mid-to-late 1970s, see reviews by Spiers[4] and by Mayneord[5].

Strength/intensity/activity: 1904

The first proposal for a gamma-ray unit was based on a comparison between the radiation emitted by the radium source and that emitted by a source of uranium. The quantity measured by this unit was variously called strength[6,7], intensity[8,9] or activity[10]. 'Radium, when pure, has about 2,000,000 times the activity of an equal weight of uranium. Its activity is therefore said to be 2,000,000' was a typical statement[11]. The strengths of clinical radium sources varied widely and the examples shown in an earlier chapter [3.16][6] were within the range 13,000–580,000. The 50 mg radium bromide applicator which was used by Williams in Boston in 1904[12] to treat his first 50 cases was strength 1,500,000, whereas the linen-based applicators of Knox[13] at the Royal Marsden Hospital, London, a decade later contained 1 cg of radium per cm^2 with strengths varying from 10,000 to 2,000,000. However, as might be expected with such a vague quantity, the published data is hardly consistent. For example, Kassabian in 1907[2] stated that 'the working unit of the most powerful and pure radium manufactured is 10 mg radium bromide strength 1,800,000', adding that it is best used in a small cell covered by a layer of mica. Radium applicators were also sometimes designated double, full, half and quarter strength, such as those of Walter in 1916[11] where full strength corresponded to an activity distribution of 10 mg radium bromide per cm^2.

Uranie: 1905

Butcher, later a President of the Röntgen Society, London, in 1905[14] when discussing the means of accurate measurement in X-ray work, defined a uranie unit as 'the quantity of radiation given out by one gram of uranium' and stated that 'a gram of radium may be roughly said to give '1,000,000 uranies' and that the output from an X-ray tube is in the range 10^8 to 10^{10} uranies. He did, however, say that the unit was chiefly of theoretical interest and he made a distinction between three groups of X-ray workers: the maker, the user and the driver (an early name for a radiographer!) of the tube.

Milligram-hour: 1909

The strength, as a quantity describing radioactivity, was not completely discarded until the early 1920s[15] although by that time many alternative units had been proposed. Notably there was the milligram-hour (mgm-hour) suggested by Turner of Edinburgh in 1909[16]. This was the product of the weight of pure salt in mg and the time in hours during which the radium source remained in contact with the tissues. It was modified in 1912 when the first International Radium Standard was prepared, so that it referred to the weight of pure radium element. This avoided confusion arising from the use of different radium salts.

Curie: 1910

When radon sources in small glass or metal containers were first used as an alternative to radium, no suitable unit had been devised to describe a 'dose of radon' analogous to the milligram-hour 'dose of radium'. At the 1910 Brussels Congress of Radiology, the curie (Ci) was first suggested as a unit and specifically defined for use with radon as 'the quantity of radon in radioactive equilibrium with 1 gm of radium'.

Rutherford: 1930

The International Radium Standard Commission in 1930 recommended extending the curie unit to include the equilibrium quantity of any decay product of radium. For example since 2.24×10 g of polonium has the same alpha particle emission rate as 1 g of radium, this mass of polonium would have an activity of 1 curie. However, the Commission expressed its opposition to the extension of the curie unit to radioactive substances outside the radium family and a rutherford (rd) unit was proposed for general use[17,18]. It was defined to be 'the amount of any radioactive isotope which disintegrates at the rate of 1,000,000 disintegrations per second'. Thus 1 rd equalled 1/37th mCi of radium, radon, polonium, etc. However, in 1953 the 7th International Congress of Radiology recommended the use of the curie as a unit of activity of any radionuclide and the rutherford was made redundant.

Millicurie destroyed: 1914

In 1914, Debierne and Regaud[19] in Paris suggested the millicurie destroyed (mcd) as a suitable unit for clinical radon dosage and this was modified in 1919 by Regaud and Ferroux[20], to become the millicurie destroyed per hour or the millicurie destroyed per cm^2. The mcd for radon and the mgm-hour for radium were the two most popular gamma-ray units in clinical practice prior to the adoption by the ICRU in 1937 of the röntgen unit for both X- and gamma-rays. Using the relationship that 1 mcd of radon is equivalent to 133 mgm-hrs of radium, Mercier in Lausanne[21] in the 1920s prepared a graph giving the dose in mcd and mgm-hrs for applicators containing initial radon activities in the range 1–100 mCi or radium activities in the range 1–30 mgm for treatment durations of 0–7 days. This graph appeared in several standard European textbooks of that time and often represented, for a radium clinic, an important part of the available physics data for planning radium and radon treatments.

Mache: 1904

The Mache unit was proposed in 1904 and was specifically defined for radon. It was expressed as a concentration of the gas in water[13,22] as 'the saturation ionisation current due to radium emanation from a litre of solution of gas expressed in electrostatic units multiplied by 1,000'. Its relationship to the curie was given as 1 Mache = 4.5 × 10 Ci. Patient prescriptions were sometimes as small as 1,000 Mache and commercially available apparatus could produce radioactive water of 5,000–10,000 Mache per litre per day [23.19]. It was also stated in 1913[23] that the natural radioactive spring waters at Gastein contain less than 250 Mache per litre. As late as 1945 this unit was still being used in Czechoslovakia for Jáchymov (formerly Joachimstal [23.21]) sources.

Biological effects: 1904–1934

The most well known of all the proposed biological units was the skin erythema dose (SED) which was used independently for both X-rays and radium for many years. They have been discussed in detail by Failla in 1921[24] for radium and by Schall in 1932[25] for X-rays, who compared a 'unit skin dose' (USD) with measurements made by other methods, such as blackening of photographic film and colour changes of chemical mixtures. The tolerance of the skin to radiation was also the limiting factor in determining the fractionation of the radiation dose and, for example, Holfelder[26] defined 1 SED = 500 r for hard X-rays (see the later section on soft and hard X-ray quality, page 177) and

Table 22.2. Relationship between dose per day and total dose (after Holfelder, and Roberts (1936)[26]).

Number of succeeding days	Dose per day (r units)	Total dose (r units)
1	500	500
2	357	714
3	275	825
4	220	880
7	165	1185
12	110	1320

produced Table 22.2 which showed that the total dose could be doubled if fractions of the total dose were given over a period of seven days rather than as a single dose.

Dose-time fractionation was recognised as an important factor very early in publications by such as Colwell and Russ in 1915[27], Regaud in 1922[28] and Juul in 1930[29] but it was not until 1944 that Strandqvist[30] produced iso-effect curves relating total dose and number of fractions in terms of skin reaction or tumour control. However, one of the first and now overlooked radiobiology pioneers was the New York surgeon, Robert Abbe, who in 1910[31] suggested the principle of afterloading. In 1904[32] he performed experiments on his own skin, varying the exposure to radium and correlating it to the degree of erythema and necrosis produced. His work with seedlings [22.2] pointed out the potential dangers of radium effects on germinative tissues.

Two other early radiobiological experiments are shown in [3.30] and [22.3] from 1918[33,34]. Krönig and Friedrich in Freiburg [3.30] used tadpoles as their biological medium but their unit of measurement was ionisation based. Russ from the Middlesex Hospital,

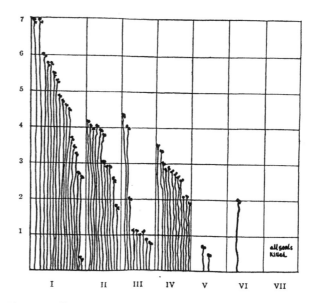

Diagrammatic representation of seed growth following exposure to radium rays; twenty seeds in each exposure; I, without exposure, 17 seeds grew; II exposure for 2 days, 11 grew; III, 3 days, 9 grew; IV, 4 days, 12 grew; V, 5 days, 2 grew; VI, 6 days, 1 grew; VII, 10 days, none grew.

[22.2] Radiobiological experiment by Robert Abbe in 1904[32] showing the quantification of time–dose factors for radium irradiation of germinative seedlings (the vertical axis is the height of seedlings in inches). (I am grateful to Dr L. L. Doss for drawing my attention to this work of Abbe.)

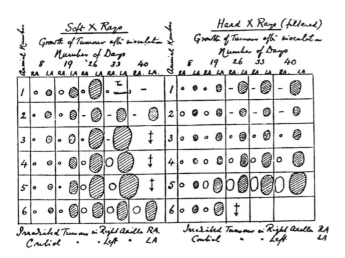

[22.3] Experimental results from the laboratory notebook of Sidney Russ using rat tumour cells, soft X-rays and hard (filtered) X-rays[34].

London, studied the biological effect of X-rays on rat tumour cells and proposed a biological unit, called the 'rad', based on his results. This is of historical interest in that the name 'rad' for a radiation unit was adopted 35 years later in 1953, with a very different concept: 1 rad = 100 ergs/g.

Since it was sometimes stated that an erythema dose of gamma-rays could be followed immediately by a full dose of X-rays without producing an excessive skin reaction, it is interesting to record that Quimby in 1927[35] studied the SED with a combination of two types of radiation: beta-rays and gamma-rays. She concluded that: (a) the formation of a skin erythema is determined not only by the quantity, but also by the quality of the radiation absorbed, and (b) if the radiation dose is measured by the erythema effect, two-thirds of a dose of soft radiation must be immediately followed by two-thirds of a dose of hard radiation, or vice-versa. This is in order to produce the same erythema as one full dose of either. It is also noted that the first publication by Quimby (1932)[36] of her linear radium source dosimetry tables, which were in advance of the Paterson and Parker Manchester radium system, was based on an erythema dose unit, although subsequent tables were quoted in röntgens[37].

The Quimby and Failla (1933)[38] definition of 'threshold erythema dose' (TED and SED are essentially the same) was 'that quantity of radiation which when delivered at a single sitting will produce in 80% of all cases tested, a feint reddening or bronzing of the skin, in from two to four weeks after irradiation, and in the remaining 20% will produce no visible effect'. The conversion from TED to radium gamma-röntgens was given by Quimby[37] as 1 TED = 1,000 gamma-röntgens.

As an indication of patient exposures given in terms of skin erythema dose (Hauteinheit dosis (HED) or unit skin dose), Solomon (1923)[39], Béclère (1927)[40] and others quote earlier data from Seitz and Wintz (1920)[41] which state the following treatment dosages. Ovarian castration dose is 34% of HED. Lethal carcinoma dose is 100%–110% of HED. Lethal sarcoma dose is 60%–70% of HED. Dose for cure of tuberculous lesions is 50% of HED.

The skin erythema dose was also used by some when considering working hours and general recommendations and the maximum tolerance dose of X-rays for peak voltages not exceeding 200 kV has been defined as:

one-thousandth of the 'erythema dose in 3 working days, under conditions when the whole body is irradiated'. Using the relationship that erythema dose for X-rays = 600 röntgens, the tolerance dose in röntgens was quoted as 0.2 r in a normal working day of seven or eight hours, or 1 r for a five-day working week by Wintz and Rump (1931)[42]. However, in the 1934 recommendations the 'safe general radiation to the whole body' is taken as 0.1 r per day for X-rays and was used as a guide in radiation protection.

It must be noted, though, that the relationship between unit skin dose (USD) and the röntgen unit (r) varies with the quality of the X-ray beam. This was recognised by many authors, including Erskine (1931)[43] who quoted the variation as 600–1,000 r and defined USD as 'enough X-rays to produce scaling of the skin and less than enough to produce blisters'. Erskine also quoted the minimum erythema dose for untanned skin as 250 r and the epilation dose for unfiltered X-rays as 450 r. There was also a definition in the literature of 1924[44] of a skin erythema unit as a function of a patient's age: 'a skin unit causes a slight erythema on flexor surfaces of the body of a young fair-skinned individual but approximately 25% more is required to produce a definite erythema on most parts of the body of an older individual with darker skin.' Other age-related dosimetry is shown in [22.4] from an adversisement of the early 1920s by Kramer of Birmingham, England[45]. The slow-radium (a mixture of radium, thorium and uranium compounds) dosimetry required not only a statement of the area of skin disease to be treated, but also the age and sex of the patient. No other information was required and the slow-radium applicator and dosimetry then posted to the 'customer'.

As a final comment, although skin erythema doses were used for many years they were really impractical since they were too indefinite and required too many tests for each measurement and apart from SED there was never any real idea of developing units from biological material by most investigators.

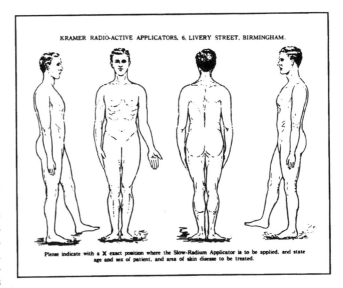

[22.4] Form for postal dosimetry for a slow-radium surface applicator in the early 1920s[45]. The exact composition of slow-radium is not clear, but Kramer's experiments using photographic film as a radiation detector indicated that the radiation dose received in 30 minutes from a pure radium bromide preparation exceeded that received in 500 hours from slow-radium. Patient treatments therefore extended over a matter of weeks rather than days.

Fluorescence: 1902–1926

The most widely used unit based on the fluorescent effect produced by X-rays was suggested by Guilleminot in 1907[46]. He used as his standard a radium source of activity 50,000 and quantified the X-ray tube output by placing the tube at a distance from the platino-cyanide screen which gave an equal illumination, when compared with that given by his radium standard behind the screen. His unit was denoted 'M' and defined as 'the quantity of X-rays falling on 1 cm² of the surface during 1 minute'. X-ray intensities at different distances were calculated using the inverse square law and as a typical example, Bordier in 1906[47] quoted a measured X-ray tube intensity of 1 M at 3 metres and calculated that it increased to 900 M at 10 cm.

The Guilleminot method was based on an earlier proposal by Courtade for a radiometer for qualitative measurements. This in turn was preceded by the radiophotometer of Contremoulins (1902)[48] which used an acetylene lamp as a standard instead of a radium source. Much later, Wintz and Rump (1926)[49] also suggested a method for determining intensity using fluorescence. In their method the fluorescence caused by the X-rays was compared to that caused by a standard lamp and measurements were made by means of a Lummer–Brodhum cube.

In addition, the 'radion' or 'radio-lux' unit of Butcher (1908)[50] was also dependent on the fluorescent effect and the fluorescence was obtained from a standard capsule of radium placed 1 cm from the screen. To define röntgen-candle and radium-candle, Butcher used the analogy of the candle power of a lamp being measured by placing the standard candle 1 metre distant from the screen and moving the lamp until the illumination of the screen from either source was equal. The square of the lamp–screen distance is the candle power and thus the square of the tube–screen distance is the röntgen-candle power. The units were defined as follows. A 'standard röntgen-candle' = 1 radium-candle = 1 mgm of radium. A 'unit of radiation' = 1 radion = the irradiation due to 1 radium-candle at a distance of 1 cm. A 'unit quantity of X-rays' = 1 radion-minute = that quantity of rays absorbed in 1 minute when the irradiation is 1 radion.

Minutes: 1916

The following anecdote was supplied by Dr Lawrence L. Doss, who when working in Santa Fe in 1986 interviewed a 76-year-old lung cancer patient regarding his past medical history. It is probably the only recorded information from a patient whose ionising radiation dose was prescribed in terms of minutes. Its interest is also enhanced when it was found that the physician was Dr Robert Abbe of New York, who is mentioned in Chapter 20 for his pioneering work in the use of afterloading in radium brachytherapy.

The patient was asked if he had received previous X-ray treatment and it was found that he had been treated in 1916 in New York for prevention of an extensive keloid scar following a severe neck and facial burn. His father had arranged a costume party for his son's sixth birthday and his clown costume had a neck ruffle which was set on fire as the boy blew out the candles on his cake. Despite surgical treatment the keloid-type scars kept recurring to the extent that his chin was pulled towards his chest. Eventually the surgeon recommended X-ray therapy followed by keloid resection. He remembered that the radiation treatment was given by Dr Robert Abbe but said that he remembered very little else as he was so frightened that he remembered more of daily visits after each day's treatment to Steeplechase and Luna Park, which were a bribe to make sure he underwent the X-ray therapy. He was initially treated on a machine which 'sparked' and scared him so much he would not lie still. Later he was treated by a quieter machine which was mounted on rails in the ceiling and could be moved from room to room where other patients had already been placed in the treatment position. After 66 treatments his skin became more pliable and he gradually recovered the full range of head and neck movement.

He later returned to the Santa Fe clinic and brought with him a record of his treatment 70 years earlier in Dr Abbe's clinic. His father had apparently not trusted the efficacy of X-ray therapy and requested a report. It was on what was by 1976, very old paper, and when unfolded there were the words: Day 1 – 6 minutes, Day 2 – 6 minutes, etc., up to Day 66 – 6 minutes.

Temperature variation: 1906–1914

A suggestion was made by Kohler of Wiesbaden that the quantity of X-rays could be measured by recording the temperature in the X-ray tube. A special X-ray tube as used by Kohler [22.5] was sold by Reiniger, Gebbert and Schall in 1914, although in 1906 Bordier, in his review of radiometric methods[47], had commented that 'we hardly need discuss this method'.

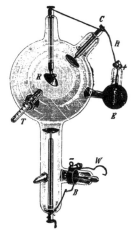

Röhren fremden Fabrikates.

„Monopol"-Oberflächen-Therapie-Röntgenröhre mit Vorrichtung zur therapeutischen Dosierung der Röntgenstrahlen nach Prof. Dr. A. Köhler, Wiesbaden.

Schutzmarke.

Diese Röhre ist besonders für die Röntgen-Oberflächentherapie bestimmt. Sie gestattet eine praktisch genügend genaue Verabreichung der für eine Sitzung erforderlichen Strahlenmenge durch bequeme direkte Ablesung an einer Thermometerskala.

[22.5] Monopol X-ray tubes were available in 1907 and some were modified to Kohler's specification by 1914. (Courtesy: Siemens AG, Erlangen.)

X-ray tube current: 1904

An ammeter was sometimes used for an indirect measurement of X-ray therapy dosage during the first decade following the discovery of X-rays. For example, an ammeter was exhibited by James G. Biddle of Philadelphia at the fifth annual meeting of the American Roentgen Ray Society in 1904 at St. Louis and the claim was made that 'by its use in therapeutic treatment an accurate standardisation of dosage becomes possible.' However, such an indirect method was clearly not satisfactory.

Photographic film blackening: 1902–1959

The ability of radiation to blacken silver bromide film was used by Kienböck in 1905[51] as the basis of his X unit for measuring X-ray doses [22.6]. Small pieces of film known as 'Kienböck strips' were exposed on the patient's skin [22.7] and the density of the developed film was compared to an arbitrary scale of blackening. Blackening of a photographic film was also proposed by Rollins in 1902[52] as a radiation protection standard. Tousey in 1921[15] suggested that film blackening could be used as a basis for a radium unit and defined a 'Tousey unit of power' as 'the photographic effect upon Kodak film by 1 candle-power incandescent electric light with a carbon filament, and with the usual brightness or whiteness'. He further defined a 'Tousey metre second' as 'the effect produced by such a light at a distance of 1 metre from Kodak film in 1 second' and quoted conversion factors as follows. A 20 mgm radium source = a radium source of 'strength' 2,000,000 exposed for 5 seconds = 1 Tousey metre second.

Film blackening as a method of dose measurements had a renaissance from the late 1940s to the late 1980s when, with small ionisation chambers designed for pocket wear, it was a main method of measuring the external radiation received by radiation workers. The film badge [22.8] consisted of a small X-ray film in an envelope wrapping, carried in a holder which in later years was made of plastic material but earlier had been made of metal. The badge holder, of which there were many different designs, incorporated several filters which enabled the energy of the photons to be distinguished approximately (e.g. orthovoltage or supervoltage X-rays) and for high energy beta radiation to be detected. Alpha-rays and low energy betas cannot be detected because they cannot penetrate the paper wrapping of the film. Typically, in the 1980s, the lowest dose equivalents that could be quoted with any degree of accuracy were 0.2 mSv for X-rays and gamma-rays and 0.8 mSv for beta-rays.

However, with photographic emulsion responding differently to radiation of different wavelengths and with the expense and time-consuming work required to set up a film badge developing and calibration laboratory, by the end of the 1980s the film badge had largely given way to the use of thermoluminescence dosimetry, in which the TLD dosimeters are also worn within a plastic badge holder.

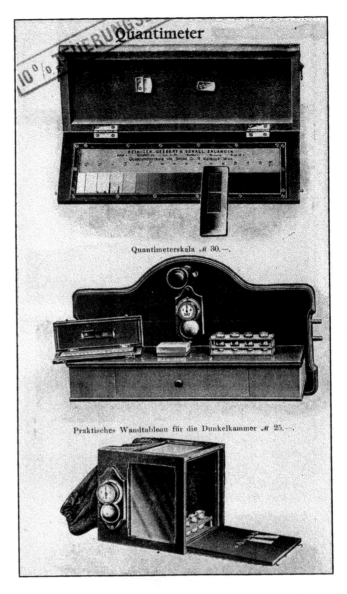

[22.6] The Kienböck quantimeter[51] as advertised in a special insert within the 1904/1905 Reiniger, Gebbert and Schall catalogue. The company overprinted the insert with an offer of a 10% price discount. The quantimeter scale in X units is seen at the top with the small rectangular metal section for holding the exposed Kienböck strip. The strip was moved along the scale to obtain a matching density. The other illustrations show the wall-mounted storage cabinet and a light-tight box for developing the Kienböck film strips. (Courtesy: Siemens AG, Erlangen.)

Thermoluminescence: 1904–1990s

Marie Curie in her 1904 doctoral thesis[54] recorded her observations on the thermoluminescence of calcium fluoride exposed to radium: 'Certain bodies such as fluorite become luminous when heated; they are thermo-luminescent. Their luminosity disappears after some time, but the capacity of becoming luminous afresh through heat is restored to them by the action of a spark, and also by the action of radium. Radium can thus restore to these bodies their thermo-luminescent property.' It was, though, not until the late 1940s that thermoluminescence was first used to make quantitative measurements of radiation exposure[55]. Several

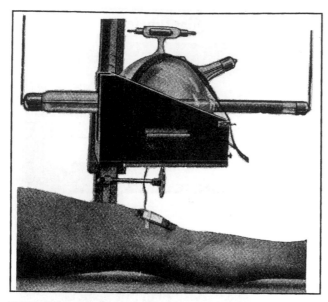

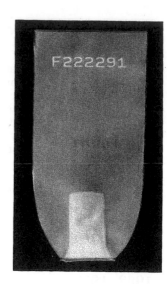

[22.9]   Finger stall containing lithium fluoride powder.

[22.7]   Shielded therapy tube of Knox in 1918[22] of the Royal Marsden Hospital, London. A Kienböck strip is seen just below the knee.

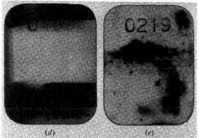

[22.8]   Film badges in 1959 after development[53]. (a) shows a film with no detectable radiation exposure. (b) is a film which has been exposed to diagnostic X-rays: the area under the filter is clear and also the blackened areas are sharply defined. (c) shows a film which has been exposed to gamma-rays. (d) was worn by a nurse in a radioactive isotope ward and shows a gamma-ray exposure (grey) as well as an X-ray exposure (black) obtained while assisting at a chest X-ray examination in the ward. It should be noted that the gamma-ray exposure of this film amounts to about four times the X-ray exposure. (e) shows splashes of radioactive material.

materials have been investigated but lithium fluoride has been found to be most suitable as a TLD phosphor.

Manufacturing methods vary between companies supplying lithium fluoride dosimeters but the material can be found as a loose powder which can be used in a finger stall to measure finger doses during nuclear medicine procedures (or during preparation of radium sources for brachytherapy before the remote after-loading era) [22.9] and can be compressed into a disc (e.g. 12 mm diameter × 4 mm thick) or square cross-section chips (e.g. 3 × 3 × 9 cm^3) or rods. It can also be

mixed with a tough substrate such as Teflon to yield a more rugged and flexible detector. The TLD personnel dosimeters which have replaced the film badge of [22.8] are in the form of a phosphor-Teflon card enclosed within a plastic badge case. Typically such a badge case would contain an open window and three filtered areas. The phosphor-Teflon can be read on either manual or automatic readers and three doses reported: low energy gamma/X/beta, high energy gamma/X and neutrons.

Chemical effects: 1904–1927

The most well known radiation unit depending upon a chemical colour change subsequent to exposure to radiation was the 'pastille unit' or 'B unit' of Sabouraud and Noiré in 1904[56] who used a small capsule of platino-barium cyanide [22.10, 22.11]. The pastille was proposed for use with X-rays but it had the relatively rare distinction of also being used in radium therapy. Also, as late as 1927 it was suggested by Codd[57] that the pastille unit should bear some relationship to a 'standard of X-ray intensity' and he commented that he 'understood that the descendants of the original Röntgen Society X-ray standard committee, appointed

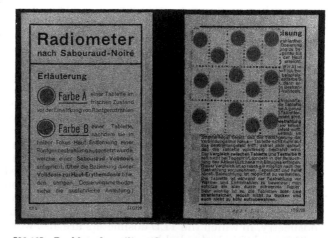

[22.10]   Booklet of pastilles. (Courtesy: Siemens AG, Erlangen.)

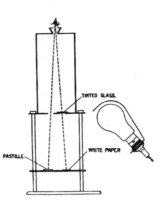

[22.11] Special holder designed for pastilles used to measure the output from an X-ray tube.

[22.12] A Lovibond tintometer for use with Sabouraud–Noiré pastilles.

in 1907, are still continuing their deliberations. No doubt the subject is a very difficult one, but until some decision is arrived at I suggest some purely arbitrary unit might be agree upon.'

The Sabouraud radiometer consisted simply of a booklet containing platino-barium cyanide pastilles and two standard tints known as tint A (unexposed) and tint B (the epilation dose), where tint B was found empirically. The instrument in [22.12] is a Lovibond tintometer which afforded a means of measuring the pastille colour in terms of fractions and multiples of tint B. The light from an 8 candle-power carbon filament lamp was allowed to fall on a sheet of white paper and on the exposed pastille. The reflected light from the white paper passed through a tinted glass before reaching the eye of the observer, whereas that from the pastille remained unchanged. A series of tinted glasses provided a range of dose values.

Another X-ray unit dependent upon a chemical colour change was the H unit of Holzknecht (1902)[58] who used a fused mixture of potassium chloride and sodium carbonate [22.13]. Tousey (1921)[15] stated that a dose of 5 H was equivalent to the dose received from a radium source of strength 7,000 after 12 hours exposure through a metal screen and that a recommended radium dose for cancer was 10 H and for lupus

was 6 H. He also stated that it was current practice to verify the H unit by self-experimentation using the forearm.

Both Holzknecht (1905)[58] and Kienböck (1904)[51] defined four degrees of exposure in terms of latency and skin reaction and both specified dosage in terms of the H unit. Holzknecht stated that only first and second degree exposures were permissible in radiotherapy and that '1 H is one-third of the dose required to determine the first signs of reaction on the face of an adult.' Alternatively, Kienböck defined a 'small normal dose' as an exposure which would produce a first degree reaction 'with a duration of about five minutes and an absorption of about 3 H units.' He further defined a 'normal dose' as that producing a slight reaction of the second degree 'with a duration of 5–10 minutes and an absorption of 3 H–5 H units'. A comparison made in 1932 by Schall[25] for filtered and unfiltered 170 kV X-rays is given in [22.14] for a unit skin dose, a pastille unit (B), Holzknecht's H unit, Kienböck's X-unit and the then röntgen unit (r) which he termed the international ionisation unit.

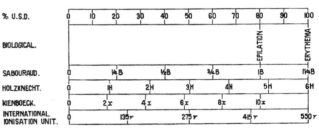

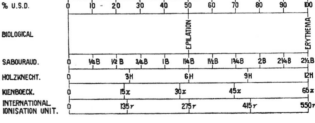

[22.14] 1932 comparison[25] of various doses for soft unfiltered 170 kV X-rays (top) and for 170 kV X-rays filtered by 0.5 mm copper + 1 mm aluminium (bottom). It was states that if erythema was produced by the 100% USD then epilation occurred at 80% USD for soft X-rays but at 50% USD for hard X-rays.

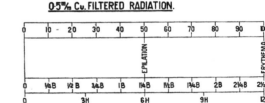

[22.13] A Holzknecht chromoradiometer. This was a combination of the Sabouraud radiometer and the Lovibond tintometer in that it was equipped with a scale of its own. Earlier models used Holzknecht's special pastille which consisted of 99.77% potassium sulphate[47]. Later models used Sabouraud–Noiré pastilles.

Other chromoradiometers were also used and that of Bordier (1906)[47] used four colour tints [22.15]. Bordier attempted to correlate his tint units with Kienböck's studies, Table 22.3. According to Bordier, tint IV was the strongest dose, should never be applied to the skin and was obtained 'after irradiation by a specimen of radium of radioactivity 100,000 for one week at a distance of 1 mm from two pastilles'.

Two other units based on chemical effects produced by X-rays were the unit proposed by Freund (1904)[59] which was dependent on the free iodine in iodoform-chloroform solution, and the 'kalom' unit proposed by Schwarz (1907)[60] which was dependent on the precipitation of calomel from a solution of ammonium oxalate and mercury bichloride. Bordier and Galimard (1906)[61]

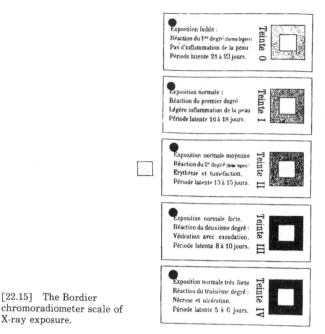

Exposition faible :
Réaction du 1ᵉʳ degré (forme legere)
Pas d'inflammation de la peau
Période latente 21 à 23 jours.

Teinte 0

Exposition normale :
Réaction du premier degré
Légère inflammation de la peau
Période latente 16 à 18 jours.

Teinte I

Exposition normale moyenne
Réaction du 2ᵉ degré (forme legere)
Erythène et tuméfaction.
Période latente 13 à 15 jours.

Teinte II

Exposition normale forte.
Réaction du deuxième degré :
Vésication avec exsudation.
Période latente 8 à 10 jours.

Teinte III

Exposition normale très forte
Réaction du troisième degré :
Nécrose et ulcération.
Période latente 5 à 6 jours.

Teinte IV

[22.15] The Bordier chromoradiometer scale of X-ray exposure.

Table 22.3. Correlation between Bordier tint units and Kienböck doses, 1906[47].

Bordier unit	Biological dose	Kienböck dose
Tint I Pale yellowish-green	Epilation 20 days after exposure. 1st degree skin reaction	Kienböck's weak normal exposure
Tint II Sulphur-yellow shade	Strong reaction: erythema tumefaction	Mild form of Kienböck's reaction of the 2nd degree
Tint III Colour of gamboge	2nd degree skin reaction, a true dermatitis	Kienböck's strong normal reaction

proposed the I unit, which was similar to that of Freund but which was more clearly defined with regard to measuring conditions. 1 I unit was found[61] to be 'approximately equal to 1 H (3.5 I = 5 H)' but the proposers commented that 'this is a pure coincidence'. Neither the I nor the kalom units were used extensively, and indeed neither was the H unit, but the reason for this was[14,47] that Holzknecht kept secret for some time the chemical composition of his dosimeter.

The only chemical dosimeters in current use in the 1990s are the ferrous sulphate systems due originally to Fricke and Morse (1927)[62], now generally known as Fricke dosimeters and used in electron beam dosimetry. They measure the oxidation of a dilute aqueous solution of ferrous sulphate to ferric sulphate on exposure to the radiation. The dose is directly proportional to the yield of ferric ions. The quantitative measurement is achieved by spectrophotometry using the absorption of ultraviolet light at 304 nm. The usual range[63] of the Fricke dosimeter is 40–400 Gy for electron beam dose rates of 0.1×10^{-2} to 40 Gy/s.

Selenium cell measurement: 1915

A dose rate unit, the 'F per minute' was proposed by Fürstenau in 1915[64]. It was based on the change in

electrical resistance of a layer of selenium caused by exposure to X-rays [22.16]. However, the response of the measuring instrument was extremely variable and depended on the quality of the radiation.

Intensimeter nach Dr. Fürstenau
D. R. P.
Nach dem Gebrauch ausschalten
Diese Seite ... rahlen aussetzen!
Nicht werfen! Zerbrechlich

[22.16] The intensimeter of Fürstenau. (Courtesy: Prof. W. Bohndorf, Strahlenklinik of the University of Würzburg.)

Idiosyncrasy and dosage: 1911

'Idiosyncrasy and dosage' is the title of a chapter in the Colwell and Russ (1915)[27] textbook *Radium, X-rays and the Living Cell.* They describe the results of an enquiry in 1911 in which 13 'radiologists of repute' were asked to explain the wide variation observed in skin erythema when similar radiation doses (albeit using different units) were given. Seven reputable opinions were that real idiosyncrasies existed in the reaction of the skin to radiation, whereas the remaining six were of the opinion that the explanation was due to 'errors in the measurement of the dose administered'. Equivalencies given by Colwell and Russ in England between different units are shown in Table 22.4, and these equivalencies were also given by Hirsch (1920)[65] in the USA, who added 3.5 kaloms for the Schwarz precipitation radiometer.

Table 22.4. Dose equivalencies, after Colwell and Russ (1915)[27].

1 skin erythema unit
Tint B of a Sabouraud–Noire pastille
Tint I of the Bordier chromoradiometer
625 M units of Guilleminot
5 H units of Holzknecht
10 X units of Kienböck
60–80 soft X-ray F units of Fürstenau

Heating effect: 1912–1953

It is interesting to note that the idea of a quantity to specify the amount of energy deposited by the radiation per unit of mass (or of volume) was proposed by Christén in 1912[66,67], but it was not until the Sixth International Congress of Radiology in 1953 that the 'rad' as a unit of absorbed dose was adopted, only to

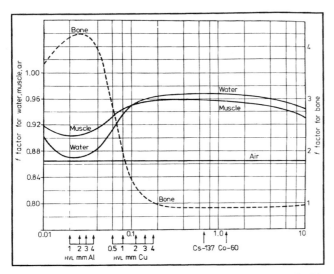

[22.17] Variation of roentgen to rad conversion factor, f with photon energy and irradiated medium.

be eventually replaced when SI units were introduced by the 'gray' (Gy) where 1 Gy = 100 rad. Absorbed dose and exposure may be related by the formula Absorbed dose = Exposure × f where f is a factor dependent upon the quality of the radiation beam and the material being irradiated [22.17].

In the period between the initial proposal in 1912 and the adoption of the rad unit some 40 years later, units of 'calories per $cm^2 \cdot mm$' were suggested in 1921 by Failla[24] for radium and of 'ergs per cm^3' for X-rays, independently by Solomon[39,86] in 1923 and by Dauvillier[68] in 1926, and in 1929 by Stahel[69] for radium. Dauvillier defined this unit as 'the energy in ergs absorbed per second in 1 cm^3 of water, supposed part of an indefinitely large volume' and called it a 'unit of power' denoted by W. The practical problems associated with making direct measurements in these units, for X-rays or for radium gamma-rays ensured that they were not widely used. In particular, the measurements of Stahel using an argon-filled ionisation chamber were subsequently found to be unreliable[70] after having produced with Murdoch of Brussels[71] dose distributions [22.18] for treatment of cancer of the cervix using the Paris system

of applicators [20.21] and dose distributions for radium brachytherapy regular-shaped and irregular-shaped surface applicators. Dose distributions for radium brachytherapy in the 1920s and 1930s (including those for the Paterson and Parker Manchester system) were therefore almost entirely based on theoretical calculations using the exposure around single or combined sources of uniform radioactive density and of simple geometry (point, line, disc, annulus, sphere)[72–74].

10 milliröntgen + FDE: 1971

In 1971, whilst undertaking a radiation protection survey of a KX-10 superficial X-ray machine sited in a hospital skin department, I was puzzled to discover some patient doses recorded by the consultant in charge as '10 mR plus FDE'. This turned out to be a placebo treatment with a sheet of lead over the applicator and the letters FDE standing for Full Dramatic Effect – consisting of a 67-year-old radiographer dressed up in the full protective armour of lead apron, gloves, and so on, and a red light on the KX-10 control panel!

Soft and hard X-ray quality: 1900–1937

The adjectives 'hard' and 'soft' were the first qualitative descriptions to be used for the penetrative quality of X-rays from a gas tube. Figure [22.19] was first published by Kienböck in 1900[75] and became the standard diagram used in the textbooks of the following decade to define the terms hard and soft.

However, although these terms remained in frequent use up to the 1930s, also being applied to Coolidge tubes, quantitative methods also became available soon after 1900 for measurement of the quality of an X-ray beam. The instruments devised for this purpose were variously known as penetrameters and quality meters. The most popular were those of Benoist (1902)[76] [22.20], Wehnelt (1904)[77], Bauer (1910)[78] [22.21], and Walter (1902)[79].

[22.18] Isodose distribution for the Paris technique (two vaginal corks each with 13.33 mg radium and an aluminium intra-uterine tube containing two radium tubes each of 6.66 mg) of brachytherapy for cancer of the cervix, as used by Murdoch and Stahel of the Brussels Radium Institute[69,71] in the late 1920s and early 1930s.

The curves are for erg/cm^3 units and it was stated that 'the uterine mucosa absorbs up to 5,000,000 ergs/ cm^3.'

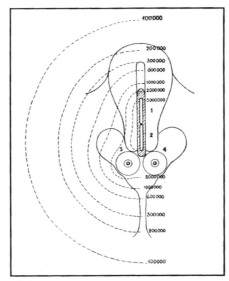

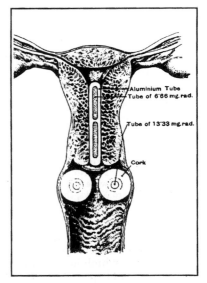

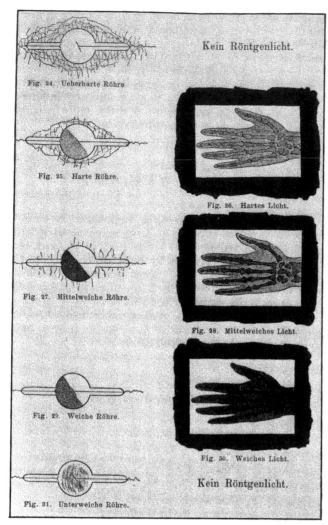

[22.19] Kienböck's 1900 illustration[75] of the terms hard, medium and soft as applied to the use of gas tubes for radiography of the bony structures and soft tissues of the hand.

Table 22.5. Material to be used for stating a half-value layer, following the ICR in 1937: after Case (1959)[80].

X-ray kilovoltage	Material
Up to 20 kV	Cellophane
20–120 kV	Aluminium
120–140 kV	Copper
400 kV	Tin

[22.21] The Bauer Qualimeter[78] was proposed in 1910 and advertised in the 1914 catalogue of Reiniger, Gebbert & Schall of Berlin, Stammhaus and Erlangen when the different instruments were compared in terms of numerical readings and the Bauer 1–10 scale was equated to the Benoist 1–10 scale. To use the qualimeter it was connected to the negative terminal of the induction coil or to the cathode of the X-ray tube. The device was a static electrometer and condenser which by indicating the potential of the cathode, gave a measurement of the quality of the X-rays. Scale reading 1 was defined to be a hardness such that the X-rays would be totally absorbed by 0.1 mm lead, whereas for scale reading 10, the X-rays would penetrate 0.9 mm of lead but be absorbed by 1.0 mm lead.

[22.20] Benoist[76] described this device in 1902 in a paper entitled 'Experimental definition of various types of X-rays by the radiochromator'. It was a thin disc of silver, 16 mm in diameter and 0.11 mm thick, surrounded by 12 aluminium steps of increasing thickness. When the penetrameter was placed behind a fluorescent screen, the luminosity of the central silver circle was compared with that of the thin steps of the aluminium ladder. For soft X-rays, steps No. 2 or No. 3 on the Benoist scale were usual, whereas for hard X-rays, the quality was No. 7 or No. 8.

[22.22] Schematic diagram of 1913 by Christén[66,67] showing the measurement of Halbertschicht or half-value layer (HVL), which is also sometimes called half-value thickness (HVT). The apparatus consists of a sheet of heavy metal (B), covered with regularly spaced small holes of such dimension that the sum of the areas of the holes is just half the area of the metal sheet. Besides this is a layer of bakelite (C). A fluorescent screen (D) was mounted at such a distance from the metal and bakelite that X-rays passing through the holes did not show the individual perforations, but produced a uniform illumination. Since half the area of the plate was open, the beam passing through it was just half of the original quantity.

Eventually though, all the penetrameters and quali-meters were superseded by the concept of half-value layer (HVL) which is generally attributed to the Swiss physician and physicist Th. Christén[66,67] in 1912 [22.22]. Later, in 1937 at the Fifth International Congress of Radiology it was agreed that for most medical purposes it was sufficient to express the quality of X-rays by HVL in a suitable material, as in Table 22.5.

Ionisation units for X-rays: 1908–1937

A unit based on the ionisation effect was used for measurements of relative intensities soon after Röntgen's discovery, but Villard in 1908[81] is generally credited as being the first to suggest a quantitative unit. There are, however, two references which predate Villard's work. In 1906 when Belot[82,83] was discussing radiotherapy and radium therapy, he remarked that 'one of the scientific methods of measurement is that derived from the ionising power of the X-rays' and 'it gives us a means of measurement more precise than any other, and has the advantage of furnishing a unit, X, which is based on the CGS system. I believe that this method will speedily find favour in radiotherapeutic practice.' Unfortunately Belot did not define his X unit.

Then in 1907 at a meeting of the American Roentgen Ray Society, C. E. S. Phillips from the Royal Cancer Hospital, London, advocated a unit based on ionisation either by X-rays or by a radioactive element. In the subsequent discussion, C. L. Leonard, one of the pioneers from Philadelphia, asked 'Why should not a unit be adopted that can be expressed in these commonly employed terms? A unit of ionisation will then be the quantity of electricity passing across a unit gap in unit time under the influences of radiation at unit distance under standard conditions of barometer and temperature.'

The Villard[81] unit was defined as 'that quantity of X-radiation which liberates by ionisation one electrostatic unit (esu) of electricity per cm^3 of air under normal conditions of temperature and pressure'. This was essentially the same as the first definition of the röntgen unit (r) in 1928. Nevertheless the Villard unit remained largely unused for several years, until it was adopted by Krönig and Friedrich in 1918[33] as the 'e unit', and later modified by Behnken in 1924[84] to become the 'R unit' or 'German unit of X-radiation' or 'German röntgen'[85].

To confuse matters on the international scene there was also the 'French röntgen' suggested by Solomon in 1925[86] and also called an 'R unit'. This was based on the ionisation produced at 2 cm distance from a 1 gm radium source. The conversion between the two R units was given by Béclère (1927)[87] as: 1 German R = 2.25 French R. Béclère also referred to the work of Grebe and Martius who measured the values of the unit skin dose (HED) in German röntgens for 27 sets of apparatus in 14 institutes. The results were very revealing and showed that not only did the HED vary among centres, but even in the same institute the HED varied from one apparatus to another. The ratio of the smallest to the largest HED was 1:3.9. If alternatively the HED is measured in different units rather than only German röntgens, then according to Holthusen in 1926[88] the picture becomes even more complex and opinions differ by a factor of 10, if, for example, the units were then converted into ergs.

However, in 1928 at the Second International Congress of Radiology, which was held in Stockholm, an absolute unit of X-ray dose was accepted internationally. This replaced the earlier e and R units, and also some less well publicised ionisation units such as the 'mega mega ion' of Szilard (1914)[89], 10^{12} ions, which could never be reproducibly measured, and the 'E unit' of Duane, which was essentially 1 German röntgen per second.

The finally acceptable international unit of 1928 was defined as the 'quantity of X-radiation which, when the secondary electrons are fully utilised and the wall effect of the chamber is avoided, produce in 1 cc of atmospheric air at 0°C and 76 cm of mercury pressure such a degree of conductivity that 1 electrostatic unit of charge is measured at saturation current'. This unit was called the 'röntgen' and was denoted by the lower case letter 'r' to avoid confusion with the German and French röntgens which were both denoted by the capital letter 'R'. Eventually, though, in the 1960s, the International Conference on Weights and Measures (or SUN Commission) decided that the use of the lower case 'r' was out of keeping with their system of terminology and the radiological professions were forced to return to the capital 'R'.

After the röntgen had been adopted in 1928 for X-rays, it was suggested by Mayneord (1931)[90] that gamma-rays could also be measured in the same units, and workers such as Coliez (1927)[91] also stressed that any international X-ray unit must also be available for gamma-ray measurements. It is interesting to note that some of the early investigators such as Cleaves (1904)[10] had referred to gamma-rays as 'radium röntgen rays', but there was never universal agreement as seen from the comment of Kassabian (1907)[2]: 'Some state that gamma and Röntgen rays are identical, though Strutt of Cambridge is of the opinion that there is a vast difference. Crookes maintains that they are actual emanations: the projection of minute particles from the radioactive body into adjacent space.'

Indeed, even in 1934 the International Commission on Radiological Units (ICRU) could still not bring itself to recommend that the röntgen be accepted as an appropriate unit for both X-rays and gamma-rays. That recommendation had to wait until 1937 when the röntgen was defined slightly differently than in 1928 as 'that amount of X or gamma radiation such that the associated corpuscular emission per 0.001293 gram of air produces in air ions carrying 1 electrostatic unit of charge of either sign'. Further revisions occurred in 1953 at the Seventh International Congress of Radiology held in Copenhagen when the röntgen was termed the special unit of a quantity called 'exposure' and the 'rad' was adopted as a unit of absorbed dose of any ionising radiation, equal to 100 ergs/gm of any absorber. Now, with the advent of SI units, the röntgen has disappeared as a special name for a radiation unit and the SI unit of exposure is 1 coulomb/kg, which is equivalent to approximately 3.876×10^3 röntgens (Table 22.1).

Ionisation units for gamma-rays: 1911–1937

The pioneer of ionisation measurements in air due to radium gamma-rays was Eve, whose papers extend back to 1904 and who, in 1911, made a good measurement of the ionisation due to various radiations as the number of ion pairs per cm^3 at 1 cm distance from a point source of radium. He used the formula $(KQ/r^2) = e^{-\mu r}$ and also realised that this was a measurement of energy absorbed, thus anticipating Bragg and Gray[92–94]. For many years the constant K was termed Eve's constant[72], later being renamed the k-factor and finally being amended by the ICRU in 1962 when it was defined relative to 1 curie of a radioactive source and called the 'specific gamma ray constant' denoted by Γ.

The ionisation effect was suggested by Failla in 1917 as a basis for a gamma-ray unit and was defined by measuring the ionisation at 1 cm from a 1 g radium source. A 'radon' unit was also defined during the same period, using the ionisation measured at 2 cm from a 1 g radium source. The 'D unit' named in honour of Dominici was proposed by Mallet and Proust[95] in 1927 at the Congress for the Advancement of the Sciences, Liege. This unit was defined to be 'the ionisation produced by 1 centigram of radium during one hour in the electroscopic chamber of an ioni-micrometer'. Mallet's ionisation instrument had 'a volume of the order of 1 cm^3 which may be compared with the same elementary volume of tissue'. Calibrations were undertaken using 10 mg radium filtered by 1 mm platinum at 2 cm distance from the electroscope[95]. Coliez (1927)[91] termed the D unit a 'decigramme hour', defined as 'the ionisation given by the 10 mg radium tube, at 2 cm from the ionisation chamber in 10 hours'. The relationship between the D unit and the röntgen was given by Cade (1948)[96] as: 1 D = 102.5 röntgens.

Other gamma-ray units based on the ionisation effect were suggested, such as the 'eve unit' of Hess (1922)[97], but only the D unit and the 'intensity-millicurie hour' (Imc) or 'cm-element hour' unit of Sievert (1932)[98] were widely used. The Imc was the 'intensity at 1 cm distance from a radium preparation containing 1 mg radium element (pure radium sulphate) in equilibrium with its breakdown products, when the source of radiation is surrounded in all directions by 0.05 cm platinum and the preparation enclosed by the filter so small that the source may be considered to be a point', Mayneord and Roberts (1937)[70]. The particular relevance of the Imc unit was in the theoretical calculations of the distributions of radiation intensity around gamma-ray sources of simple geometrical shape, since these were assumed to consist of an infinite number of point sources[72–74].

Ionisation measuring instruments: 1896–1990s

An enormous number of different designs of ionisation measuring instruments have been developed over the years for the measurement of X-rays and gamma-rays. This section illustrates only a small fraction of what has been available to medical physicists from the year 1896.

One of the most comprehensive reviews of ionisation measuring instrument development has been given by Peter Pychlau of PTW-Freiburg (1983)[99] and the following chronology to the 1920s is taken from his publication.

'It was reported in *The Electrician* of February 7th 1896 that an electrically charged gold leaf electroscope and an electrically charged plate had been discharged by exposure to X-rays. This was probably the first measurement of the ionisation effect although it was not directly recognised as such. J. Perrin in Paris in November 1896 published a drawing of the principle of a free-in-air chamber and listed all the essential elements of an experimental arrangement.

'However, it was to be many years before the goal of a practical free-air chamber was achieved. Measurement of photons using ionisation chambers means measuring small electrical charges and to illustrate the difficulties encountered in making the necessary electrical measuring instruments, it has to be remembered that in the

[22.23] The gold leaf electroscope was one of the standard radiation measuring instruments in many hospitals prior to the availability of sturdy ionisation chambers and of Geiger counters. From a 1929 description[100] 'it consists of a small gold leaf strip about 2 cm long and 1–2 mm wide attached at one end to a rigid metallic support from which it normally hung down freely. If, however, this arrangement is placed inside an earthed metallic case to shield it from draughts and electrostatic influences, and the gold leaf system insulated, then on giving the leaf and its support a charge, the former will stand out away from the latter by an amount depending on the charge given. Having given it a charge, if the insulation is perfect the position of the leaf is permanent. However, if X-rays or gamma-rays fall on the air in the metallic box the charge leaks away and the leaf returns to its uncharged position.' The upper illustration shows a 1904 electroscope[8] both discharged (left) and charged (right). These devices were still commercially available in 1950 as shown by the illustration (right) from a catalogue of the Central Scientific Company of Chicago. The overall height is 23 cm and the thickness 6 cm. In the 1929 diagram[21] (far right) the metal rod T, supported by a perfect insulator I (amber, paraffin, wax or orca), carries a very thin aluminium or gold leaf F, a few mm in width. In this design the graduated scale was recommended for use in 'measuring the unknown content of a radium preparation' after calibration using a radium source of known content, such as 5 mg. A comparison of the rate of fall of the gold leaf enabled the unknown mg radium content to be determined.

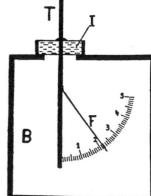

early years of this century it was just not possible to go to a shop and buy resistors, capacitors, plugs, switches or voltage sources. The change in this situation in Germany, occurred on December 22nd 1920 when the main radio station near Berlin broadcast the first wireless concert. From then on, radio sets were manufactured commercially and broadcasting created a market demand for radio-electrical components. Before then each laboratory had to manufacture its own components.

'The first dosimeters used electrostatic methods and an electroscope or electrometer was charged with a frictional electricity machine and then connected to an ionisation chamber, which was always a thimble chamber. Irradiation of the chamber caused the electrometer to discharge (as reported by Perrin in 1896). X-ray outputs were measured on the assumption of a constant dose rate, which was not at all correct for the 1920s. Also, with these early devices the electrometer often discharged even when the chamber was not irradiated because of poor insulation. This situation was improved with the development of condenser chambers such as those of Glasser in Freiburg, Sievert in Stockholm and by the Victoreen Company in the USA. The operational technique was to irradiate for a specified time and then calculate the dose rate from the measured dose and the irradiation time.'

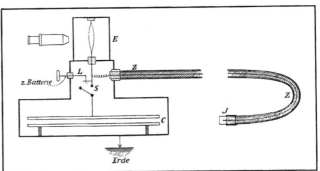

[22.24] Ionisation chamber and Wulf electrometer used by Krönig and Friedrich of Freiburg in 1918[33]. Their choice of instrument was determined by the following consideration: 'As it was necessary for our biological experiments to measure the intensity at each location within the irradiated medium, we had to design the apparatus in a way that allows us to insert the ionisation chamber like a probe into the medium. The method of measurement was the discharge of an electrometer combined with a capacitor. We used the well known two-fibre electrometer of Wulf. The amount of charge can be determined by the spreading of two conducting quartz-fibres under a microscope'.

[22.25] Compensation electroscope described by the Union Minière du Haut Katanga (1929)[21] for the measurement of radium preparations. It consisted of three parts: the electroscope proper and two spherical ionisation chambers. The filament is a platinum wire 0.004–0.006 mm in diameter, attached to a rod. It hangs freely between the two plates of the electrometer. The radium preparation is placed near one chamber with the other chamber and the electroscope protected by a thick lead screen. The rate of movement of the filament, expressed in divisions/second is proportional to the gamma-ray dose rate of the radium preparation.

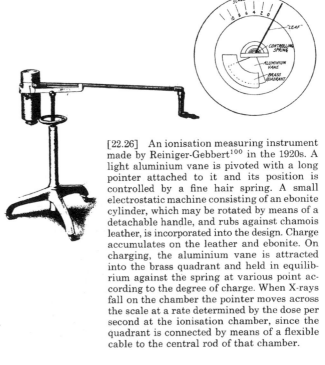

[22.26] An ionisation measuring instrument made by Reiniger-Gebbert[100] in the 1920s. A light aluminium vane is pivoted with a long pointer attached to it and its position is controlled by a fine hair spring. A small electrostatic machine consisting of an ebonite cylinder, which may be rotated by means of a detachable handle, and rubs against chamois leather, is incorporated into the design. Charge accumulates on the leather and ebonite. On charging, the aluminium vane is attracted into the brass quadrant and held in equilibrium against the spring at various point according to the degree of charge. When X-rays fall on the chamber the pointer moves across the scale at a rate determined by the dose per second at the ionisation chamber, since the quadrant is connected by means of a flexible cable to the central rod of that chamber.

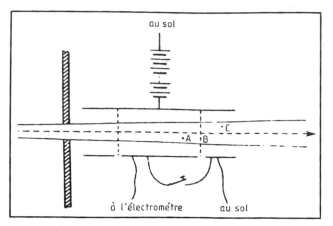

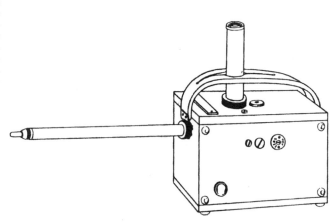

[22.27] Illustrations of the principle of the 'free-air' ionisation chamber. Left: diagram after Perrin[99] dated November 1896 with the legend 'Luftkondensator zur Bestimmung des durch Ionisation verursachten Elektrizitätsverlustes und damit der Intensität der ionisierenden Röntgenstrahlen'. Right: in a free-air chamber (1945)[101] the X-ray beam entering the chamber is limited by a system of diaphragms and so prevented from striking the chamber walls. The complete chamber is quite elaborate and very special attention has to be given to the definition of the X-ray beam, to the determination of the volume of air actually ionised, to the saturation potential of the chamber and to the spacing of the electrodes, which must be sufficient to ensure complete utilisation of the secondary electrons produced by the passage of the X-ray beam. If a charge Q esu is collected in time t seconds and the ionised volume of air is v cm^3 the dose rate is Q/fvt esu/cm^3/sec (röntgen/sec) where f is the factor which reduces the atmospheric pressure and temperature to those stated in the definition. Kaye and Binks (1936)[102] at the National Physical Laboratory used a chamber some 3.5 metres high with a plate separation which could be varied between 25 cm and 3 metres. When the ionising source was radium it was placed several metres from the chamber[101].

[22.28] The Victoreen r-meter of the 1930s. It consists of a small ionisation chamber rigidly connected to a string electrometer, the scale of which is calibrated in röntgens. The electrometer, of dimensions 6 × 4 × 2.5 inches^3, is charged by a small built-in static charger. The chamber tube is 7 inches in length. The ionisation chamber is placed at the point of measurement and exposed for one minute for a scale reading in röntgens/minute. By the 1960s, although still in use, it had been largely superseded for routine radiotherapy dosimetry on the output of therapy machines by the Baldwin–Farmer dosemeters.

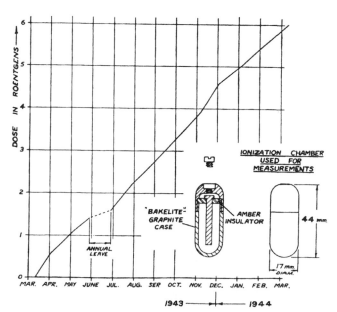

[22.29] Schematic diagram of a condenser chamber used in the 1940s for personnel monitoring[103]. The sensitivity was about 450 volts/röntgen. They were usually charged to about 100 volts and worn for 7–10 days. The graph shows the dose received in a year by one radiographer working an eight-hour day, five days per week, on both radium bomb teletherapy and orthovoltage X-ray machines. The total dose received is 6 röntgen but 1 röntgen has to be subtracted from this figure to take into account the natural leakage of the chambers over a year.

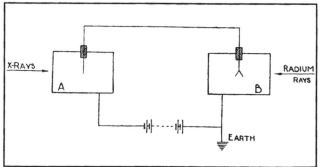

[22.30] This device, the X-ray/radium balance, was designed by Sidney Russ. The ionisation due to the X-ray beam is balanced against that which is due to a fixed radium standard incorporated in the instrument. The X-ray intensity is then expressed in terms of a known quantity of radium. A and B are the two ionisation chambers. Schall[25] stated that the critical part of the instrument was the ionisation chamber, constructed with wall material such that its effective atomic number is equal to that of air.

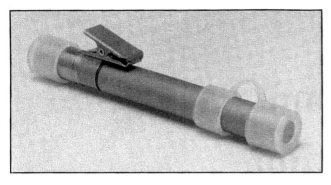

[22.31] Pocket pen-type ionisation chamber dosimeter of the 1980s. These instruments are used not only in hospitals but also in the nuclear power industry, fire service, armed forces and in civil defence. Right: reading the radiation exposure. (Courtesy: R. A. Stephen & Co. Ltd, Mitcham, UK.)

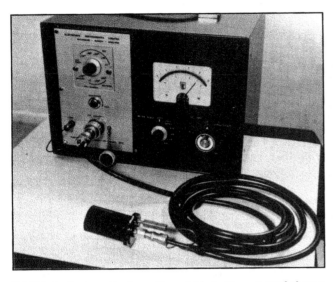

[22.32] Ionisation chamber with portable battery-operated electrometer for measurements on diagnostic X-ray equipment, 1970[104]. The chamber volume is approximately 40 cm^3. Full-scale deflections are designed within the range 0.45 milliröntgen to 150 röntgen and for exposure rates within the range 0.045 mr/sec to 15 r/sec.

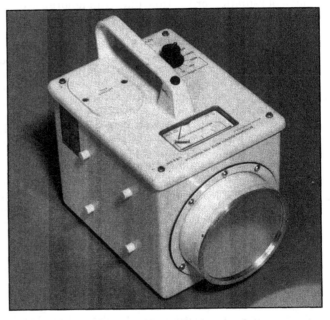

[22.33] Portable EMI Electronics ionisation chamber survey meter of the 1980s known as a beta-gamma dose rate meter. Its size is approximately $25 \times 15 \times 20$ cm^3. The chamber protrudes about 4 cm from the front of the monitor and normally has a black plastic cap protecting the front of the chamber, except (as shown here) when beta-rays are to be measured: since they could not penetrate the plastic end cap.

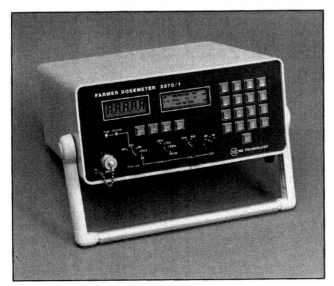

[22.34] A Farmer dosemeter of the 1990s. This modern dosimeter includes a digital readout (top left), information display (top centre) and a range of buttons (right) which include programmable correction factors for temperature and pressure, preset dose and preset time. Ionisation chambers which can be used with this dosemeter include a 0.03 cm^3 parallel plate chamber for measurement of low energy X-rays, 0.6 cm^3 thimble chambers for therapy measurements and 35 cm^3 and 600 cm^3 chambers for diagnostic radiology and radiation protection measurements. Strontium-90 reference sources are available for chamber stability checks. (Courtesy: NE Technology.)

Chapter 23

Radiation Risks and Radiation Protection

In the first two to three years following the medical use of X-rays and radium, there was almost a complete lack of radiation protection, due to an ignorance of the hazards involved. However, when the injurious effects started to become apparent, as in [3.33], described in 1904 as a Röntgen light burn of the second degree[1], some national attempts began to be made in some, but not all countries. For example, the Röntgen Society's proposals in 1898 for a Committee on X-ray Injuries has already been described in Chapter 3, and in America in 1903 a Protection Committee was proposed within the American Roentgen Ray Society by S. H. Monell. Table 23.1 lists some of the other milestones in the development of radiation protection standards[2, 3] over the 50-year period between the adoption of an international radium standard and the publication of the first ICRP report. Later milestones include the publications of several other ICRP reports, such as *ICRP Report 26*[4] in 1977, and the establishment of SI units in radiological medicine[5] in the mid-1970s.

From the early years of this century, [23.1] is a unique set of photographs[6] in chronological sequence which document the various stages of radiation injury from dermatitis to cancer to amputation of the left hand of an X-ray engineer, 1910–1913. The degrees of Röntgen-ray dermatitis were also defined by Beck[1] in 1904, see Table 23.2. The treatment of X-ray burns, also as of 1904, was described by Pusey and Caldwell[7] who began their commentary by stating 'It is often necessary to carry the reaction to the point of producing some degree of burn in order to get the desired result when giving X-ray exposures for therapeutic purposes.' They continued by stating that the treatment is 'along ordinary medical and surgical lines' and suggest lanolin as a salve base, following a proposal by Schiff and Freund in 1898. For reactions where there was no weeping a calamine and zinc oxide lotion was sometimes used and the authors had also found that a lead and opium wash sometimes gave comfort to the patient. Their most radical option was given as skin grafting.

Radium burns to the hands also occurred due to lack of proper precautions when handling sources and those who exposed their hands for diagnostic purposes of a radiumgraph, as distinct from an X-ray radiograph, would have received a significant radiation dose. Figure [23.3] shows a direct comparison in 1911 with the same hand, coin and compass used for demonstration[8], see also [3.8].

A major advance in fluoroscopic and radiographic

Table 23.1. Some of the milestones in the development of radiation protection standards during the 50 years 1910–1960.

1911	Adoption of an international radium standard and the curie as a unit of activity.
1915	Röntgen Society in London adopts protection recommendations.
1921	British X-ray and Radium Protection Committee adopts radiation protection recommendations.
1921	Maximum Tolerance Dose principle stated for X-rays as 'A sort of grand average of the protective measures could be gleaned from the working conditions of a number of experienced radiologists who had escaped injury and still enjoyed normal health'.
1922	American Roentgen Ray Society adopts radiation protection rules.
1925	The first tolerance dose proposed: 0.01 skin erythema dose per month.
1926	Dutch Board of Health adopts the first regulatory exposure limit: 1 skin erythema dose per 90,000 working hours.
1928	Röntgen unit (R) adopted for exposure to X-rays.
1928	International X-ray and Radium Committee (ICXRP, now ICRP) formed and its first recommendations published.
1929	U.S. Advisory Committee on X-ray and Radium Protection (USACXRP) formed, which later (1946) is reorganised into the National Committee on Radiation Protection (NCRP).
1931	USACXRP recommends exposure limit of 0.2 R per day.
1931	A League of Nations report recommends a limit of 10^{-5} R per second for 8 hours.
1934	ICXRP recommends a limit of 0.2 R per day.
1934	USACXRP recommends a separate limit of 5 R per day for hands and in 1936 reduces the recommended exposure limit from 0.2 to 0.1 R per day.
1941	USACXRP recommends adoption of a maximum body burden of 0.1 μCi for radium.
1944	Maximum Permissible Dose (MPD) concept for inhaled radioactivity introduced, together with the rem and the rep, by H. M. Parker.
1948	British Radioactive Substances Act passed in Parliament.
1949	NCRP lowers the basic MPD for radiation workers to 0.3 R per week and introduces the risk-benefit concept.
1950	ICRP adopts a basic occupational MPD of 0.3 R per week.
1953	ICRU introduces the concept of absorbed dose.
1954	NCRP puts forward the ALARA concept: 'as low as reasonably achievable'.
1955	NCRP recommends 5 rem per year as the basic MPD for occupational exposure.
1956	United National Scientific Committee on the Effects of Atomic Radiation (UNSCEAR) organised.
1959	*ICRP Publication 1* issued, stating ALARA concept and recommending a limit of genetically significant dose to the general population of 5 rem in 30 years.

safety for X-ray workers was the invention of test objects for assessing the hardness of the X-ray beam, such as the chiroscope in 1903 and the very similar osteoscope[1] in 1904 [23.4, 23.5]. These were the first devices to ensure both good image quality and good radiation protection.

Table 23.2. Degrees of Röntgen ray burns: after Beck (1904)[1].

First degree	Characterised by the symptoms of hyperaemia. The most pronounced subjective symptom is a tormenting itching of the skin.
Second degree	Main feature is the formation of blisters containing clear or yellowish contents. Inflammatory signs are well pronounced. Pain is intense: see [3.33].
Third degree	This is the gravest degree and is characterised by the escharotic destruction of the irradiated tissues. They show signs of dry gangrene and their appearance is brownish-black. The necrotic area should be surgically removed. An ulcer remains which may take months to heal.

The introduction of radiation units of measurement (Chapter 22) with a more scientific basis than a subjective opinion of a skin erythema dose was another advance which directly improved radiation protection for both operator (over-exposure) and patients (under- or over-exposure), although even until the end of the 1930s it was often difficult to actually measure the proposed radiation units in practice.

Some examples of protective wear from 1910 have been shown in [3.34] but the apparatus itself could also be shielded. A shielded X-ray tube[9] of 1907 is illustrated in [23.8] together with short tubes of lead glass (or 'heavy glass') used for 'sharpening the image', and a shielding design of 1918 is shown in [22.7]. The operator could be shielded by a mobile screen of lead and lead glass [23.9], as used with the 'Blitzapparatus' of 1909, or by a permanent lead wall [23.10] as in Kassabian's 1907 diagnostic X-ray room[10]. Protection for the eyes was afforded by various designs of X-ray goggles [23.6] incorporating lead glass lenses.

With procedures to ensure safety with radium brachytherapy (radium and cobalt-60 teletherapy are discussed in Chapter 19) the situation is entirely different from that with X-ray machines for diagnostic radiology or for teletherapy (see Chapter 18) where a major aspect is room design, incorporating lead plywood shielding on walls and doors and a shielded cubicle for the operator for diagnostics, and a room with thick concrete walls and a maze entrance, and patient viewing using closed circuit TV for therapy. Thus the United States Bureau of Standards 1938 publication[11] *Radium Protection* included the following major section headings: storage, manipulation and preparation; radon plants; transportation within an institution; and transportation by common carriers.

One design of radium storage safe, having eight drawers which can be pulled out using a pistol-grip device, is shown in [23.13]. At the rear of the open drawer is a cavity in which radium sources can be stored, and in front of this cavity is solid lead. Manipulation and preparation of the sources was always undertaken behind a shielded bench such as in [23.11] which incorporated lead block shielding in front of the chest of the operator and on the base of the bench to reduce the exposure to the gonads. Sources were handled using long-handled forceps or tweezers and devices were used to drop the sources into applicators so that handling could be made easier [23.12]. Thus for example, a needle could be dropped into a cavity with the pointed end first, and only the eyelet end remaining above lead, and the thread could then be passed through the eyelet and tied so that it would not become loose.

(Patients have been known to have tried to pull out radium sources and in some cases were successful.)

Manual afterloading obviated the need for many of these procedures and hence significantly reduced the exposure dose to personnel. Some form of lead shielding was still necessary, however, when nursing procedures were undertaken while the radionuclide sources were in position within the patient, for a low dose rate treatment such as for cancer of the cervix [23.15]. With remote afterloading the radiation hazard is eliminated except for remote afterloading machines where a small amount of exposure is obtained due to preparation of iridium-192 seed or ribbon sources, such as with the microSelectron-LDR remote afterloader.

Two hazards of radium brachytherapy were 'lost sources' and mishandling of the radium sources, which if fractured could lead to serious contamination and the escape of radon gas. In the early 1930s there was a relatively large amount of literature on radium source mishandling and on the mechanical life of such sources. Since the half-life of radium was 1670 years there was an advantage in dosimetry calculations in that the activity per source remained constant, but this was offset by the disadvantage of the limited mechanical life of a source. Statistics of radium needle damage[12] from St. Bartholomew's Hospital, London, are given in Table 23.3. Stories of lost radium needles and subsequent searches are anecdotal history for most hospitals which ever used radium tubes and needles. Indeed, in 1938 a complete book[13] was devoted to 'Radium lost and found' and included humorous illustrations, such as [23.16], of radium losses in the USA.

With the development of radon brachytherapy in the 1920s [20.13], it became advantageous for hospitals to have their own radon production plants [23.17] and many designs were invented for what was always a potential radiation hazard [23.18, 23.19, 23.21].

Radon, also known as radium 'emanation', was also used for less scientifically sound medical purposes and was recommended as a general health cure. As 'radio-active drinking water' [23.19, see also 20.58] it was prescribed for gout, rheumatism and a variety of other complaints. Other remedial treatments included injection of petroleum jelly impregnated with radon (1909) and application of radioactive mud poultices (1913)[3]. Such treatments flourished until the late 1920s and one physician[14] stated in 1921 that since 1913 he had given over 7,000 injections of radium element in doses ranging from 10 to 100 microgram. One of the ailments treated by this physician was hypertension. Other diseases treated orally and parenterally (with either radium, mesothorium or polonium) included those as diverse as syphilis, gout, infectious polyarthritis, rheumatism, leukaemia, pernicious and other anaemias, epilepsy and multiple sclerosis[15]. In addition

Table 23.3. Radium needle damage statistics[12].

Year	No. of needles	No. of damages	Percentage damages
1927	220	108	49
1928	226	127	56

Needle lengths were from 1.8 to 6.0 cm and long needles were found more vulnerable than short ones. There is a marked gain in mechanical strength through using 0.6 mm platinum walls instead of 0.5 mm. Damages included complete fracture, ruptured eyelet, indented and bent eyelets and slight and severe bending.

there was a whole body radium irradiation technique which was used to treat migraine[16].

Another therapeutic use of radium emanation was inhalation of the gas, using small portable devices [23.18, 23.22] or within large closed chambers known as 'emanatoriums' or 'inhalatoriums' such as those at the Joachimsthal health spa [23.21][17]. Radium baths were also recommended for their therapeutic properties, being combined at Joachimsthal with a simultaneous inhalation treatment [23.22].

Radon was also the sources of an unusual case of metal contamination[18] which occurred in New York in the 1950s. Spent radon sources, or more precisely the gold used for their encapsulation, found their way to the jewellery industry. Rings and other jewellery were manufactured from this gold without proper decontamination. In 1981, a total of 170 pieces of contaminated jewellery were identified, and nine persons were found to have skin lesions. These persons had on average worn their jewellery for 17 years. Also, as late as 1989 three new cases of contaminated gold rings were discovered, two of which had resulted in skin cancer on the wearer's finger[18]. In one of those cases a radon plant technician at Memorial Hospital, New York, had previously made a wedding ring for his wife, the final outcome being amputation of her affected finger[19].

Not only radon but also radium was used for treatment of non-malignant conditions, in some cases purely for cosmetic purposes, and many of these patients eventually suffered from radiation-induced cancers. A series of these case histories was collected by Sir Stanford Cade[20] in 1957. The latent periods between irradiation and diagnosis of the induced cancer varied widely and have been reported as long as 56 years [23.23], although the mean value for the Cade data is 30 years.

Many of the cases documented by Cade related to the use of X-rays or radium creams from a 'beauty specialist' or a 'commercial firm' (the first of whom may well have been the unsuccessful M. Gaudoin of 1896 [3.1]). This chapter closes with a series of advertisements for these extremely dangerous preparations, all sold with the promise of improvement of facial beauty—all that is, except the last one for Zoé Atomic Soda [23.28] from the archives of the Marie Curie Museum in the Institut-Curie, Paris. It is claimed that Zoé 'Gives infinite energy, just like an atomic pile.' If only I could have obtained a bottle when I started writing this book it would have been complete much quicker!

[23.1] These photographs of progressive radiation injury are reproduced from the 1930 textbook[6] by Flaskamp and Wintz of Erlangen and I am grateful to Dr Gregor Bruggmoser for this translation from the original German text.

186

Radiation injury to an X-ray engineer: Erlangen, 1904–1913

A
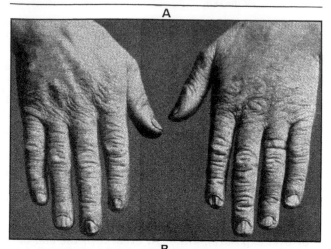

B
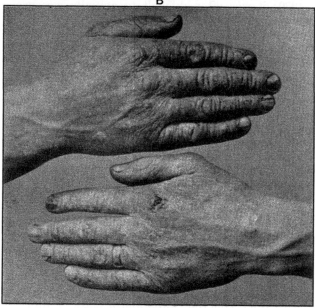

C
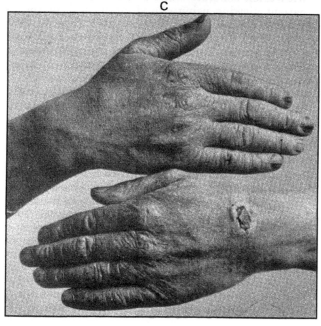

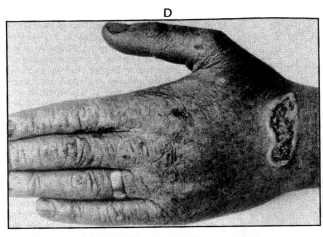

D

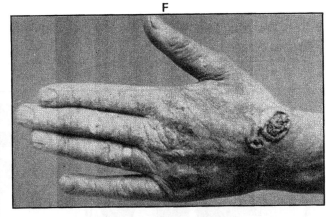

F

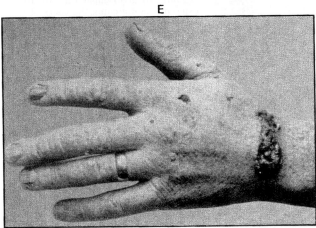

E

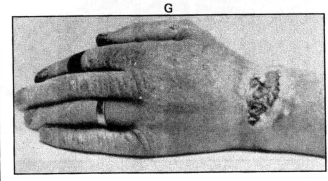

G

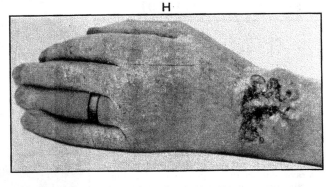

H

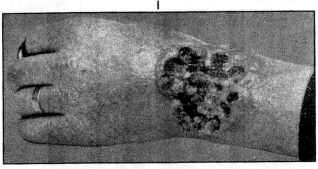

I

The photographs show the typical course of a professional X-ray damage, beginning with an 'X-ray hand' and ending with an ulcer and a carcinoma. They are of the hands of an X-ray engineer who has been working with X-rays since 1904. He worked in the Diagnostics Department making 300–400 films per month. After six months (in the autumn of 1904), he noticed a reddening of the skin and a slight swelling of both hands. He thought that this was an effect of the chemicals used for the film development. In spite of treatment, there was no recovery. The fingernails became rough and the skin dry and cracked. He protected himself with frequent treatments with ointment, and succeeded in keeping the state of the hands stationary until 1910.

The photograph (A) in April 1910 is typical of Röntgen X-ray hands. A swelling first occurred above the wrist of the left hand, looking like a small carbuncle. After a couple of days a small plug came out of this carbuncle. The patient thought it was an irritation caused by his cuffs. The cuffs rubbed the skin off, especially during bicycle riding. All treatment modalities had no success: moist treatments, ointment bandages, transplantations. (B) and (C) are from March and July 1911 and (D) from November 1911 shows an ulcer on the wrist of the left hand. Until the end of 1912 this ulcer decayed in combination with a serious necrosis. (E)–(I) show the deterioration during 1912 and in February 1913 the lower one-third of the left forearm was amputated (J). This cured the cancer.

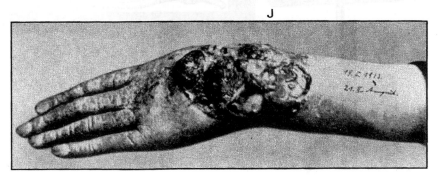

J

William Morton's New York X-ray laboratory: 1896

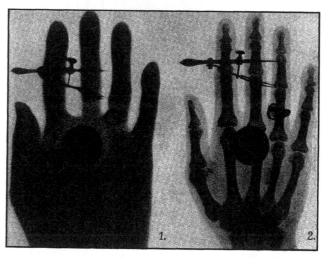

X-ray quality assurance: 1904–1911

[23.3] Comparison of 1911 images of a hand using radium (left) and X-rays (right) as the imaging source[8]. The radiation dose received during a radiumgraph was much greater, but the image far less distinct, than when using X-rays, due to the high energy of radium gamma-rays. Consequently very few radium-graphs were ever published; other examples are shown in [2.16, 2.17, 3.7–3.9].

[23.2] William Morton's X-ray laboratory in New York. This photograph is reproduced from Morton's 1896 book[21] where it is described as 'representing the apparatus as arranged for actual taking of an X-ray picture. The devices here shown are now being assembled into more compact form for greater convenience and portability.' Elsewhere in the book, Morton advises that: 'It is presumed that the careful operator has seen that his machinery is well oiled, that his break-wheel has been polished with sandpaper and that all the electrical connections are properly made and that the Crookes tube is in position over the object. It must be observed that the wires leading from the induction coil to the Crookes tube do not touch the glass walls of the tube, for fear of the electrical discharge perforating the glass . . . We would advise the beginner to give ample time to all of his exposures by watching the bones of his own hand in a fluoroscope and guided also by previous experience in developing his X-ray negatives.' Morton's apparatus is set up for simultaneous radiography and fluoroscopy and the standing operator, identifiable as Morton himself, is, as he advises, viewing his own hand during radiography of the patient's hand.

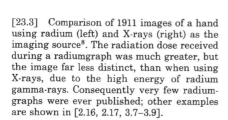

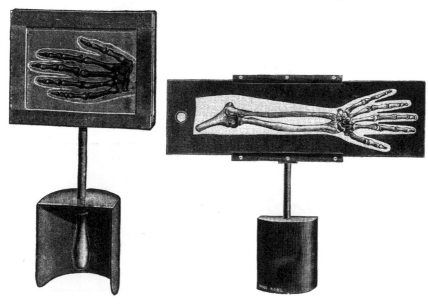

[23.4] The chiroscope (left) was demonstrated at a meeting of The Röntgen Society, London, in May 1903. 'Almost every worker with X-rays has to use the fluorescent screen very frequently in order to test the condition of his tube by reference to a shadow of his own hand. This process has so often to be gone through that most workers have had to suffer from more or less acute inflammation of the skin of their hands. To obviate this is the function of the chiroscope. This instrument consists of an articulated skeleton hand suitably mounted behind a small fluorescent screen, which is to serve as a test object. The fleshy parts of the hand are represented by suitably cut-out tin foil, and the whole is mounted on a holder, the construction of which affords protection to the hand of the operator.' A similar instrument (right) was designed by Max Kohl of the Chemnitz Company using a bony skeleton hand, forearm and elbow joint as described by Carl Beck[1] in 1904. He termed it an osteoscope and hoped that its use would limit the number of 'wrinkled and shrivelled Röntgen hands of physicians'. Both devices were mounted on metal-shielded handles to afford further protection to the user.

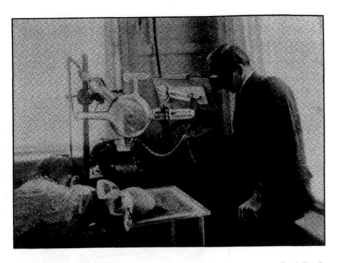

[23.5] The osteoscope of Max Kohl as used in 1904 by Carl Beck of New York. This photograph was entitled by Beck[1] 'Controlling the vacuum by the osteoscope during exposure'.

Spectacles: 1910–1939 (and 1980!)

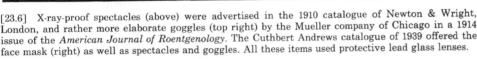

X-RAY-PROOF SPECTACLES.

Spectacles to protect the eyes are very necessary to X-Ray operators. Having had great experience in this department we have taken pains to make our spectacles as light and effective as possible.

10230   **X-Ray-Proof Spectacles**, in case, per pair
(Fig. 10230) £0   5   6

[23.6] X-ray-proof spectacles (above) were advertised in the 1910 catalogue of Newton & Wright, London, and rather more elaborate goggles (top right) by the Mueller company of Chicago in a 1914 issue of the *American Journal of Roentgenology*. The Cuthbert Andrews catalogue of 1939 offered the face mask (right) as well as spectacles and goggles. All these items used protective lead glass lenses.

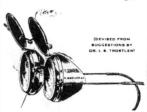

FACE MASK		
Carefully padded and with adjustable straps, PROTEX glass eyepieces. £3	7	6
SPECTACLES .. per pair 12	6	
GOGGLES .. ,, ,, 17	6	

[23.7] These rather peculiar for 'X-ray Gogs' offering 'instant X-ray sight' were advertised in a British newspaper in 1980. The underlying physical concept of this party piece was a small aperture, at the centre of the cardboard lens, containing a piece of nylon stocking material. This acted as a diffraction grating which blurred fingers when viewed through these 'Gogs'. The blur was supposed to represent soft tissue and that which was not blurred represented bone!

Lead protection features, X-rays: 1907–1909

Lead protection features, radium: 1920–1960

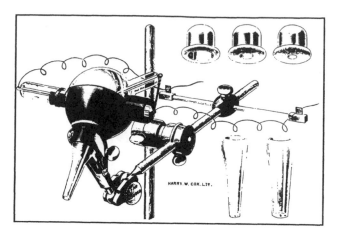

[23.8]   X-ray tube shielding and beam focusing in 1907[9].

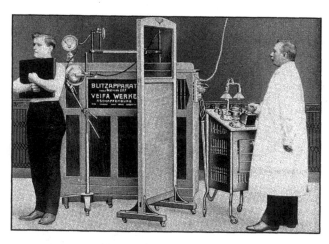

[23.9]   A 1909 diagnostic X-ray apparatus designed by Friedrich Dessauer and termed the Blitz Apparatus. The photograph shows one of the early mobile lead shields incorporating a lead glass window. No cassette holders were available at this time and it was standard practice for the patient to hold their chest film.

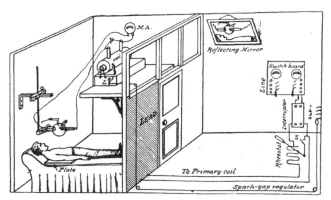

[23.10]   Protection features in the diagnostic X-ray room of Mihran Kassabian of Philadelphia, who was one of the early American radiological pioneers and the author of a standard textbook of 1907[10].

[23.11]   Handling radium sources behind a specially designed lead-protected bench in the 1920s. (Courtesy: Institut-Curie, Paris.)

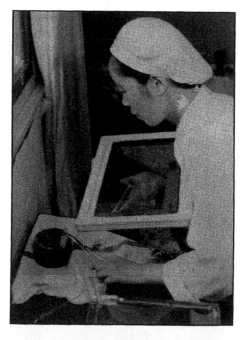

[23.12]   Radium needle handling by a nurse at the Cancer Hospital of the Shanghai Medical University in the 1950s. She is preparing radium needles for a tongue implant. The needles are in a lead container and are being viewed through a lead glass window.

[23.13] Example of a design of lead safe for the storage of radium sources.

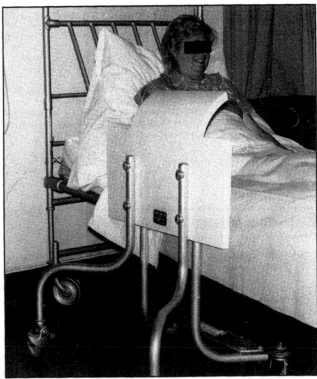

[23.15] Lead bed shield in use for a patient receiving radium brachytherapy for cancer of the cervix in the 1960s. The curved top enables the nurse to lean over the patient more easily and the two cutaway portions at the top of the shields are for her arms to reach round to the patient.

[23.14] Radiation protection during production of radium needles and tubes at the Union Minière factory at Oolen, Belgium [2.20], reproduced from the company's *Radium* textbook[28] of 1929. The upper photograph shows a lead glass glovebox for filling the tubes. The technicians in the lower photograph are partly protected, as in [23.11], by lead body shields mounted on the edge of the workbench.

Lost radium: USA, 1938

[23.16] 'Lost' radium sources usually ended up in incinerators, trash waste or down drains but one missing sources was recorded by Taft[13] as being located inside a pig, and the episode being summarised as 'Complete recovery of radium, complete loss of pig'.

Radon production: Paris, 1920s

[23.17]   Radon production plant at the Institut-Curie, Paris, in the 1920s. A readily available supply of the gas was required by those hospitals providing radon brachytherapy treatment, as discussed in [20.13]. The technician in this photograph is wearing protective goggles similar to those advertised in [23.6].

Radon therapy: 1913–1920s

RADIUM THERAPY

The only scientific apparatus for the preparation of radio-active water in the hospital or in the patient's own home.

This apparatus gives a high and measured dosage of radio-active drinking water for the treatment of gout, rheumatism, arthritis, neuralgia, sciatica, tabes dorsalis, catarrh of the antrum and frontal sinus, arterio-sclerosis, diabetes and glycosuria, and nephritis, as described in Dr. Saubermann's lecture before the Roentgen Society, printed in this number of the "Archives."

DESCRIPTION.

The perforated earthenware "activator" in the glass jar contains an insoluble preparation impregnated with radium. It continuously emits radium emanation at a fixed rate, and keeps the water in the jar always charged to a fixed and measureable strength, from 5,000 to 10,000 Maché units per litre per diem.

SUPPLIED BY

RADIUM LIMITED,
93, MORTIMER STREET, LONDON, W.

[23.18]   An emanation apparatus, the Aktivator, designed by a Dr H. Staumm of Villingen/ Schwarzwald, and used for inhalation purposes. The accompanying documentation states: 'This certificate guarantees the strength stated above [30,000 millistats per 24 hours] under the conditions that the above production number [2491] agrees with that stamped on the apparatus itself and that the seal fixed to the inside of the apparatus is unbroken.' (Courtesy: Prof. W. Bohndorf, Strahlenklinik, University of Würzburg.)

[23.19]   Radium emanation (radon) apparatus for the production of radioactive drinking water, as recommended for a wide variety of ailments. This advertisement appeared in a 1913 issue of the *Archives of the Roentgen Ray* and refers to a lecture[22] given by Dr J. Saubermann of Berlin to the Röntgen Society in London on April 1st 1913 on progress in radium emanation therapy. His dosimetry (as in the advertisement) was given in terms of the Mache unit (Chapter 22).

Radium compress: Strasbourg, 1920s

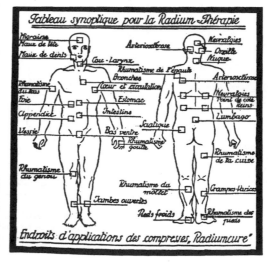

[23.20]   The Radium Compress from the Laboratories Pierre Koehren (Pharmacists) of Strasbourg made major claims for its Radium Cure and, as seen from the Radium Therapy Summary Table from the sales literature, a wide spectrum of conditions could be treated including migraine, arteriosclerosis and appendicitis.

Taking a bath at Joachimsthal: 1918

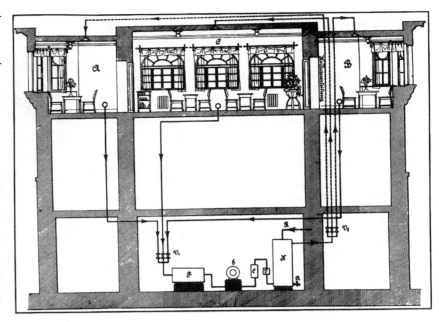

[23.21] It is of particular interest to record the radon apparatus used in the spa at Joachimsthal[17] from where Marie Curie obtained her earliest supplies of radium (Chapter 2). This figure shows the architectural drawings of the radon inhalation rooms at Joachimsthal, complete with chairs and tables and flower vases. The solid lines on the left represent the air tubes from the radon rooms back to the gas exchanger tank and the dotted lines on the right represent the air tubes from the gas tank to the room. $\mathfrak{R}$ is the radon water supply, $\mathfrak{F}$ an air filter, $\mathfrak{E}$ an electric motor for the compressor $\mathfrak{C}$, $\mathfrak{K}$ the gas exchanger tank and $\mathfrak{A}$ the exit tube for 'exhausted' water.

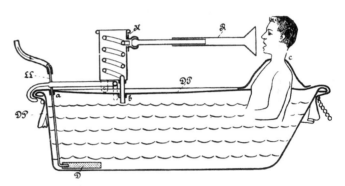

[23.22] Treatment could be given in various ways at Joachimsthal. This inhalation apparatus was connected by a hose to the jar system of [23.17] and incorporated a special breathing valve and a nose clip. The two valves in the glass cylinder prevented the breathing in of outside air.

In the Joachimsthal bath, air is pumped into the bath tub which contains radium and is diffused through a porous stone on the bottom left. The air pushes the radon gas into the system for inhalation. The cover on the bath is designed with three holes: for the air tube, the air cooler and the patients's head. On top of the bath towards the left end is the cooling coil for cooling the air from the bath tub such that the condensed water flows back into the tub. The radon gas is 'transmitted' to the patient through the trumpet-like pipe and the entire treatment is described as 'radium bath with inhalation'.

Radiation-induced cancer: 42-year latency

[23.23] Carcinoma of the thyroid 42 years after irradiation in 1912 for enlarged cervical nodes[20].

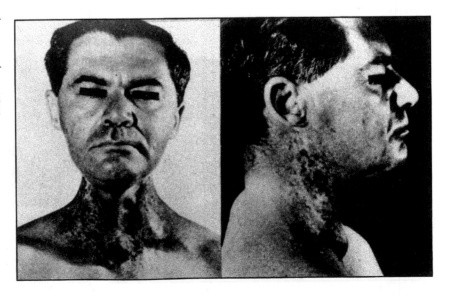

Hiroshima, 1945 and Chernobyl, 1986

[23.24] No chapter on radiation risks would be complete without at least a passing reference to the Chernobyl accident of April 1986, and given below are details of radiation injuries suffered in the aftermath of the explosion from information[23] given by the Russian Delegation to IAEA, Vienna, in August 1986 and that obtained when I visited Chernobyl in December 1987. The long-term effects of the accident have still to appear but already, by 1993, more workers who assisted in the clean-up operation, including some helicopter pilots, have died, and the TASS photographer who returns regularly to the site to record the situation for posterity has not long to live. With the break-up of the Soviet Union and the current economic situation in the Ukraine, public and government interest in Chernobyl has waned and adequate funds have not been made available to guarantee that no further disaster will occur at the power station. For example, the protective concrete building, the 'sarcophagus', is now crumbling and the radioactive mass beneath is critically dangerous.

The previous pictures of radiation burns in this book are due to X-rays and are localised to one portion of the body, such as the chest or abdomen [3.33, 8.6, 8.7] or the hands and wrist [3.35, 23.1], whereas those suffered at Chernobyl by the firemen and power plant technicians who initially fought the blaze were much more extensive and earlier to appear, partly because of the much higher radiation doses and also because they were beta-radiation burns, although two of the victims had very severe thermal burns.

56 patients were able to be evacuated to Moscow where they were treated by Professor Angelina Guskova who classified the group as follows. The 20 terminal cases were defined as having between 40% and 100% body surface burns, whereas the remaining 36 cases had between 1% and 40% body surface burns.

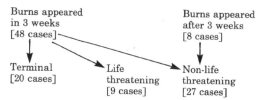

Burns appeared in 3 weeks [48 cases] → Terminal [20 cases], Life threatening [9 cases], Non-life threatening [27 cases]

Burns appeared after 3 weeks [8 cases] → Non-life threatening [27 cases]

As in the early days of the century when various degrees of radiation burn were defined, so did Guskova also define a scale of radiation syndrome, as a function of dose. In the fourth degree the primary reactions are early, in the first 15–30 minutes, and by the 7th to 9th day there is vomiting and damage to the digestive tract and fever. In this group of patients the first death occurred on the 9th day and all patients were dead by the 28th day.

Degree	Radiation dose
First	<1 Gy
Second	1– 4 Gy
Third	4– 6 Gy
Fourth	6–16 Gy

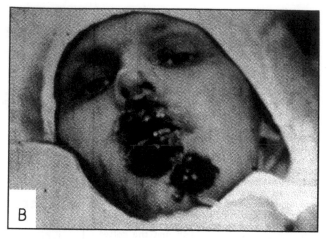

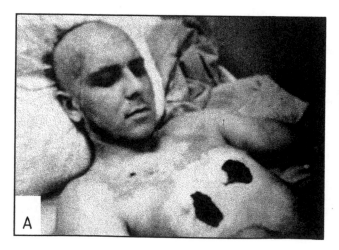

If the colour videotape at the IAEA Post-Chernobyl Conference in August 1986, showing the reactor's burning graphite moderator, was dramatic then so also were a series of some six or seven slides (two of which are reproduced as here) of firemen victims in their early twenties at various stages of their medical history. The original slides were returned to Moscow and have previously been reproduced only once[23] after they were obtained when a colleague photographed the projected slides. My tape recordings of Guskova's commentary on (A) was 'Epilation of the scalp, and blue skin, where there was total ulceration', and on (B) was 'Characteristic of skin where you have infection as well as radiation burns. This has spread to the membrane of the nose and it causes the patient the most terrible pain. Explosions of this sickness from viruses is something we had in about 20 patients.' The viral infections were herpes simplex of the facial skin, lips and oral mucosa and were treated with Acylovir.

In terms of injuries, the experience of the Chernobyl victims has sometimes been likened to the experience of the victims of Hiroshima. Figure (C) shows[24] the back of a 17-year-old female with keloids in an early stage combined with ulcers. She was at a distance of 1.6 kilometres from the hypocentre of the bomb (August 1945). This comparison is obviously false in terms of casualties and the few accounts show that however bad Chernobyl was in terms of the cost to individuals the experience of many of the Hiroshima victims was much worse: as recorded by John Hersey[25], a journalist sent out by the *New Yorker* in May 1946, and by Anne Chisholm[26] in her 1985 follow-up of young girls who survived the explosion, recalling the personal memory[26] of a 13-year-old girl who was in a street 1.6 km from the hypocentre: '. . . whole face burned, . . . no eyebrows . . . mother had to pull my eyes open . . . my skin came off when they tried to remove my burnt clothes . . . 4 days later the burned skin was peeled off my face, it was all black, underneath was full of pus . . .'

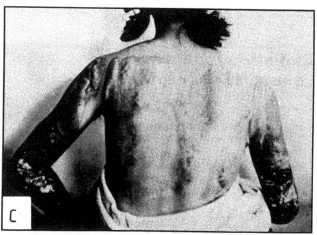

Additionally it should be noted that the large number of Hiroshima burns were of thermal origin and not from beta-radiation fallout. They were of two types, flash burns from the direct heat of the bomb fireball and flame burns from burning clothes and ignited buildings. There are many differences between casualties from a nuclear power plant accident and from nuclear war.

X-ray treatment and sterility: New York 1925–1958

[23.25] There are well documented records by Ira Kaplan of New York[27] on the X-ray treatment of menstrual dysfunction and sterility. Kaplan started in 1925 to apply low doses for the treatment of female sterility and in 1954 published a third generation follow-up of these patients with detailed histories of the grandchildren born to these women. In his report of 1937 he described his 'stimulation therapy' technique using 200 kV X-rays and a four-field pelvic technique. 75–100 röntgen were given per field per treatment to a total of three weekly treatments. The ovarian dose was stated to be 10%–12% of the dose to produce a full skin erythema. In addition, one or two treatments of 75–100 röntgen were given to the forehead over the pituitary area, simultaneously with the pelvic treatment. The patient workload and follow-up data are given in the table below as of 1959.

No. of married women irradiated	800
No. lost to follow-up	156/800 (19.5%)
No. of women who conceived	351/644 (54.5%)
No. of conceptions in the subgroup of 351	688
No. of children born alive and normal	543/688 (78.9%)
No. of children born dead or died sometime after birth	20/688 (2.9%)
No. of children born alive with abnormalities	3/688 (0.4%)
No. of children died by accident	2/688
No. of married children producing grandchildren	35
No. of grandchildren born	45

Kaplan (1959) concludes that '(a) despite pronouncements that with respect to genetic damage there is no safe dose of irradiation, no increase in genetic damage was found in the group of 644 women given approximately 65 röntgen to their ovaries and followed-up for 1–33 years, and (b) the incidence of genetic damage to the children and grandchildren of this group is less than that in the normal population'. His final comment is '(c) some have stated that the series studied is too small to yield significant data. This may be true, but the results are more impressive than the geneticists' calculations based on theory.'

Radium beauty creams

[23.26] Tho-Radia creme was advertised in 1933. With a radium and thorium base it was stated that it was the formula of 'Dr Alfred Curie': which gentleman never existed and was certainly not related to Marie Curie. It was described as 'a revolution in the art of beautifying the face' by Dr F. Tixier of the rue des Capucines, Paris.

[23.27] Crème Activa dates from 1919 and was advertised from the rue d'Amsterdam, Paris, claiming to promote youthful skin and to make wrinkles disappear.

Zoé Atomic Soda

[23.28] Zoé Atomic Soda: if only it worked!

Bibliography

Selected Books for Further Reading

The following books are recommended for further reading since they represent the major historical works published to date.

The Trail of the Invisible Light by Grigg in 1965 has the most illustrations including a large number of photographs of pioneer radiologists, radiotherapists and equipment manufacturers and of logos of the early manufacturers. In addition, brief biographical details of the pioneers are also given by Grigg. The two-volume book on *Classical Descriptions in Diagnostic Roentgenology* by Bruwer in 1964 is the best publication of its kind for diagnostic work.

The 1969 book by the Brechers is an excellent review of radiology in USA and Canada although it does not contain many illustrations. An earlier review for North America is the *Science of Radiology* edited by Glasser in 1933 which is extremely well referenced. For a review of British radiology, that by Burrows (1986) provides the most detailed reference work, including well researched biographies of pioneers in the United Kingdom.

Glasser also wrote in 1931 an illustrated biography of Röntgen which is still the standard text today. The best biographies of Marie Curie, although they contain only a very few photographs, are by her daughter Eve Curie in 1938 and by Robert Reid in 1974.

For extensive reviews of radium therapy see *Radium Therapy* by Wickham and Degrais published in 1910 and for a later period the well illustrated book of some 340 pages entitled *Radium* and published in 1929 by the Union Minière du Haut Katanga, the then major supplier of radium. This was a standard working text in many hospitals even until the early 1940s and was unusual for being published by a manufacturer. No similar book was ever published by an X-ray equipment manufacturer. This was perhaps because of the wealth of books written on X-ray applications by physicians, the most exhaustive of which for the first two decades of this century are those by Williams (1902), Kassabian (1907) and Knox (1915).

The most extensive review of nuclear medicine is the chronology (1600–1989) written by Marshall Brucer in 1990. The format subdivides references into sections on scientific and technical background; organisations and internal politics; instruments, units of measurement and drugs; radioiodine, phosphorous-32 and radioactive therapy; scans and other tests of function; radiation hysteria and health physics.

Recommended books in languages other than English include the two French publications: on radiology by Pallardy *et al* (1989) and on Marie Curie by the Université Libre de Bruxelles (1990). In German, the recently published *Strahlentherapie* journal issue by Scherer (1992) on the history of radiotherapy during 1900–1925 is, although not illustrated, an excellent reference source. A second issue for 1925–1960 is scheduled for publication but is not yet available.

Brecher R & Brecher E, *The rays: a history of radiology in the United States and Canada*, Williams & Wilkins: Baltimore, Maryland, 1969.

Brucer M, *A chronology of nuclear medicine*, Heritage Publications, St. Louis, 1990.

Bruwer A, *Classical descriptions in diagnostic roentgenology*, 2 volumes, Charles C Thomas: Springfield, Illinois, 1964.

Burrows EH, *Pioneers and early years: a history of British radiology*, Colophon: Alderney, 1986.

Curie E, *Madame Curie*, first published 1938. English edns: Reprint Society: London, 1942; Doubleday: New York, 1949. Dutch edn: Leopolds Uitgeversmij: Den Haag, 1948.

Glasser O, *Wilhelm Conrad Röntgen and the early history of roentgen rays*, Julius Springer: Berlin, 1931. English edns: John Bale Sons & Danielsson: London, 1933; Charles C Thomas: Springfield, Illinois, 1934.

Glasser O (ed), *The science of radiology*, Charles C Thomas: Springfield, Illinois, 1933; Baillière Tindall & Cox: London, 1933.

Grigg ERN, *The trail of the invisible light*, Charles C Thomas: Springfield, Illinois, 1965.

Kassabian MK, *Röntgen rays and electro-therapeutics with chapters on radium and phototherapy*, Lippincott: Philadelphia, 1907.

Knox R, *Radiography, X-ray therapeutics and radium therapy*, A & C Black: London, 1915; 2nd edn published in 2 volumes as: *Radiography and radio-therapeutics*, Part I *Radiography* (1917) and Part II *Radio-therapeutics* (1918); 3rd edn published 1919; 4th edn of Part I published 1923.

Pallardy G, Pallardy M-J & Wackenheim A, *Histoire illustrée de la radiologie*, Roger Dacosta: Paris, 1989.

Reid R, *Marie Curie*, first published 1974. Camair Press: London, 1984.

Scherer E, *Chronik der Strahlentherapie: Ausgewählte Kapitel aus der deutschsprächigen radioonkologischen Literatur*, Band 1: 1900–1925, Urban & Vogel: München, 1992.

Union Minière du Haut Katanga, *Radium: production, general properties, its applications in therapeutics*, Union Minière du Haut Katanga: Brussels, 1929.

Université Libre de Bruxelles, *Marie Sklodowska Curie et la Belgique*, ULB: Brussels, 1990.

Wickham L & Degrais P, *Radium therapy*, English edn Cassell: London, 1910.

Williams FH, *The Roentgen rays in medicine and surgery as an aid to diagnosis and as a therapeutic agent*, Macmillan: New York, 1902.

References

Chapters 1–23

References: Chapter 1

1 Glasser O, *Wilhelm Conrad Röntgen and the early history of roentgen rays*, Julius Springer: Berlin, 1931. English edns: John Bale Sons & Danielsson: London, 1933; Charles C Thomas: Springfield, Illinois, 1934.

2 Bellgarths E, *The new photography and vivisection*, London Saturday Review, February 29th 1896.

3 Wenz W, Elke M & Wackenheim A, *Radiologie am Oberrhein 1895 bis heute*, Schering Aktiengesellschaft: Berlin, 1987.

4 *Röntgen*, Röntgen-Gedachtnisstätte: Würzburg, 1985.

5 Comroe JH, *Retrospectroscope: insights into medical discovery*, Von Gehr Press: Menlo Park, California, 1977.

6 Harder D, *Röntgen's discovery—how and why it happened*, Int. J. Radiation Biology, **51**, 815, 1987.

7 Eder JM & Valenta E, *Versuche uber photographie mittelst der Rontgen'schen strahlen von regierungsrath*, Lechner: Vienna, 1896.

8 Kaye GWC, *X-rays: an introduction to the study of röntgen rays*, appendix 1, 218, Longmans Green: London, 1914.

9 Röntgen WC, *Opera Selecta*, Deutsches Röntgen-Museum: Remscheid-Lennep, 1979.

10 Thompson EP, *Roentgen rays and phenomena of the anode and cathode: principles, applications and theories*, Van Nostrand: New York, 1896.

11 Niewenglowski GN, *La photographie de l'invisible au moyen des rayons X, ultra-violets, de la phosphorescence et de l'effluve électrique: Historique théorie pratique* (pamphlet), Société d'Editions Scientifiques: Paris, 1896.

12 Peters PE, Hennig U & Scharmann A, *A new kind of rays— W. C. Röntgen 1895—First reactions in the United States*, Exhibition panel for the Congress of the Radiological Society of North America, Chicago, 1989.

13 Isenthal AW, *The Isenthal bequest of historical photographs and papers*, deposited at the British Institute of Radiology, London.

14 Kassabian MK, *Röntgen rays and electro-therapeutics with chapters on radium and phototherapy*, Lippincott: Philadelphia, 1907.

References: Chapter 2

1 Becquerel H, *Emission de radiations nouvelles par l'uranium métallique*, Comptes Rendus de l'Académie des Sciences, Paris, **122**, 1086, 1896.

2 Curie M, *Rayons émis par les composés de l'uranium et du thorium*, Comptes Rendus de l'Académie des Sciences, Paris, **126**, 1101, 1898.

3 Becquerel H & Curie P, *Action physiologique des rayons au radium*, Comptes Rendus de l'Académie des Sciences, Paris, **132**, 1289, 1901.

4 Curie P, Curie M & Bemont MG, *Sur une nouvelle substance fortement radio-active, contenue dans la pechblende*, Comptes Rendus Hebdomadaires des Séances de l'Académie des Sciences, Paris, **127**, 1215, 1898.

5 Turner D, *Radium, its physics and therapeutics*, Baillière Tindall: London, 1911.

6 Levy LA & Willis HG, *Radium and other radioactive elements: a popular account treated experimentally*, Percival Marshall: London, 1904.

7 Walter AE, *A practical guide to X-rays, electrotherapeutics and radium therapy for students and practitioners*, Thacker Spink: Calcutta, 1916.

8 Union Minière du Haut Katanga, *Radium: production, general properties, its applications in therapeutics*, Union Minière du Haut Katanga: Brussels, 1929.

9 Landa ER, *Buried treasure to buried waste: the rise and fall of the radium industry*, Colorado School of Mines Quarterly, **82**, 17, Colorado School of Mines Press: Golden, Colorado, 1987.

10 Landa ER, *The first nuclear industry*, Scientific American, **247**, 180, 1982.

11 Tousey S, *Medical electricity, röntgen rays and radium*, 3rd edn, Saunders: Philadelphia, 1921.

12 Kassabian MK, *Röntgen rays and electro-therapeutics with chapters on radium and phototherapy*, Lippincott: Philadelphia, 1907.

13 Curie E, *Madame Curie*, (first published 1938), English edn, Reprint Society: 1942; Dutch edn, Leopolds Uitgeversmij: Den Haag, 1948.

14 Reid R, *Marie Curie*, (first published 1974), Camair Press: London, 1984.

15 Université Libre de Bruxelles, *Marie Sklodowska Curie et la Belgique*, Université Libre de Bruxelles: Brussels, 1990.

16 Danne J, *Le radium sa préparation et ses propriétés*, Librairie Polytechnique Ch. Béranger: Paris, 1904.

17 Curie M, *Traité de radioactivité*, Gauthier-Villars Imprimeur-Librairie du Bureau des Longitudes, de l'Ecole Polytechnique: Paris, 1910.

18 Curie M, *La radiologie et la guerre*, Librairie Félix Alcan: Paris, 1921.

19 Balladur J, *Le parti architectural: construire un nouvel hôpital*, Journal de l'Institut Curie: Comprendre & Gir, 1ᵉʳ trimestre 1988.

20 Gricouroff G, *De l'Institut du Radium à l'Institut Curie: 70 ans de l'histoire de la lutte contre cancer*, Rapport 1982 sur la recherche et le traitement du cancer, 4, Institut-Curie: Paris, 1982.

21 Joliot F & Curie I, *Artificial production of a new kind of radio-element*, Nature, **201**, February 10th 1934 (reprinted in: J. Nuclear Medicine, **25**, 201, 1984).

22 Becquerel J & Matout L, *Die Strahlung der radioaktiven Substanzen*, 1–12, in: *Handbuch der radium-biologie und therapie einschliesslich der anderen radioaktiven elemente*, Lazarus P (ed), Bergmann: Wiesbaden, 1918.

23 Curie Mme S, *Untersuchungen über die Radioaktiven Substanzen*, Die Wissenschaft, Sammlung naturwissenschaftlicher und mathematischer Monographien, **1**, Kaufmann W (translator), Druck und Verlag von Friedrich Vieweg und Sohn: Braunschweig, 1904. Also published in 1904 by Mme Sklodowska Curie were *Recherches sur les substances radioactives*, Thèse présentée à la Faculté des Sciences de Paris pour obtenir le grade de Docteur des Sciences Physiques, 2nd edn, Gauthier-Villars: Paris (English translation: *Radioactive substances*, Philosophical Library: New York, 1961); and in Polish, *Badanie cial radioaktywynych*, Sklad Glowny: Warsaw.

24 Davis KS, *The history of radium* (lecture delivered in April 1923 to the Osler Medical Historical Society in Rochester, USA, on the 25th anniversary of the discovery of radium), Radiology, **2**, 331–342, 1924.

References: Chapter 3

1 Dittmar A, *Prof. Rontgen's "X" rays and their applications in the new photography* (pamphlet), F. Bauermeister: Glasgow, 1896.

2 Morton WJ, *The X-ray or photography of the invisible and its value in surgery*, American Technical Book Co: New York, 1896; Simpkin Marshall Hamilton Kent: London, 1896.

3 Holland CT, *X-rays in 1896*, Liverpool Medico-Chirurgical Journal, **45**, 61, 1937.

4 Grubbé E, *Priority in therapeutic use of X-rays*, Radiology, **31**, 156, 1933.

5 del Regato JA, *Wilhelm Conrad Röntgen*, Int. J. Radiation Oncology, Biology & Physics, **1**, 133, 1976.

6 Glasser O, *Wilhelm Conrad Röntgen and the early history of roentgen rays*, Julius Springer: Berlin, 1931. (English edns: John Bale Sons & Danielsson: London, 1933; Charles C Thomas: Springfield, Illinois, 1934.)

7 Despeignes V, *Observations on a case of cancer of the stomach treated by röntgen rays*, Lyon Medical Journal, **82**, 428, 1896.

8 Freund L, *Elements of general radiotherapy for practitioners*, Rebman: New York, 1904.

9 Schiff E, *The therapeutics of röntgen rays*, Rebman: London, 1901.

10 Daniel J, *The X-rays*, Medical Record, **49**, 17, 1896.

11 Edison T, Morton WJ, Swinton AAC & Stanton E, *The effect of X-rays upon the eyes*, Nature, **53**, 421, 1896.

12 Stevens LG, *Injurious effects on the skin*, Brit. Medical Journal, **1**, 998, 1896.

13 Gilchrist TC, *A case of dermatitis due to the X-rays*, Bulletin Johns Hopkins Hospital, **7**, 71, 1897.

14 Scott NS, *X-ray injuries*, Amer. X-ray Journal, **1**, 57, 1897.

15 Codman EA, *A study of cases of accidental X-ray burns hitherto recorded*, Philadelphia Medical Journal, **7**, 15, 1902.

16 Butcher WD, *Protection in X-ray work*, Archives of the Roentgen Ray, **10**, 38, 1905.

17 Franklin M, *The dangers of the X-ray*, Archives of the Roentgen Ray, **9**, 161, 1905.

18 Lebon H, *Moyens de protection contre l'action nocive des rayons Röntgen*, Clinique (Paris), **2**, 35, 1907.

19 Mutscheller A, *Physical standards of protection against roentgen-ray dangers*, Amer. J. Roentgenology, **13**, 65, 1925.

20 Barclay AE & Cox S, *The radiation risks of the roentgenologist. An attempt to measure the quantity of roentgen rays used in diagnosis and to assess the dangers*, Amer. J. Roentgenology, **19**, 551, 1928.

21 Glasser O, *First observations on the physiological effects of roentgen rays on the human skin*, Amer. J. Roentgenology, **28**, 75, 1932.

22 Meyer H, *Ehrenbuch der Röntgenologen und Radiologen aller Nationen*, Urban & Schwarzenberg: Berlin, 1937.

23 9th International Congress for Radiology, *Congress Handbook*, ICR: Munich, 1959.

24 Tousey S, *Medical electricity, röntgen rays and radium*, 3rd edn, Saunders: Philadelphia, 1921.

25 Kassabian MK, *Röntgen rays and electro-therapeutics with chapters on radium and phototherapy*, Lippincott: Philadelphia, 1907.

26 Giesel F, *Ueber Radium und Polonium*, Physikalische Zeitschrift, **1**, 16, 1899.

27 Himstedt F & Nagel WA, *Ueber die Einwirkung der Becquerel- und der Röntgenstrahlen auf das Auge*, Physikalische Zeitschrift, **2**, 362, 1901.

28 London ES, *Physio-pathological value of radium rays: their relation to the domain of vision*, Arkhiv. Biologiceskikh Nauk, **10**, 191, 1903.

29 Javal LE & Curie P, Wien Medizinische Wochenschrift, **52**, 2055, 1902.

30 MacKee GM, *X-rays and radium in the treatment of diseases of the skin*, 2nd edn, Henry Kimpton: London, 1927.

31 Wickham L & Degrais P, *Radium therapy*, English edn, Cassell: London, 1910.

32 Bergonié J & Tribondeau L, *Interprétation de quelques résultats de la radiothérapie*, Comptes Rendus de l'Académie des Sciences, **143**, 983, 1906.

33 Lawrence H, *Radium: how and when to use*, Stillwell: Melbourne, 1911.

34 Fowler JH, *The radiobiology of brachytherapy*, in: *Brachytherapy HDR and LDR*, Martinez AA, Orton CG & Mould RF (eds), 121–137, Nucletron: Leersum, 1990.

35 Brenner DJ & Hall EJ, *Fractionated high dose rate versus low dose rate regimes for brachytherapy of the cervix: a non-mathematical guide for the perplexed*, Activity Selectron Brachytherapy Journal, Supplement 2, 11, 1991: see also: Brenner DJ & Hall EJ, *The radiobiological basis of recent innovations in brachytherapy*, Activity Selectron Brachytherapy Journal, **5**, 91, 1991.

36 Dale RG, *The application of the linear-quadratic dose-effect equation to fractionated radiotherapy and protracted radiotherapy*, Brit. J. Radiology, **58**, 515, 1985: see also: Dale RG, *The use of small fraction numbers in high dose rate gynaecological afterloading: some radiobiological considerations*, Brit. J. Radiology, **63**, 290, 1990.

37 Orton CG, *Recent developments in time-dose modelling*, Australasian Physics & Engineering Science in Medicine, **14**, 57, 1991.

38 Krönig B & Friedrich W, *Physikalische und biologische Grundlagen der Strahlentherapie*, Urban & Schwarzenburg: Berlin, 1918.

39 del Regato JA, *Claudius Regaud*, Int. J. Radiation Oncology, Biology & Physics, **1**, 993, 1976.

40 Regaud C, *The influence of the duration of irradiation on the changes produced in the testicle by radium*, Int. J. Radiation Oncology, Biology & Physics, **2**, 565, 1977 (English translation of: Regaud C, Comptes Rendus de Société Biologique, **86**, 787, 1922).

41 Regaud C, *The alternating rhythm of cellular mitoses and the radiosensitivity of the testis*, Int. J. Radiation Oncology, Biology & Physics, **2**, 569, 1977 (English translation of: Regaud C, Comptes Rendus de Société Biologique, **86**, 822, 1922).

42 Packard C, *Biologic effects of roentgen rays and radium*, in: *The science of radiology*, Glasser O (ed), 319, Baillière-Tindall & Cox: London, 1933.

43 Czerny V, *Bemerkungen über die Injektion von Radiumpräparaten bei malignen Tumoren*, Deutschen Medizinischen Wochenschrift, no. 51, 1–2, 1909.

44 Clerk A, Personal communication, 1979.

45 Hall-Edwards J, *Bullets and the billets: experiences with X-rays in South Africa*, Archives of the Roentgen Ray, **6**, 31, 1902.

46 Beck C, *Röntgen ray diagnosis and therapy*, Appleton & Lange: New York, 1904.

47 Levy LA & Willis HG, *Radium and other radioactive elements: a popular account treated experimentally*, Percival Marshall: London, 1904.

48 Cleaves MA, *Light energy*, Rebman: New York, 1904.

49 Fuchs G & Hofbauer J, *Das Spätresultat einer vor 70 Jahren durchgeführten Röntgenbestrahlung*, Strahlentherapie, **130**, 161–166, 1966.

50 Freund L, *Ein mit Röntgenstrahlen behandelter Fall von Naevus pigmentosus piliferus Demonstrations in der Gesellschaft der Artze in Wien*, 15.1.1897, Wien klin. Wschr., 147, 1937.

51 Tonta I, *Raggi di Röntgen e loro pratiche applicazioni*, Ulrico Hoepli: Milan, 1898.

52 Kaye GWC, *X-rays*, Longman Green: London, 1923.

53 Turner D, *Radium, its physics and therapeutics*, Baillière-Tindall: London, 1911.

54 Riddell JR, *Handbook of medical electricity and radiology*, Livingstone: Edinburgh, 1926.

55 Bouchacourt L, *L'exploration des organes internes à l'aide de la lumière éclairante et non éclairante, endoscopie par les rayons de Röntgen*, University of Paris Doctorat en Médecine, 1898.

56 Ramsey LJ, *British X-ray patents 1896–1900*, Radiography, **35**, 285–291, 1969.

57 Hope A, *Why didn't I think of it first? A collection of unusual patents from the British Patent Office*, David & Charles: Newton Abbot, 1973.

58 Galitzin Prince B & Karnojitzky AV, *Ueber die Ausgangspunkte und Polarisation der X-Strahlen*, Report to Imperial Academy of Science, St. Petersburg, March 6th 1896 (see also: Dibner B, *The new rays of Professor Röntgen*, Burndy Library: Norwalk, 1963).

59 Firth I, New Scientist, December 25th 1969. Wood's own account of his exposure of the N-ray fraud is reproduced in: Seabrook W, *Dr Wood, Modern Wizard of the Laboratory*, Harcourt Brace: New York, 1941.

60 Hall-Edwards J, *On X-ray dermatitis and its prevention*, Proc. Roy. Soc. Med., **2**, 11–34, 1908.

References: Chapter 4

1 Wolfenden RN, *Radiography in marine zoology*, Archives of the Roentgen Ray, Supplement 2, 1897/1898.

2 Mould RF, *Norris Wolfenden and X-rays*, J. Laryngology & Otology, **103**, 1020, 1989.

3 Morton WJ, *The X-ray or photography of the invisible and its value in surgery*, American Technical Book Co: New York, 1896; Simpkin Marshall Hamilton Kent: London, 1896.

4 Holland CT, *X-rays in 1896*, Liverpool Medico-Chirurgical Journal, **45**, 61, 1937.

5 Walsh D, *The Röntgen rays in medical work*, Baillière Tindall & Cox: London, 1897.

6 Mould RF, *Self-induced excess iron therapy*, in: *Mould's medical anecdotes*, 36, Adam Hilger: Bristol, 1984.

7 Skinner EH, *Radiologic societies and literature*, in: *The science of radiology*, Glasser O (ed), 375–384, Charles C Thomas: Springfield, Illinois, 1933; Baillière Tindall & Cox: London, 1933.

8 Bishop PJ, *Evolution of the British Journal of Radiology*, Brit. J. Radiology, **46**, 833–836, 1973.

9 Brecher R & Brecher E, *The rays: a history of radiology in the United States and Canada*, Williams & Wilkins: Baltimore, Maryland, 1969.

References: Chapter 5

1 Glasser O, *Wilhelm Conrad Röntgen and the early history of roentgen rays*, Julius Springer: Berlin, 1931. English edns: John Bale Sons & Danielsson: London, 1933; Charles C Thomas, Springfield, Illinois, 1934.

2 Hennig U, *Deutsches Röntgen-Museum Remscheid-Lennep*, Westermann: Braunschweig, 1989.

3 Trevert E, *Something about X-rays for everybody*, Bubier Publishing: Lynn, 1898 (reprint, Medical Physics Publishing Corporation: Madison, WI, 1988).

4 Kassabian MK, *Röntgen rays and electro-therapeutics with chapters on radium and phototherapy*, Lippincott: Philadelphia, 1907.

5 Albers-Schönberg H, *Die Röntgentechnik*, Lucas Gräfe & Sillem: Hamburg, 1903.

6 Beck C, *Röntgen ray diagnosis and therapy*, Appleton & Lange: New York, 1904.

7 Williams FH, *The Roentgen rays in medicine and surgery as an aid in diagnosis and as a therapeutic agent*, 2nd edn, Macmillan: New York, 1902. (See also: Williams FH, *Reminiscences of a pioneer in roentgenology and radium therapy, with reports of some recent observations*, Amer. J. Roentgenology, **13**, 253, 1925.)

References: Chapter 6

1 Ward HS, *Practical radiography. A handbook of the applications of the X-rays*, published for The Photogram Ltd, Dawbarn & Ward: London, 1896.

2 Morton WJ, *The X-ray or photography of the invisible and its value in surgery*, American Technical Book Co: New York, 1896; Simpkin Marshall Hamilton Kent: London, 1896.

3 Mould RF, *Spark coils and spy scares*, in: *Mould's medical anecdotes*, Adam Hilger: Bristol, 1984.

4 Tyler AF, *Roentgenotherapy*, Henry Kimpton: London, 1919.

5 Thompson EP, *Roentgen rays and phenomena of the anode and cathode: principles, applications and theories*, Van Nostrand: New York, 1896.

6 Schall WE, *X-rays: their origin, dosage and practical application*, John Wright: Bristol, 1932.

7 Kassabian MK, *Röntgen rays and electro-therapeutics with chapters on radium and phototherapy*, Lippincott: Philadelphia, 1907.

References: Chapter 7

1 Coolidge WD, *A powerful roentgen ray tube with a pure electron discharge*, Physical Review, **2**, 409, 1913.

2 Coolidge WD, *A new radiator type of hot cathode roentgen ray tube*, General Electric Review, **21**, 55, 1918.

3 U.S. Army Division of Roentgenology, *United States Army X-ray manual*, Hoeber: New York, 1918 edn.

4 Kaye GWC, *X-rays*, Longmans Green: London, 1923.

5 Coolidge WD, *Summary of physical investigation work in progress on tubes and accessories*, Amer. J. Roentgenology, December 1915.

6 van der Plaats GJ, *Medical X-ray technique*, Philips Technical Library: Eindhoven, 1969.

7 Seram E, *X-ray imaging equipment*, Charles C Thomas: Springfield, Illinois, 1985.

8 Hill DR (ed), *Principles of diagnostic X-ray apparatus*, Philips Technical Library: Eindhoven, 1975.

9 Coolidge WD & Charlton EE, *Roentgen ray tubes*, in: *The science of radiology*, Glasser O (ed), 77, Charles C Thomas: Springfield, Illinois, 1933; Baillière Tindall & Cox: London, 1933.

References: Chapter 8

1 Reynolds L, *The history of the use of the roentgen ray in warfare*, Amer. J. Roentgenology, 54, 649, 1945.

2 Arthur D & Muir J, *A manual of practical X-ray work*, Rebman: London, 1909.

3 Battersby J, *The present position of the Roentgen rays in military surgery*, Archives of the Roentgen Ray, 3, 74, 1899.

4 Borden WC, *The use of the Röntgen ray by the Medical Department of the United States Army in the war with Spain (1898)*, House of Representatives, 56th Congress, 1st Session, Document No. 729, Government Printing Office: Washington, DC, 1900.

5 Curie M, *La radiologie et la guerre*, Librairie Félix Alcan: Paris, 1921.

6 Skjei E, *Military sends CTs to Saudi field hospitals*, Diagnostic Imaging, 113–112, March 1991.

7 Christie AC, *Mobile roentgen ray apparatus*, Amer. J. Roentgenology, 6, 358, 1919.

8 U.S. Army Division of Roentgenology, *United States Army X-ray manual*, Hoeber: New York, 1918 edn.

9 Hall-Edwards J, *Bullets and their billets: experiences with X-rays in South Africa*, Archives of the Roentgen Ray, 6, 31, 1902.

References: Chapter 9

1 Rodman GH, *The historical collection of X-ray tubes*, Journal of the Röntgen Society, 5, 85, 1909.

2 Miller CW, *Historical X-ray tubes*, Brit. Inst. of Radiology Bulletin, 1, 3, 1975.

3 Mould RF, *The BIR historical collection of X-ray tubes, lantern slides, journals and books*, Brit. Inst. of Radiology Bulletin, 5, 3, 1979.

4 Trevert E, *Something about X-rays for everybody*, Bubier Publishing: Lynn, 1898 (reprint, Medical Physics Publishing Corporation: Madison, WI, 1988).

5 Holland CT, *X-rays in 1896*, Liverpool Medico-Chirurgical Journal, 45, 61, 1937.

6 Porter AW, *The new photography*, The Strand Magazine, 12, 107, 1896.

7 Londe A, *Radiographie et radioscopie*, Gauthier-Villars: Paris, 1898.

8 Eder JM & Valenta E, *Versuche über photographie mittelst der Rontgen'schen strahlen von regierungsrath*, Lechner: Vienna, 1896.

9 Isenthal AW & Ward HS, *Practical radiography. A handbook of the applications of the X-rays*, 2nd edn, published for The Photogram Ltd, Dawbarn & Ward: London, 1898.

10 Wright L, *The induction coil in practical work including Röntgen X-rays*, Macmillan: London, 1897.

References: Chapter 10

1 Thompson EP, *Roentgen rays and phenomena of the anode and cathode: principles, applications and theories*, Van Nostrand: New York, 1896.

2 U.S. Army Division of Roentgenology, *United States Army X-ray manual*, Hoeber: New York, 1915 edn.

3 Williams FH, *The Roentgen rays in medicine and surgery as an aid in diagnosis and as a therapeutic agent*, 2nd edn, Macmillan: New York, 1902. (See also: Williams FH, *Reminiscences of a pioneer in roentgenology and radium therapy, with reports of some recent observations*, Amer. J. Roentgenology, 13, 253, 1925.)

4 Russell JGB, *Pioneer discovery of rare earth fluorescence*, Brit. J. Radiology, 62, 765, 1989.

5 Knox R, *Radiography and radio-therapeutics*, Part 1: *Radiography*, 4th edn, A & C Black: London, 1923.

6 Ramsey LJ, *Luminescence and intensifying screens in the early days of radiography*, Radiography, 42, 245, 1976.

7 Morgan RH & Lewis I, *Notes on the use of protective glass in photofluorographic equipment*, Amer. J. Roentgenology, 54, 403, 1945.

8 Londe A, *Radiographie et de radioscopie*, Gauthier-Villars: Paris, 1898.

9 Walsh D, *The Röntgen rays in medical work*, Baillière Tindall & Cox: London, 1897.

10 Hildebrand H, Scholz W & Wieting, *Das Arteriensystem des Menschen im stereoskopischen Röntgenbild*, Sammlung von Stereoskopischen Röntgenbildern I, Neuen Allgemeinen Krankenhaus Hamburg-Eppendorf, 1901.

11 Grigg ERN, *The trail of the invisible light*, Charles C Thomas: Springfield, Illinois, 1965.

12 Battermann JJ, Szabó BG, van Rossum HJM & Mould RF, *Brachytherapy Teaching Project: an Interactive Multimedia Programme*. (The COMETT Technical Assistance Office is part of the European Center for Strategic Management of Universities.) in: *International Brachytherapy*, Mould RF (ed), 527–535, Nucletron: Veenendaal, 1992.

13 Pallardy G, Pallardy M-J & Wackenheim A, *Histoire illustrée de la radiologie*, Roger Dacosta: Paris, 1989.

References: Chapter 11

1 Strauss H, *Beitrag zur Wurdigung der Diagnostischen Bedeutung der Röntgendurchleuchtung*, Deutsch. med. Wchnschr, 22, 161, 1896.

2 Hemmeter JC, *Photography of the human stomach by the roentgen method: a suggestion*, Boston Medical and Surgical Journal, 134, 609, 1896.

3 Knox R, *Radiography and radio-therapeutics*, Part 1: *Radiography*, 4th edn, A & C Black: London, 1923.

4 Cerné A & Delaforge, *La radioscopie clinique de l'estomac: normal et pathologique*, Baillière: Paris, 1908.

5 Case JT, *50 years of roentgen rays in gastroenterology*, Amer. J. Roentgenology, 54, 607, 1945.

6 Hunter D, *The diseases of occupations*, 6th edn, Hodder & Stoughton: London, 1978.

7 Taylor DM, Mays CW, Gerber GB & Thomas RG (eds), *Risks from radium and thorotrast*, Report 21, British Institute of Radiology: London, 1989.

8 Bruwer AJ, *Classic descriptions in diagnostic roentgenology*, 2 vols, Charles C Thomas: Springfield, 1964.

9 Brecher R & Brecher E, *The rays: a history of radiology in the United States and Canada*, Williams & Wilkins: Baltimore, Maryland, 1969.

10 Phillips CES, *Bibliography of X-ray literature and research*, 1896–1897, The Electrician Printing & Publishing Company: London, 1898.

References: Chapter 12

1 Knox R, *Radiography and radio-therapeutics*, Part 1: *Radiography*, 4th edn, A & C Black: London, 1923.

2 Albers-Schönberg H, *Die Röntgentechnik*, Lucas Gräfe & Sillem: Hamburg, 1903.

3 Shanks CS, *Radiology in the twenties*, Brit. J. Radiology, 46, 766, 1973.

4 Kirklin BR, *Background and beginning of cholecystectomy*, Amer. J. Roentgenology, 54, 637, 1945.

5 Holbeach CH, *The Potter–Bucky diaphragm*, Journal of the Röntgen Society, **15**, 179, 1921.

6 Pallardy G, Pallardy M-J & Wackenheim A, *Histoire illustrée de la radiologie*, Roger Dacosta: Paris, 1989.

7 Morton WJ, *The X-ray or photography of the invisible and its value in surgery*, American Technical Book Co: New York, 1896; Simpkin Marshall Hamilton Kent: London, 1896.

8 Kassabian MK, *Röntgen rays and electro-therapeutics with chapters on radium and phototherapy*, Lippincott: Philadelphia, 1907.

References: Chapter 13

1 Bruwer AJ, *Classic descriptions in diagnostic roentgenology*, 2 vols, Charles C Thomas: Springfield, Illinois, 1964.

2 Tousey S, *Medical electricity, röntgen rays and radium*, 1st edn, Chapter: *Examples of the value of the X-ray in the study of anatomy*, Saunders: Philadelphia, 1915.

3 Glasser O, *Wilhelm Conrad Röntgen and the early history of roentgen rays*, Julius Springer: Berlin, 1931. English edns: John Bale Sons & Danielsson: London, 1933; Charles C Thomas, Springfield, Illinois, 1934.

4 Thompson EP, *Roentgen rays and phenomena of the anode and cathode: principles, applications and theories*, Van Nostrand: New York, 1896.

5 Isenthal AW & Ward HS, *Practical radiography. A handbook of the applications of the X-rays*, 2nd edn, published for The Photogram Ltd, Dawbarn & Ward: London, 1898.

6 Ward HS, *Practical radiography. A handbook of the applications of the X-rays*, published for The Photogram Ltd, Dawbarn & Ward: London, 1896.

7 Isenthal AW, *The Isenthal bequest of historical photographs and papers*, deposited at the British Institute of Radiology, London.

8 Isenthal AW & Ward HS, *Practical Radiography. A handbook for physicians, surgeons and other users of X-rays*, 3rd edn, published for The Photogram Ltd, Dawbarn & Ward: London, 1901.

References: Chapter 14

1 Morton WJ, *The X-ray or photography of the invisible and its value in surgery*, American Technical Book Co: New York, 1896; Simpkin Marshall Hamilton Kent: London, 1896.

2 Williams FH, *The Roentgen rays in medicine and surgery as an aid to diagnosis and as a therapeutic agent*, Macmillan: New York, 1902. (See also: Williams FH, *Reminiscences of a pioneer in roentgenology and radium therapy, with reports of some recent observations*, Amer. J. Roentgenology, **13**, 253, 1925.)

3 Mould RF, *Radiation protection in hospitals*, Adam Hilger: Bristol, 1985.

4 Kodak Ltd, *X-rays in dentistry*, Kodak: London, 1969.

5 Tousey S, *Medical electricity, Röntgen rays and radium*, Saunders: Philadelphia, 1915.

References: Chapter 15

1 Grossmann G, *Lung tomography*, Brit. J. Radiology, **8**, 733, 1935.

2 Andrews JR, *Planigraphy. I: Introduction and history*, Amer. J. Roentgenology, **36**, 575, 1936.

3 Andrews JR & Stava RJ, *Planigraphy. II: Mathematical analyses of the methods, description of apparatus, and experimental proof*, Amer. J. Roentgenology, **38**, 145, 1937.

4 Andrews JR & Turek RO, *Planigraphy. III: An evaluation of the method in the diagnosis of cancer of the lower respiratory tract*, Amer. J. Roentgenology, **58**, 173, 1947.

5 Howes WE, *Sectional roentgenography of the larynx*, Radiology, **33**, 586, 1939.

6 Camp JD & Allen EP, *Microtia and congenital atresia of the external auditory canal*, Amer. J. Roentgenology, **43**, 201, 1940.

7 Twining EW, *Tomography, by means of a simple attachment to the Potter–Bucky couch*, Brit. J. Radiology, **10**, 332, 1937.

8 Bricker JD, *Tomography*, in: *Classic descriptions in diagnostic roentenology*, Bruwer AJ (ed), 1406, Charles C Thomas: Springfield, Illinois, 1964.

9 Webb S (ed), *The physics of medical imaging*, Adam Hilger: Bristol, 1988. (See also: Webb S, *From the watching of shadows: the origins of radiological tomography*, Adam Hilger: Bristol, 1990.)

10 Ambrose J & Hounsfield G, *Computerised transverse axial tomography*, Brit. J. Radiology, **46**, 148, 1972. (See also: *X-ray diagnosis peers inside the brain*, New Scientist, April 27th 1972; and Higson GR, *The beginning of CT scanning: a personal recollection*, British Institute of Radiology Bulletin, **5**, 3, 1979.)

11 Hounsfield G, *Computerised transverse axial scanning (tomography), Part 1: Description of system*, Brit. J. Radiology, **46**, 1016, 1973.

12 Ambrose J & Hounsfield G, *Computerised transverse axial scanning (tomography), Part 2: Clinical applications*, Brit. J. Radiology, **46**, 1023, 1973.

13 Radon J, *Ueber die Bestimmung von Funktionen durch ihre Integralwerte langs gewisser Mannigfaltigkeiten*, Ber. Verh. Sachs. Akad. Wiss. Leipzig Math. Phys. Kl., **69**, 262, 1917.

14 Oldendorf WH, *Isolated flying spot detection of radiodensity discontinuities: displaying the internal structural pattern of a complex object*, IRE Trans. Biomedical Electronics, **BME-8**, 68, 1961.

15 Cormack AM, *Early two-dimensional reconstruction and recent topics stemming from it* (Nobel Prize lecture), Science, **209**, 1482, 1980. (See also: Hounsfield G, *Computed medical imaging* (Nobel Prize lecture), Science, **210**, 22, 1980.)

16 Dümmling K, *10 years of computed tomography: a retrospective review*, Siemens Electromedia, **52**, 13, 1984.

17 General Electric, *Computed tomography: do patient benefits justify the investment?*, 1979.

18 Zaklad H, *Computerized multiple X-rays give a new view of the body's interior*, Electronics, 89, October 1976.

19 Brooks RA & Di Chiro G, *Principles of computer assisted tomography (CAT) in radiographic and radioisotopic imaging*, Physics in Medicine and Biology, **21**, 689, 1976.

20 Pullan BR, *The scientific basis of computerised tomography*, in: *Recent advances in radiology and medical imaging*, Lodge T & Steiner RE (eds), **6**, 1, Churchill Livingstone: Edinburgh, 1979.

21 Kreel L, *Computed axial tomography in the diagnosis and treatment of malignancy*, Cancer Treatment Reviews, **5**, 117, 1978.

References: Chapter 16

1 Glasser O, *Wilhelm Conrad Röntgen and the early history of Röntgen rays*, Julius Springer: Berlin, 1931. English edns: John Bale Sons & Danielsson: London, 1933; Charles C Thomas: Springfield, Illinois, 1934.

2 van Puyvelde L, *X-rays and picture research*, Brit. J. Radiology, **4**, 136, 1930.

3 Rawlings FIG, *X-rays in the study of pictures*, Brit. J. Radiology, **12**, 239, 1939.

4 Elliott WJ, *The use of the roentgen ray in the scientific examination of paintings*, Amer. J. Roentgenology, **50**, 779, 1943.

5   Macht SH & Etchison B, *Roentgen examination of paintings*, Amer. J. Roentgenology, 84, 958, 1960.

6   van der Wetering E, *The invisible Rembrandt: the results of technical and scientific research*, in: *Rembradt: The master and his workshop: paintings*, 90, Brown C, Kelch J & van Thiel P (eds), Yale University Press and National Gallery Publications: London, 1991.

7   König W, *14 Photographien mit Röntgen-Strahlen, aufgenommen im Physikalische Verein zu Frankfurt a. M.*, J. A. Barth: Leipzig, 1896.

8   Isenthal AW & Ward HS, *Practical Radiography. A handbook of the applications of the X-rays*, 2nd edn, published for The Photogram Ltd, Dawbarn & Ward: London, 1898.

9   Fleming SJ, *Authenticity in art: the scientific detection of forgery*, Institute of Physics: Bristol, 1975.

10 Londe A, *Radiographie et de radioscopie*, Gauthier-Villars: Paris, 1898.

11 Corney GM, *Radiography of La Pietà*, Medical Radiography and Photography, 41, 1, 1965.

12 Zwicker C, Hosten N, Wildung D & Felix R, *Nofretete Analyse der Büste mit dreidimensionaler CT*, Deutsches Arzteblott-Arztliche Mitteilungen, 89, B1753–B1755, 1992.

References: Chapter 17

1   Isenthal AW & Ward HS, *Practical radiography. A handbook for physicians, surgeons and other users of X-rays*, 3rd edn, published for The Photogram Ltd, Dawbarn & Ward: London, 1901.

2   Isenthal AW & Ward HS, *Practical radiography. A handbook of the applications of the X-rays*, 2nd edn, published for The Photogram Ltd, Dawbarn & Ward: London, 1898.

3   Kaye GWC, *X-rays and the war*, Journal of the Röntgen Society, 14, 2, 1918.

4   *Exhibition of radiographic prints*, Journal of the Röntgen Society, 16, 75, 1920.

5   St. John A & Isenburger HR, *Industrial radiography*, Wiley: New York, 1934.

6   General Electric X-Ray Corporation, *Radiographs by the mile on Hoover Dam penstocks*, Ideas, 5, May 15th 1933.

7   Bibergal AV, Margulis YJ & Vorobev EI, *Radiation protection from gamma-rays*, Medgiz: Moscow, 1960.

8   Crowther JA, *Handbook of industrial radiology*, Arnold: London, 1944. (The subsequent edition of this book is: Wiltshire WJ, *A further handbook of industrial radiology*, Arnold: London, 1957.)

9   Porter AW, *The new photography*, Strand Magazine, 12, 107, 1896.

10 Londe A, *Radiographie et de radioscopie*, Gauthier-Villars: Paris, 1898.

11 Glasser O, *Wilhelm Conrad Röntgen and the early history of roentgen rays*, Julius Springer: Berlin, 1931. (English edns: John Bale & Sons & Danielsson: London, 1933; Charles C Thomas: Springfield, Illinois, 1934.)

12 Hempelmann LH, *Potential dangers in the uncontrolled use of shoe-fitting fluoroscopes*, New England J. Medicine, 241, 335–337, 1949.

13 Williams CR, *Radiation exposures from the use of shoe-fitting fluoroscopes*, New England J. Medicine, 241, 333–335, 1949.

References: Chapter 18

1   Haagensen CD, *An exhibit of important books, papers and memorabilia illustrating the evolution of the knowledge of cancer*, Amer. J. Cancer, 18, 42, 1933.

2   Lennmalm F (ed), *Förhandlingar vid Svenska Lakare-Sallskapets Sammankomster ar 1899*, 205–209, Isaac Marcus: Stockholm, 1900.

3   Berven E, *The development and organisation of therapeutic radiology in Sweden*, Radiology, 79, 829–841, 1962.

4   Sequeira JH & Morton ER, *The Light, X-ray and Electrical Departments at the London Hospital*, Archives of the Roentgen Ray, 10, 274, 1905.

5   Wetterer J, *Handbuch der Röntgentherapie nebst Anhang: die Radium therapie*, Otto Nemich: Leipzig, 1908; and *The homogeneous radiation of Dessauer*, Fortschritte a.d. Gebiete d. Röntgenstrahlen, 13, 189, 1908.

6   Voltz F, *Dosage tables for deep-therapy*, Heinemann: London, 1922.

7   British Institute of Radiology, *Central axis depth dose data for use in radiotherapy*, suppl. no. 17, 1983.

8   Mayneord WV, *The physics of X-ray theory*, Churchill: London, 1929.

9   Hirsch IS, *The principles and practice of roentgen therapy: with dosage formulae and dosage tables by Guido Holzknecht*, American X-ray Publishing Company: New York, 1925.

10 Pilon H, *Protection in radiotherapy*, Brit. J. Radiology (Röntgen Society section), 23, 164, 1927.

11 Holfelder H, *Die Röntgen-Tiefentherapie*, Georg Thieme: Leipzig, 1938.

12 Phillips R & Innes GS, *Supervoltage X-ray therapy: a report for the years 1937–1942 on The Mozelle Sassoon Supervoltage X-ray Therapy Department, St. Bartholomew's Hospital*, HK Lewis: London, 1944.

13 Charlton EE, Westendorp WF, Dempster LE & Hotaling G, *A million-volt X-ray unit*, Radiology, 35, 585, 1940.

14 Friedman M, Brucer M & Anderson E (eds), *Roentgens, rads and riddles: a symposium on supervoltage radiation therapy*, United States Atomic Energy Commission, 1959.

15 Kerst DW, *The Betatron*, Radiology, 40, 115, 1943.

16 Wideröe R, *Über ein neues Prinzip zur Herstellung höher Spannungen*, Arkiv Elekrotech, 21, 387, 1928.

17 Ising G, *Prinzip einer Methode zur Herstellung von Kanalstrahlen höher Voltzahl*, Arkiv Mat. Astr. Fys., 18, 1, 1925.

18 Karzmark CJ & Pering NC, *Electron linear accelerators for radiation therapy: history, principles and contemporary developments*, Physics in Medicine and Biology, 18, 321, 1973.

19 Sloan DH & Lawrence EO, *The production of heavy high speed ions without the use of high voltages*, Physics Review, 38, 2021, 1931.

20 Alvarez LW, *The design of a proton linear accelerator*, Physics Review, 70, 799, 1946. (See also: Alvarez et al, *Berkeley proton linear accelerator*, Rev. Sci. Instrum., 26, 111, 1955.)

21 Fry DW, Harvie RB, Mullett LB & Walkinshaw W, *Travelling wave linear accelerator for electrons*, Nature, 160, 351, 1947. (See also: Fry DW et al, *A travelling wave linear accelerator for 4 MeV electrons*, Nature, 162, 859, 1948; and: Frey DW, *The linear accelerator*, Philips Technical Review, 14, 1, 52.)

22 Ginzton EL, Hansen WW & Kennedy WR, *A linear electron accelerator*, Rev. Sci. Instrum., 19, 89, 1948.

23 Meredith WJ & Massey JB, *Fundamental physics of radiology*, 2nd edn, John Wright: Bristol, 1971.

24 Greene D, *Linear accelerators for radiation therapy*, Medical Physics Handbook 17, Adam Hilger: Bristol, 1986.

25 Klevenhagen SC, *Physics of electron beam therapy*, Medical Physics Handbook 13, Adam Hilger: Bristol, 1985.

26 Beck C, *Röntgen ray diagnosis and therapy*, Appleton and Lange: New York, 1904.

27 Mould RF, *Radiotherapy treatment planning*, 2nd edn, Medical Physics Handbook 14, Adam Hilger: Bristol, 1985.

28 Schmitz H, *Technique and statistics in the treatment of carcinoma of the uterus and contiguous organs with the combined use of radium and X-rays*, Amer. J. Roentgenology, 9, 662, 1922.

29 Rowlands JJ, *Van de Graaff electrostatic generators*, The Technology Review (Massachusetts Institute of Technology), 63, 1, 1961.

30 Johns HE & Cunningham JR (eds), *The physics of radiology*, 3rd edn, Charles C Thomas: Springfield, Illinois, 1969.

31 Mullard Ltd, *World's first gantry-mounted linear accelerator installed at the General Hospital, Newcastle upon Tyne*, Radiography, **30**, 206, 1954.

References: Chapter 19

1 Lederman M, *The early history of radiotherapy: 1895–1939*, Int. J. Radiation Oncology, Biology & Physics, **7**, 639, 1981.

2 Bailey H & Quimby E, *Radium report of the Memorial Hospital, New York*, 2nd Report, 1917–1922, Hoeber: New York, 1923. (See also: Janeway HH, *Radium therapy at the Memorial Hospital, New York*, 1st Report, 1915–1916, Hoeber: New York, 1917.)

3 Medical Research Council, *Medical uses of radium*, MRC Special Report Series 62, MRC: London, 1922.

4 Lysholm E, *Apparatus for the production of a narrow beam of gamma-rays in treatment by radium at a distance*, Acta Radiologica, **2**, 516, 1923.

5 Mallet L & Coliez R, *Techniques de curiethérapie profonde*, Brit. J. Radiology (BIR Section), **31**, 393, 1926.

6 Union Minière de Haut Katanga, *Radium: production, general properties, its applications in therapeutics*, Union Minière du Haut Katanga: Brussels, 1929.

7 Laborde S, *La technique de la curiethérapie*, Gauthier-Villars: Paris, 1933.

8 Berven EGE, *Development of technique and results of treatment of tumours of the oral and nasal cavities*, Amer. J. Roentgenology, **28**, 332, 1932.

9 Johns HE & Watson TA, *The cobalt-60 story*, Theratronics Information Sheet, ASTRO Meeting, Washington DC, November 1991. (See also: Johns HE, Bates IM & Watson TA, *1000 curie cobalt units for radiation therapy. 1. The Saskatchewan cobalt-60 unit*, Brit. J. Radiology, **25**, 296, 1952, and: Johns HE, Epp ER, Cormack DV & Fedoruk SO, *Depth dose data and diaphragm design for the Saskatchewan 1000 curie cobalt unit*, Brit. J. Radiology, **25**, 302, 1952.)

10 Friedman M, Brucer M & Anderson E (eds), *Roentgens, rads and riddles: a symposium on supervoltage radiation therapy*, United States Atomic Energy Commission, 1959.

11 Myers WG, *Some medical applications of atomic energy*, Proceedings of the 12th Annual Conference of State & Local Trade Associations, Ohio State University, 46, 1948.

12 Freundlich HF, Haybittle JL & Quick RS, *Radio-iridium teletherapy*, Acta Radiologica, **34**, 115, 1950.

13 Marshall E, *Juarez: an unprecedented radiation accident*, Science, **223**, 1152, 1984. (See also: Anderson I, *Radioactive scrap ends up in table legs*, New Scientist, February 23rd 1984.)

14 International Atomic Energy Agency, *The radiological accident in Goiania*, IAEA: Vienna, 1988.

15 Flint HT, *Radium teletherapy: latest modification of the Westminster apparatus and its use*, Brit. Medical Journal, **1**, 653, 1934.

16 Lederman M, *The historical development of laryngectomy. VI. History of radiotherapy in the treatment of cancer of the larynx 1896–1939*, Laryngoscope, **85**, 333, 1975.

17 Flint HT & Grimmett L, *Measurement of the distribution of gamma rays around a 4 gram mass of radium*, Brit. Medical Journal, **2**, 321, 1930. (See also: Rock Carling E, *Radium teletherapy: note on the apparatus at present in use at the Westminster Hospital with 4 gram of radium*, Brit. Medical Journal, **1**, 232, 1930.)

18 Failla G, *Design of a well-protected radium pack*, Amer. J. Roentgenology, **20**, 128, 1928.

19 Sluys F & Kessler E, *Gammathérapie*, Le Cancer, **2**, 88 & 150, 1925.

20 du Mesnil de Rochemont R, *Einführung in die Strahlenheilkunde ein Lehrbuch für Studierende und Arzte*, Urban & Schwarzenberg: Berlin, 1937.

21 Watson Ltd, *Cobalt-60: Canada's contribution to cancer treatment*, Watson X-ray News, No. 5, 1957.

22 Lederman M & Mould RF, *Radiation treatment of cancer of the pharynx: with special reference to telecobalt therapy*, Brit. J. Radiology, **41**, 251–274, 1968.

23 Wheatley BM, *Physical aspects of the use of caesium fission products in teletherapy*, Brit. J. Radiology, **33**, 246, 1960.

24 Wheatley BM, Jones JC & Sinclair TC, *A caesium beam therapy unit*, Brit. J. Radiology, **33**, 251, 1960.

References: Chapter 20

1 Goldberg SW & London ES, *Zur Frage der Beziehungen zwischen Becquerelstrahlen und Hautaffectionen*, Dermatologische Zeitschrift, **10**, 457, 1903.

2 Wickham L & Degrais P, *Radium therapy*, English edn, Cassell: London, 1910.

3 Meredith WJ (ed), *Radium dosage: the Manchester system*, 2nd edn, Livingstone: Edinburgh, 1967. (See also the original papers of: Paterson R & Parker HM, *A dosage system for gamma-ray therapy*, Brit. J. Radiology, **7**, 592, 1934; Paterson R, Parker HM & Spiers FW, *A system of dosage for cylindrical distributions of radium*, Brit. J. Radiology, **9**, 487, 1936; Paterson R & Parker HM, *A dosage system for interstitial radium therapy*, Brit. J. Radiology, **11**, 252 & 313, 1938; Tod M & Meredith WJ, *A dosage system for use in the treatment of cancer of the uterine cervix*, Brit. J. Radiology, **11**, 809, 1938.)

4 Pierquin B & Dutreix A, *Pour une nouvelle méthodologie en curie thérapie; le Système de Paris (endo et plésio-radiothérapie avec préparation non radioactive)*, Note préliminaire, Annals Radiologie, **9**, 757, 1966. (See also: Pierquin B, Dutreix A, Paine CH, Chassagne D, Marinello G & Ash D, *The Paris System in interstitial radiation therapy*, Acta Radiologica Oncology, **17**, 33, 1978.)

5 Pierquin B, Wilson JF & Chassagne D, *Modern brachytherapy*, Masson: New York, 1987. (See also: Dutreix A, Marinello G & Wambersie A, *Dosimétrie en curiethérapie*, Masson: Paris, 1982.)

6 Pernot M, Hoffstetter S & Forçard JJ, *Interstitial low dose rate curiethérapy for head and neck cancers in 1991*, Activity Selectron Brachytherapy Journal, **5**, 122, 1991. (See also: Pernot M, *Brachytherapy in head and neck tumours*, in: *Brachytherapy 2*, Mould RF (ed), 364, Nucletron: Leersum, 1989.)

7 Hurdon E, *Cancer of the uterus*, Oxford University Press: London, 1942.

8 Regaud C, *Services de Curiethérapie*, in: *Radiophysiologie et radiothérapie recuil de travaux biologiques, techniques et thérapeutiques*, **2**, Regaud C, Lacassagne A & Ferroux R (eds), 218, Archives de l'Institut du Radium de L'Université de Paris et de la Fondation Curie, Les Presses Universitaires de France: Paris, 1922.

9 Fletcher GH, Stovall M & Sampiere V, in: *Carcinoma of the uterine cervix endometrium and ovary*, 69, Year Book Medical Publishers: Chicago, 1962. (See also: Fletcher GH, *Textbook of radiotherapy*, 2nd edn, Lea & Febiger: Philadelphia, 1973.)

10 Bell AG, *The uses of radium*, Amer. Medicine, **6**, 261, 1903.

11 Strebel H, *Vorschlaege zur Radiumtherapie*, Deutsche Medizinal Zeitung, **24**, 11, 1903.

12 Abbe R, *Radium in surgery*, J. Amer. Medical Association, **47**, 183, 1906. (See also: Abbe R, News item concerning 'a very ingenious method of introducing radium into the substance of a tumour', Archives of the Roentgen Ray, **15**, 74, 1910.)

13 Glaser FH, *Zur historischen Entwicklung der Brachytherapie*, Deutsches Ärzteblatt, **87**, B-2123, 1990.

14 Senate Committee on Mines and Mining, *A bill to provide for and encourage prospecting, mining, and treatment of radium-bearing ores in lands belonging to the United States, for the purpose of securing an adequate supply of radium for Government and other hospitals in the United States and for other purposes*, (63rd Congress, 2nd Session), Government Printing Office: Washington, DC, 1914.

15 Henschke UK, Hilaris BS & Mahan GD, *Intracavitary radiation therapy of the uterine cancer by remote afterloading with cycling sources*, Amer. J. Roentgenology, **96**, 45, 1966. (See also: Henschke UK, Hilaris BS & Mahan DG, *Remote afterloading with intracavitary applicators*, Radiology, **83**, 344, 1964.)

16 Joslin CA, Smith CW & Mallik A, *The treatment of cervix cancer using high activity cobalt-60 sources*, Brit. J. Radiology, **45**, 257, 1972. (See also: Joslin CA, *Afterloading methods in radiotherapy*, in: *Recent advances in cancer and radiotherapeutics*, Halnan KE (ed), 353, 1971, and: Bates TD & Berry RJ (eds), *High dose rate afterloading in the treatment of cancer of the uterus*, Brit. J. Radiology Special Report No. 17, 1980, and: O'Connell D, Howard N, Joslin CAF, Ramsey NW & Liversage WE, *A new remotely controlled unit for the treatment of uterine cancer*, Lancet, **18**, 570, 1965, and: Taina E (ed), *High versus low dose rate intracavitary radiotherapy in the treatment of carcinoma of the uterus: a comparative study of afterloading cobalt-60 [Cathetron] and conventional [radium] treatment*, Acta Obstetricia et Gynecologica Scandinavica, suppl. 103, 1981.)

17 Joslin CA, *A place for high dose rate brachytherapy in gynecological oncology: fact or friction*, Activity Selectron Brachytherapy Journal, suppl. no. 2, *Gynaecological high dose rate brachytherapy*, 1991.

18 Mould RF (ed), *Pulmonary brachytherapy*, Activity Selectron Brachytherapy Journal, suppl. no. 1, 1990.

19 Lawrence H, *Radium: how and when to use*, Stillwell: Melbourne, 1911.

20 Union Minière du Haut Katanga, *Radium: production, general properties, its applications in therapeutics*, Union Minière du Haut Katanga: Brussels, 1929.

21 Cade S, *Radium treatment of cancer*, Churchill: London, 1929.

22 Wilson CW, *Radium therapy: its physical aspects*, Chapman & Hall: London, 1945.

23 Stevenson WC, *Preliminary clinical report on a new and economical method of radium therapy by means of emanation needles*, Brit. Medical Journal, **2**, 9–10, 1914. (See also: Stevenson WC, Medical Press & Circular, March 11th 1914, and: Joly J, *On the local application of radium in therapeutics*, Proceedings of the 24 March scientific meeting of the Royal Dublin Society, May 8th 1914.)

24 Fowler JF & Mould RF, *High dose rate (HDR) consequences of the tilt of the uterine tube*, Activity Selectron Brachytherapy Journal, supplement 2, *Gynaecological HDR brachytherapy*, 19, 1991. (See also: Mould RF & Hobday PA, *Radiation dosimetry for the Amersham caesium-137 manual afterloading system for gynaecological brachytherapy*, Amersham International: Amersham, 1984.)

25 Mould RF, *The calculation of the distribution of absorbed energy from gamma ray sources in media of low atomic number*, MSc thesis, London University, 1966. (See also: Mould RF, *Röntgen archeology 1895–1937*, Brit. Inst. Radiology Bulletin, **2**, 8, 1976.)

26 Quimby EH, *The grouping of radium tubes in packs and plaques to produce the desired distribution of radiation*, Amer. J. Roentgenology, **27**, 18, 1932. (See also: Quimby EH, *Dosage table for linear radium sources*, Radiology, **43**, 572, 1944; Quimby EH, *Fifty years of radium*, Amer. J. Roentgenology, **60**, 723, 1948; Quimby EH, *The development of dosimetry in radium therapy*, in: *Afterloading: 20 years of experience 1955–1975*, Hilaris BS (ed), 1, Memorial Sloan-Kettering Cancer Center: New York, 1975.)

27 Sievert R, *Die Intensitaetsverteilung der primaren Gamma-Strahlung in der Nache medizinischen Radiumpraeparate*, Acta Radiologica, **1**, 89, 1921. (See also: Sievert R, *Eine Methode zur Messung von Roentgen-, Radium-, und Ultra-Strahlung nebst einigen Untersuchungen ueber die Anwendbarweit derselben in der Physik und der Medizin*, Acta Radiologica, suppl. 14, 1932.)

28 Jennings WA & Russ S, *Radon: its technique and use*, Middlesex Hospital Press: London, 1948.

29 Hodt HJ, Sinclair WK & Smithers DW, *A gun for interstitial implantation of radioactive gold grains*, Brit. J. Radiology, **25**, 419, 1952.

30 Wallace DM, Stapleton JE & Turner RC, *Radioactive tantalum wire implantation as a method of treatment for early carcinoma of the bladder*, Brit. J. Radiology, **25**, 421, 1952.

31 Hilaris BS (ed), *Afterloading: 20 years of experience, 1955–1975*, Memorial Sloan-Kettering Cancer Center: New York, 1975. (See also: Hilaris BS, Nori D & Anderson LL, *Atlas of brachytherapy*, Macmillan: New York, 1988, and: Nori D, *The VIII brachytherapy update 1988*, Memorial Sloan-Kettering Cancer Center: New York, 1988.)

32 Rotte K & Sauer O, *From radium to remote afterloading: German gynaecological experience 1903–1992 with special reference to Würzburg*, in: *International brachytherapy*, Mould RF (ed), 581–587, Nucletron: Veenendaal, 1992.

33 von Seuffert E, *Die Radiumbehandlung maligner Neubildungen in der Gynakologie*, in: *Lehrbuch der Strahlentherapie: Gynakologie*, Meyer H (ed), 940, Urban & Schwarzenberg: Berlin, 1929.

34 Dessauer F, *Radium, Mesothorium und Harte X-Strahlung und die Grundlagen ihrer medizinischen Anwendung*, Otto Nemnich: Leipzig, 1914. (See also: Gauss CJ & Lembcke H, *Röntgentiefentherapie ihre theoretischen Grundlagen, ihre praktische Anwendung und ihre klinischen Erfolge*, Urban & Schwarzenberg: Berlin, 1912.)

35 Heyman J, Reuterwall O & Benner S, *The Radium hemmet experience with radiotherapy in cancer of the corpus of the uterus: classification, method and treatment results*, Acta Radiologica, **22**, 31, 1941.

36 Simon N, Silverstone SM & Roach LC, *Afterloading Heyman applicators*, Acta Radiologica, **10**, 231, 1971. (See also: Simon N & Silverstone SM, *Intracavitary radiotherapy of endometrial cancer by afterloading*, Gynaecologic Oncology, **1**, 13, 1972, and: Simon N, *Intracavitary radiation of endometrial cancer*, in: *Cancer of the uterus in developing countries: report of an International Working Party & Conference in Brazil 1973*, Simon N & Snelling M (eds), 302, 1974.)

37 Blomfield GW, *Malignant tumours of the ovary, uterus, vagina and vulva*, in: *Cancer, 5, Radiotherapy*, Raven RW (ed), 243, Butterworth: London, 1959.

38 Battermann JJ, *Breast conserving therapy: changing of indications and techniques*, Activity Selectron Brachytherapy Journal, **4**, 46, 1990.

39 Mate TP, Kwiatkowski TM & Hatton JW, *Remote HDR afterloading brachytherapy of the prostate: a preliminary report*, Activity Selectron Brachytherapy Journal, **4**, 65, 1990.

40 Kassabian MK, *Röntgen rays and electro-therapeutics, with chapters on radium and phototherapy*, Lippincott: Philadelphia, 1907.

41 Finzi NS, *Radium therapeutics*, Frowde: London, 1913. (See also: Mould RF, *Oesophageal applicators before remote afterloading*, Activity Selectron Brachytherapy Journal, **6**, 44, 1992.)

42 Souttar HS, *Radium and its surgical applications*, Heinemann: London, 1929. (See also: Mould RF, *Brachytherapy for brain tumours in 1929*, Activity Selectron Brachytherapy Journal, **5**, 162, 1991.)

43 Warszawski N, Pfreunder L, Bohndorf W, Bratengeier K & Krone A, *Interstitial brachytherapy with flexible catheters in brain tumours*, Activity Selectron Brachytherapy Journal, **6**, 28, 1992.

44 Yankauer S, *Two cases of lung tumour treated brochoscopically*, New York Medical Journal, **115**, 741, 1922. (See also: Mould RF, *Lung cancer brachytherapy in the 1920s*, Activity Selectron Brachytherapy Journal, Suppl. No. 1, 30, 1990.)

45 Pancoast HK, *Superior pulmonary sulcus tumour*, J. Amer. Medical Association, **99**, 1391, 1932.

46 Knox R, *Radiography, X-ray therapeutics and radium therapy*, A & C Black: London, 1915.

47 Ward WR & Smith AJD, *Recent advances in radium*, Churchill: London, 1933.

48 Pavlou WJ, Vikram B, Urban MS, Siegel JH & Aubrey R, *The use of high dose rate remote afterloading brachytherapy in the treatment of malignant biliary obstruction*, Activity Selectron Brachytherapy Journal, 5, 13, 1991.

49 Philips TL, Gutin P & Sneed PK, *Brachytherapy and hyperthermia for brain tumours: the UCSF experience*, Activity Selectron Brachytherapy Journal, 5, 102–106, 1991.

50 Walstam R, *History of brachytherapy in Sweden including the origins of the Stockholm and Heyman techniques*, in: *Proceedings of the 1st Nordic Brachytherapy Working Conference*, May 1992, Linköping, Mould RF (ed), 34, Nucletron: Veenendaal, 1992. (See also: Forsell G, *Radiumbehandling av maligna tumörer i kvinnliga genitalia*, Hygiea, **74**, 1912, and: Heyman J, *The so-called Stockholm method and the results of treatment of uterine cancer at Radiumhemmet*, Acta Radiologica, **16**, 129, 1935, and: Heyman J, *The Radiumhemmet experience with radiotherapy in cancer of the corpus of the uterus*, Acta Radiologica, **22**, 11, 1941, and: Walstam R, *Remotely-controlled afterloading radiotherapy apparatus*, Physics in Medicine and Biology, **7**, 225, 1962, and: Walstam R, *A remotely controlled afterloading apparatus*, Acta Radiologica, Suppl. 236, 84, 1965, and: Walstam R, *Studies on short-distance and intracavitary gamma beam techniques: physical considerations with special reference to radiation protection*, Acta Radiologica, Suppl. 236, 1965.)

51 Okeanova N, *Treatment of endometrical carcinoma using a combination of preoperative brachytherapy and surgery*, Activity Selectron Brachytherapy Journal, **6**, 18, 1992.

52 ICRU, *Dose and volume specification for reporting intracavitary therapy in gynaecology*, ICRU Report no. 38, ICRU: Bethesda, 1985.

53 Ellis F & Taylor CBG, *The American caesium-137 afterloading system for gynaecological brachytherapy*, Amersham International, 1982, and: Mould RF & Hobday PA, *Radiation dosimetry for the Amersham caesium-137 manual afterloading system for gynaecological brachytherapy*, Amersham International, 1984.

54 Flamant F, Chassagne D, Cosset JM, Gerbaulet A & Lamerle J, *Embryonal rhabdomyosarcoma of the vagina in children*, Eur. J. Cancer, **15**, 527–532, 1979; Gerbaulet A, Panis X, Flamant F & Chassagne D, *Iridium afterloading curietherapy in the treatment of pediatric malignancies*, Cancer, **56**, 1274–1279, 1985; Gerbaulet A, *Pediatric neoplasms*, in: *Modern brachytherapy*, Pierquin B, Wilson JF & Chassagne D (eds), 315–316, Masson: New York, 1987; Gerbaulet A, Esche B, Haie C, Castaigne D, Flamant F & Chassagne D, *Conservative treatment for lower gynaecological tract maglignancies in children and adolescents: the Institut Gustave-Roussy experience*, Int. J. Radiation Oncology Biology & Physics, **17**, 655–658, 1989; Flamant F, Gerbaulet A, Nihoul-Fekete C, Valteau-Couanet D, Chassagne D & Lemerle J, *Long-term sequelae of conservative treatment by surgery, brachytherapy and chemotherapy for vulvar and vaginal rhabdomyosarcoma in children*, J. Clinical Oncology, **8**, 1847–1853, 1990; Gerbaulet A, Habrand JL, Haie C et al, *The role of brachytherapy in the conservative treatment of pediatric malignancies: experience of the Institut Gustave-Roussy*, Activity Selectron Brachytherapy Journal, **5**, 85–90, 1991.

55 *Une première médicale européenne. Distilbène, la revanche de la fécondité*, Le Figaro, 3 December 1992; *Grâce à une nouvelle méthode de curiethérapie. Une 'fille du distilbène' a pu accoucher*, Le Monde, 4 December 1992.

References: Chapter 21

1 *Availability of radioactive isotopes (Manhattan Project)*, Science, **103**, 348, 1946, and: *Availability of radioactive isotopes (Manhattan Project), Corrected tables*, Rev. Scientific Instruments, 17, 348, 1946.

2 Beck C, *Röntgen ray diagnosis and therapy*, Appleton & Lange: New York, 1904.

3 Blumgart HL & Weiss S, *Studies on the velocity of blood flow*, J. Clinical Investigation, 4, 15–31, 1927.

4 Joliot F & Curie I, *Artificial production of a new kind of radio-element*, Nature, 201, February 10th 1934.

5 de Hevesy GC, *Marie Curie and her contemporaries: the Becquerel–Curie memorial lecture*, J. Nuclear Medicine, 2, 167, 1961, see also: *Biography of George de Hevesy* & reprint of *Marie Curie and her contemporaries*, J. Nuclear Medicine, 25, 116–131, 1984.

6 Mayneord WV, *Some applications of nuclear physics to medicine*, Brit. J. Radiology, suppl. no. 2, 1950.

7 International Atomic Energy Agency, *Isotopes in day to day life*, IAEA: Vienna, 1981.

8 Pochin E, *Nuclear radiation: risks and benefits*, Clarendon Press: Oxford, 1983.

9 Anderson EC, *Scintillation counters*, Brit. J. Radiology, suppl. no. 7, 27, 1957, and: Oberhausen E, *Liquid scintillation whole-body counters*, in: *Clinical uses of whole-body counting*, 3–4, IAEA: Vienna, 1966.

10 Anderson EC, Hayes FN & Hiebert RD, *Walk-in human counter*, Nucleonics, 16, 106, 1958.

11 Tauxe WN & Orvis AL, *The Mayo Clinic whole body counter*, Mayo Clinic Proceedings, 41, 18–23, 1966, and: Orvis AL, *Whole-body counting*, in: *Medical radionuclides—radiation dose and effects*, AEC Symposium Series 20, Cloutier RJ, Edwards CL & Snyder WS (eds), 115–132, US Atomic Energy Commission: Oak Ridge, 1970.

12 IAEA, *The radiological accident in Goiania*, IAEA: Vienna, 1988.

13 Mayneord WV, Turner RC, Newbery SP & Hodt HJ, *A method of making visible the distribution of activity in a source of ionising radiation*, Nature, **168**, 762–765, 1951.

14 Mayneord WV & Newbery SP, *An automatic method of studying the distribution of activity in a source of ionising radiation*, Brit. J. Radiology, 25, 589–596, 1952.

15 Mayneord WV, *Radiological research*, Brit. J. Radiology, 27, 309–317, 1954.

16 Cassen B, Curtis L, Reed CW & Libby R, *Instrumentation for iodine-131 in medical use*, Nucleonics, 9, 46–49, 1951.

17 Curtis L & Cassen B, *Speeding up and improving contrast of thyroid scintigrams*, Nucleonics, 10, 58–59, 1950.

18 Beck RN, *Radioisotope scanning systems*, Instrument Society of America Trans., 5, 335–348, 1966.

19 Mallard JR, *Medical radioisotope scanning*, Physics in Medicine and Biology, 10, 309–334, 1965.

20 Leach KG, *The physical aspects of radioisotope organ imaging*, Teaching Booklet No. 1, British Institute of Radiology: London, 1976.

21 Mallard JR & Peachey CJ, *A quantitative automatic body scanner for the localisation of radioisotopes in vivo*, Brit. J. Radiology, 32, 652, 1959, and: *A scanning machine for presenting pictorially and quantitatively the distribution of radioisotopes in vivo*, Proceedings of the 3rd International Conference on Medical Electronics, 511, 1960.

22 Kuhl DE, Chamberlain RH, Hale TH & Gorson RG, *A high contrast photographic recorder for scintillation counter scanning*, Radiology, 66, 730, 1956, see also: Herring CE, *A universal photorecording system for radioisotope area scanners*, Journal of Nuclear Medicine, 1, 83–101, 1960.

23 Anger HO, *Scintillation camera*, Rev. Scientific Instruments, **29**, 27–33, 1958.

24 Maisey M, *Nuclear medicine, a clinical introduction*, Update Books: London, 1980.

25 McCready VR, *Clinical radioisotope scanning*, Brit. J. Radiology, 40, 401–423, 1967, see also: Wagner HN, *Principles of nuclear medicine*, W B Saunders: Philadelphia, 1968, and: Gottschalk A & Beck RN, *Fundamental problems in scanning*, Charles C Thomas: Springfield, Illinois, 1968, and: Britton KE, *Radionuclides: clinical uses*, in: McAinsh TF (ed), *Physics in medicine and biology encyclopedia, medical physics, bioengineering and biophysics*, 691–698, Pergamon: Oxford, 1986.

26 Harper PV, Beck R, Charleston D & Lathrop KA, *Optimisation of a scanning method using technetium-99m*, Nucleonics, 22, 50–54, 1964, see also: McAfee JG, Fueger CF, Stern HS, Wagner HN & Migita T, *Technetium-99m pertechnetate for brain scanning*, Journal of Nuclear Medicine, 5, 811–827, 1965.

A Century of X-rays and Radioactivity in Medicine

27 International Atomic Energy Agency, *Quality control of nuclear medicine instruments*, IAEA-TECDOC-317, IAEA: Vienna, 1984.

28 Mould RF, *Phantoms in nuclear medicine*, in: McAinsh TF (ed), *Physics in medicine and biology encyclopedia, medical physics, bioengineering and biophysics*, 567–572, Pergamon: Oxford, 1986.

29 Goddard BA, Gregory C, Keeling DH, McCready VR, Mould RF, Potter DC, Rogers RT, Williams ES & Williams HS, *A survey of images of a phantom produced by radioisotope scanners and cameras*, British Institute of Radiology Special Report No. 9, British Institute of Radiology: London, 1976.

30 Mould RF, *An investigation of the variations in the normal liver shape*, Brit. J. Radiology, 45, 586–590, 1972.

31 Mould RF, *A liver phantom for evaluating camera and scanner performance in clinical practice*, Brit. J. Radiology, 44, 810–811, 1971, and: Goddard BA, Gregory C, Keeling DH, McCready VR, Mould RF, Potter DC, Rogers RT, Williams ES & Williams HS, *A survey of some radionuclide imaging equipment with an anthropomorphic phantom*, Department of Health and Social Security Report STB/3/78, DHSS: London, 1978, and: Mould RF, *The London liver phantom*, J. Nuclear Medicine, 24, 974–975, 1933, and: Volodin V, Souchkevitch G, Racoveanu N, Bergmann H, Busemann-Sokole E, Delaloye B, Georgescu G, Herrera N, Jasinski W, Kasatkin Y, Paras P & Mould RF, *World Health Organisation interlaboratory comparison in 12 countries on quality performance of nuclear medicine devices*, European J. Nuclear Medicine, 10, 193–197, 1985.

32 Herrera NE, Herman GA, Hauser W & Paras P, *College of American Pathologists series X survey program*, in: *Medical radionuclide imaging*, Proceedings of a symposium organised by IAEA in cooperation with WHO, September 1980, Heidelberg, 177–187, 1981, and: Souchkevitch GN, Asikainen M, Bauml A, Bergmann H, Busemann-Sokole E, Carlsson S, Delaloye B, Dermetzoglou F, Herrera N, Jasinski W, Karanfilski B, Mester J, Oppelt A, Perry J, Skretting A, van Herk G, Volodin V & Mould RF, *World Health Organisation and International Atomic Energy Agency 2nd interlaboratory comparison study in 16 countries on quality performance of nuclear medicine imaging devices*, European J. Nuclear Medicine, 13, 495–501, 1988.

33 Haerten RL & Hernandez T, *Single photon emission computed tomography (SPECT): principles, technical implementation and clinical application*, Electromedica, 52, 66–80, 1984.

34 Alavi A & Hirsch LJ, *Studies of central nervous system disorders with SPECT and PET: evolution over the past two decades*, Seminars in Nuclear Medicine, 21, 58–81, 1991.

35 Anger HO & Rosenthal DJ, *Scintillation camera and positron camera*, in: *Medical radioisotope scanning*, Proceedings of an IAEA/WHO seminar, February 1959, Vienna, 59–82, IAEA: Vienna, 1959.

36 Sweet WH, Mealey J, Brownell GL & Aronow S, *Coincidence scanning with positron-emitting arsenic or copper in the diagnosis of focal intracranial disease*, in: *Medical radioisotope scanning*, Proceedings of an IAEA/WHO seminar, February 1959, Vienna, 163–188, IAEA: Vienna, 1959.

37 Freeman LM & Blaufox MD, *Editorial*, in: *Positron emission tomography*, Part I, Seminars in Nuclear Medicine, 22, 139, 1992.

38 Ter-Pogossian MM, *The origins of positron emission tomography*, Seminars in Nuclear Medicine, 22, 140–149, 1992, see also: Maisey M & Jeffery P, *Clinical applications of PET*, Brit. J. Clinical Practice, 45, 265–272, 1991.

39 Greig WR, Boyle IT & Boyle JA, *Diagnosis of thyroid disorders using radioactive iodine*, Medical Monograph 5, Radiochemical Centre: Amersham, 1967.

40 Ansell G & Rotblat J, *Radioactive iodine as a diagnostic aid for intrathoracic goitre*, Brit. J. Radiology, 21, 552–558, 1948, see also: McAlister J, *The development of radioisotope scanning techniques*, Brit. J. Radiology, 46, 889–898, 1973.

41 Meredith WJ & Massey JB, *Fundamental physics of radiology*, 2nd edn, John Wright: Bristol, 1971.

42 Rassow G, *Basic information on routine diagnosis in nuclear medicine*, Siemens Aktiengesellschaft: Erlangen, 1970.

43 Kniseley RM, *Training for clinical use of radioisotopes: qualification courses*, J. Nuclear Medicine, 1, 239–250, 1960.

44 Pircher FJ, Sitterson BW & Andrews GA, *The ORINS linear scanner in diagnosis and treatment of thyroid carcinoma with iodine*, J. Nuclear Medicine, 1, 251–261, 1960.

45 Woldring MG, *Technetium atlas*, Philips-Duphar: Petten, 1973.

46 Harmer CL, Burns JE, Sams A & Spittle M, *Clinical evaluation of bone scanning with fluorine*, in: *Radioactive isotopes in the localisation of tumours*, McCready VR, Trott NG et al (eds), Proceedings of an International Nuclear Medicine Symposium, Imperial College, London, September 1967, 117, Heinemann: London, 1968.

47 Myers WG, *The Anger scintillation camera becomes of age*, J. Nuclear Medicine, 20, 565–567, 1979.

48 Doss LL, Ho T & Lange B, *High dose rate brachytherapy for osteolytic metastases in previously irradiated sites*, in: *International brachytherapy*, Mould RF (ed), 412–415, Nucleton: Veenendaal, 1992.

49 Donaldson RM & Jarritt PH, *Nuclear cardiology*, Brit. J. Clinical Equipment, 4, 136–143, 1979.

50 Kuhl DE, *A clinical radioisotope scanner for cylindrical and section scanning*, in: *Medical radioisotope scanning*, Proceedings of a symposium, April 1964, Athens, vol. I, 273–289, IAEA: Vienna, 1964, and: Kuhl DE, *The current status of tomographic scanning*, in: *Fundamental problems in scanning*, Gottschalk A & Beck RN (eds), 179–188, Charles C Thomas: Springfield, Illinois, 1968.

51 Waldemar G, Hasselbalch SG, Andersen AR, Delecluse F, Petersen P, Johnsen A & Paulson OB, ^{99m}Tc-d, l-HMPAO and SPECT of the brain in normal aging, J. Cerebral Blood Flow and Metabolism, 11, 508–521, 1991.

52 Wilcke O, *Results of positron scanning in 1200 cases for diagnosis of intracranial lesions*, in: *Medical radioisotope scanning*, Proceedings of a symposium, April 1964, Athens, Vol. II, 95–107, IAEA: Vienna, 1964.

References: Chapter 22

1 Pullin VE & Wiltshire WJ, *X-rays past and present*, Ernest Benn: London, 1927.

2 Kassabian MK, *Röntgen rays and electro-therapeutics, with chapters on radium and phototherapy*, Lippincott: Philadelphia, 1907.

3 Jennings WA & Axton EJ, *Radiation quantities: measurements*, in: *Physics in medicine and biology encyclopedia*, McAinsh TF (ed), 638–642, Pergamon: Oxford, 1986, and: Jennings WA, *Radiation quantities and units*, in: *Physics in medicine and biology encyclopedia*, McAinsh TF (ed), 634–638, Pergamon: Oxford, 1986.

4 Spiers FW, *SI units in radiology*, Radiological Protection Bulletin, No. 18, 4, 1977.

5 Mayneord WV, *SI units in medical physics*, Contemporary Physics, 17, 1, 1976.

6 Wickham L & Degrais P, *Radium therapy*, English edn, Cassell: London, 1910.

7 Turner D, *Radium, its physics and therapeutics*, Baillière Tindall: London, 1911.

8 Levy LA & Willis HG, *Radium and other radioactive elements: a popular account treated experimentally*, Percival Marshall: London, 1904.

9 King LV, *Absorption problems in radioactivity*, London, Edinburgh and Dublin Philosophical Magazine and Journal of Science, 23, 242, 1912.

10 Cleaves MA, *Light energy*, Rebman: New York, 1904.

11 Walter AE, *A practical guide to X-rays, electrotherapeutics and radium therapy for students and practitioners*, Thacker Spink: Calcutta, 1916.

12 Williams FH, *Reminiscences of a pioneer in roentgenology and radium therapy, with reports of some recent observations*, Amer. J. Roentgenology, 13, 253, 1925. (See also: Williams FH, *The Roentgen rays in medicine and surgery as an aid to diagnosis and as a therapeutic agent*, 2nd edn, Macmillan: New York, 1902.)

13 Knox R, *Radiography, X-ray therapeutics and radium therapy*, A & C Black: London, 1915.

14 Butcher WD, *The means of accurate measurement in X-ray work*, Journal of the Röntgen Society, **1**, 74, 1905, see also: *Protection in X-ray work*, Archives of the Roentgen Ray, **10**, 38, 1905.

15 Tousey S, *Medical electricity, Röntgen rays and radium*, 3rd edn, Saunders: Philadelphia, 1921.

16 Turner D, *Remarks on the effects and use of radium*, Report of a paper read at the December 1st 1909 meeting of the Edinburgh Medico-Chirurgical Society, Lancet, **2**, 1873, 1909.

17 Curie M, Debierne A, Eve AS, Geiger H, Hahn O, Lind SC, Meyer ST, Rutherford E & Schweidler E, *Radioactive constants as of 1930*, J. Amer. Chemical Society, **53**, 2437, 1931.

18 Condon EU & Curtiss LF, *New units for measurement of radioactivity*, Science, **103**, 712, 1946.

19 Debierne A & Regaud C, *Sur l'emploi de l'émanation du radium condensée en tubes clos à la place des composés radifères et sur le dosage (en millicuries d'émanation détruite) de l'énergies depensées pendant les applications radioactives locales*, Comptes Rendus de l'Académie des Sciences, Paris, **161**, 422, 1915.

20 Regaud C & Ferroux R, *Durée d'application en radium-thérapie: procédés, notation et de calcul; table pour l'emploi de l'émanation du radium*, Journal de Radiologie et Electrologie, **3**, 481, 1919.

21 Union Minière du Haut Katanga, *Radium: production, general properties, its applications in therapeutics*, Union Minière du Haut Katanga: Brussels, 1929.

22 Knox R, *Radiography and radio-therapeutics*, 2nd edn, Part II: *Radio-therapeutics*, A & C Black: London, 1918.

23 Finzi NS, *Radium therapeutics*, Frowde: London, 1913. (See also: Mould RF, *Oesophageal applicators before remote afterloading*, Activity Selectron Brachytherapy Journal, **6**, 44, 1992.)

24 Failla G, *Dosage in radium therapy*, Amer. J. Roentgenology, **8**, 674, 1921.

25 Schall WE, *X-rays: their origin, dosage and practical application*, 4th edn, John Wright: Bristol, 1932.

26 Roberts F, *The principles and practice of X-ray therapy*, H K Lewis: London, 1936.

27 Collwell HA & Russ S, *Radium, X-rays and the living cell*, Bell: London, 1915.

28 Regaud C, *The influence of the duration of irradiation on the changes produced in the testicle by radium*, Int. J. Radiation Oncology Biology & Physics, **2**, 565, 1977 (English translation of: Regaud C, Comptes Rendus de Société Biologique, **86**, 787, 1922).

29 Juul J, *Significance of the time factor in radium irradiation*, Acta Radiologica, **11**, 226, 1930.

30 Strandqvist M, *Studien über die kumulative Wirkung der Röntgensstrahlung bei Fraktionierung*, Acta Radiologica, Suppl. 55, 1944.

31 Abbe R, News item concerning '*a very ingenious method of introducing radium into the substance of a tumour*', Archives of the Roentgen Ray, **15**, 74, 1910.

32 Abbe R, *Report of the Practitioner's Society of New York*, New York Medical Record, **65**, 938, 1904, and: *The subtle power of radium*, New York Medical Record, **66**, 321, 1904, and: *The status of radium*, New York Medical Record, **65**, 356, 1904, and: Doss LL, *The history of radium*, Missouri Medicine, **75**, 594–599, 1978.

33 Krönig B & Friedrich W, *Physikalische und biologische Grundlagen der Strahlentherapie*, Urban & Schwarzenberg: Berlin, 1918.

34 Russ S, *A suggestion for a new X-ray unit in radiotherapy*, Archives of Radiology and Electrotherapy, **23**, 226, 1918.

35 Quimby EH, *The skin erythema dose with a combination of two types of radiation*, Amer. J. Roentgenology, **17**, 612, 1927.

36 Quimby EH, *The grouping of radium tubes in packs and plaques to produce the desired distribution of radiation*, Amer. J. Roentgenology, **27**, 18, 1932.

37 Quimby EH, *Dosage tables for linear radium sources*, Radiology, **43**, 572, 1944.

38 Quimby EH & Failla G, *Radium dosimetry*, in Glasser O (ed), *The science of radiology*, ch. XIV, Charles C Thomas: Springfield, Illinois, 1933; Baillière Tindall & Cox: London, 1933.

39 Solomon I, *La radiothérapie profonde*, Masson: Paris, 1923.

40 Béclère A, *On international standardisation of measures in röntgentherapy*, Brit. J. Radiology (Röntgen Society Section), **23**, 66, 1927.

41 Seitz L & Wintz H, *Unsere Methode der Röntgen Tiefentherapie*, Urban & Schwarzenberg: Berlin, 1920.

42 Wintz H & Rump W, *Protective measures against dangers resulting from the use of radium, roentgen and ultraviolet rays*, League of Nations, CH 1054, League of Nations: Geneva, 1931.

43 Erskine AW, *Practical X-ray treatment*, Bruce: Minneapolis, 1931.

44 Christie AC, *Röntgen diagnosis and therapy*, Lippincott: Philadelphia, 1924, see also: *Roentgenology in North America: a historical sketch*, Acta Radiologica, **6**, 281, 1926.

45 Kramer JB, *Radiations from slow-radium*, Baillière Tindall & Cox: London, 1921.

46 Guilleminot H, *Nouveau quantitometer pour rayons X*, Comptes Rendus de l'Académie des Sciences, Paris, **145**, 711, 1907.

47 Bordier H, *Radiometric methods*, Archives of the Roentgen Ray, **11**, 6, 1906.

48 Contremoulins G, *Recherche d'une unité mesuré pour la force de pénétration des rayons X et pour leur quantité*, Comptes Rendus de l'Académie des Sciences, Paris, **134**, 169, 1902.

49 Wintz H & Rump W, *Das Roentgenphotometer*, Strahlentherapie, **22**, 451, 1926.

50 Butcher WD, *The measurement of X-rays*, Journal of the Röntgen Society, **4**, 36, 1908.

51 Kienböck R, *Ueber Dosimeter und das quantrimetrische Verfaren*, Fortschritte a. d. Gebiete d. Röntgenstrahlen, **9**, 276, 1905. (See also: *Radiotherapie. Ihre biologischen Grundlagen, Anwendungmethoden und Indikationen*, Ferdinand Enke: Stuttgart, 1907.)

52 Rollins W, *Notes on X-ray light: vacuum tube burns*, Boston Medical and Surgical Journal, **146**, 39, 1902.

53 Osborn SB, *Radiation protection*, in: *Cancer*, vol. 5, Raven RW (ed), 117–133, Butterworth: London, 1959.

54 Curie M (1904), *Recherches sur les substances radioactives, Thèse présentée à la Faculté de Paris pour obtenir le grade de Docteur des Sciences Physiques*, 2nd edn, Gauthier-Villars: Paris (English translation: *Radioactive substances*, Philosophical Library: New York, 1961).

55 McKinlay AF, *Thermoluminescence dosimetry*, Adam Hilger: Bristol, 1981.

56 Sabouraud R & Noiré H, *Traitement des teignes tondantes par les rayons X*, La Presse Mèdicale, **12**, 825, 1904.

57 Codd MA, *Standardisation and the X-ray industry*, Brit. J. Radiology (Röntgen Society Section), **23**, 221, 1927.

58 Holzknecht G, *Das Chromoradiometer*, Congrès International d'Electrologie et Radiologie Mèdical, Vol. 2, 377, 1902. (See also: *Roentgen-therapy as practiced in the Röntgen Laboratory of the Allgemeine Krankenhaus of Vienna*, Archives of the Roentgen Ray, **9**, 262, 1905.)

59 Freund L, *Ein Neues Radiometrisches Verfahren*, Wien Klinische Wochenschrift, **17**, 1904. (See also: *Elements of general radio-therapy for practitioners*, Rebman: New York, 1904.)

60 Schwarz G, *Ueber die Einwirkung der Roentgenstrahlung auf Ammonium-oxalat-Sublimatloesung*, Fortschritte a. d. Gebiete d. Roentgenstrahlen, **11**, 114, 1907.

61 Bordier H & Galimard J, *A new unit for measuring X-rays*, Archives of the Roentgen Ray, **11**, 164, 1906.

62 Fricke H & Morse S, *The chemical action of roentgen rays on dilute ferrosulphate solutions as a measure of dose*, Amer. J. Roentgenology, **18**, 430, 1927.

63 Klevenhagen SC, *Physics of electron beam therapy*, Adam Hilger: Bristol, 1985.

64   Fürstenau R, *Ueber die Verwendbarkeit des Selens zu Roentgenstrahlenenergiemessungen*, Physikalische Zeitung, **16**, 276, 1915.

65   Hirsch IS, *The principles and practice of roentgenological technique*, American X-ray Publishing Co: New York, 1920.

66   Christén Th, *Messung und Dosierung der Röntgenstrahlen*, in: *Archiv und Atlas der normalen und pathologischen Anatomie in typischen Röntgenbildern*, Albers-Schönberg H (ed), Fortschritte auf dem Gebiete der Röntgenstrahlen, Band 28, Lucas Gräfe & Sillem: Hamburg, 1913.

67   Christén Th, *Absolute measurement for quality for Röntgen rays and its use in Röntgen therapy*, Verhandlungen der Deutschen Röntgengesellschaft, **13**, 119, 1912.

68   Dauvillier A, *X-ray dosage*, Brit. J. Radiology (Röntgen Society Section), **22**, 115, 1926.

69   Stahel E, *Bestimmung der bei Gamma- und Röntgen-strahlenbehandlung von Gewebe absorbierten Energiemengen*, Strahlentherapie, **33**, 296, 1929. (See also: Stahel E, *Eine Mikro-ionisationskammer für Röntgen- und Radiumstrahlen*, Strahlentherapie, **31**, 582, 1929.)

70   Mayneord WV & Roberts JE, *An attempt at precision measurements of gamma rays*, Brit. J. Radiology, **10**, 365, 1937.

71   Murdoch J, *Dosage in radium therapy*, Brit. J. Radiology, **4**, 256, 1931.

72   Meyer H & Schweidler E, *Radioactivität*, Teubner: Berlin, 1916. (See also: Hess V, *On the gamma-ray action of extensive flat radium preparations at different distances, with and without absorbing materials*, Physical Review, **19**, 73, 1922, and: Souttar HS, *On fields of radiation from radon seeds*, Brit. J. Radiology, **4**, 681, 1931, and: Mayneord WV, *The distribution of radiation around simple radioactive sources*, Brit. J. Radiology, **5**, 677, 1932, and: Mayneord WV, *The plane surface applicator*, Acta Radiologica, **14**, 95, 1933.)

73   Sievert R, *Die Intensitaetsverteilung der primaren Gamma-Strahlung in der Nache medizinischer Radiumpraeparate*, Acta Radiologica, **1**, 89, 1921. (See also: Sievert R, *Eine Methode zur Messung von Roentgen-, Radium-, und Ultra-Strahlung nebst einigen Untersuchungen ueber die Anwendbarweit derselben in der Physik und der Medizin*, Acta Radiologica, Suppl. 14, 1932.)

74   Mould RF, *Calculation of the distribution of absorbed energy from gamma-ray source in media of low atomic number*, MSc thesis, London University, 1966.

75   Kienböck R, *Ueber die Einwirkung des Röntgen-Lichtes auf die Haut*, Wiener klinische Wochenschrift, **13**, 1153, 1900. (See also: *Radiotherapie. Ihre biologischen Grundlagen, Anwendungsmethoden und Indikationen*, Ferdinand Enke: Stuttgart, 1907.)

76   Benoist L, *Experimental definition of various types of X-rays by the radiochromator*, Comptes Rendus de l'Académie des Sciences, Paris, **134**, 225, 1902.

77   Wehnelt A, *Über eine Röntgenröhre mit veranderlichen Härtegrade und über einen neuen Härtmesser*, in: *Festschrift Ludwig Boltzmann gewidment zum sechzigsten Geburtstage*, Meyer S (ed), 160, Barth: Leipzig, 1904.

78   Bauer H, *Über einen objectiven Härtmesser mit Ziegerausschlag für den Röntgenbetrieb*, Deutsche Medizinische Wochenschrift, **36**, 2099, 1910.

79   Walter B, *Zwei Härtskalen für Röntgenröhren*, Fortschritte auf dem Gebiete der Röntgenstrahlen, **6**, 2, 1902.

80   Case JT, *The early history of radium therapy and the American Radium Society*, 1959 Janeway Lecture, Amer. J. Roentgenology, **82**, 574, 1959.

81   Villard P, *The radiosclerometer*, Archives d'èlectricitè médicale, Bordeaux, **14**, 692, 1908.

82   Belot J, *The principal factors in radiotherapy and radiumtherapy*, Archives of the Roentgen Ray, **11**, 36, 1906.

83   Belot J, *Radiotherapy in skin disease*, Rebman: New York, 1905.

84   Behnken H, *Die Eichung von Dosismessern in der Physikalische-Technischen Reichanstalt*, Fortschritte a. d. Gebiete d. Röntgenstrahlen, **31**, 479, 1924.

85   Behnken H, *The German unit of X-radiation*, Brit. J. Radiology (Röntgen Society Section), **23**, 72, 1927.

86   Solomon I, *Über die Wahl einer Quantimetrischen Einheit*, Strahlentherapie, **20**, 642, 1925.

87   Béclère A, *On international standardisation of measures in röntgentherapy*, Brit. J. Radiology (Röntgen Society Section), **23**, 66, 1927.

88   Holthusen H, *A unit of X-ray dosage*, Brit. J. Radiology (Röntgen Society Section), **22**, 42, 1926.

89   Szilard B, *Absolute measurement of röntgen and gamma rays in biology*, Archives of the Roentgen Ray, **19**, 3, 1914.

90   Mayneord WV, *The measurement in units of 'r' of the gamma rays from radium*, Brit. J. Radiology, **4**, 693, 1931.

91   Coliez R, *Discussion on international units and standards*, Brit. J. Radiology (Röntgen Society Section), **23**, 100, 1927.

92   Bragg WH, *Studies in radioactivity*, Macmillan: New York, 1912.

93   Fricke H & Glasser O, *Eine theoretische und experimentelle Untersuchung der kleinen Ionisationskammer*, Fortshritte a. d. Gebiete d. Roentgenstrahlen, **33**, 329, 1925, and: *A theoretical and experimental study of the small ionisation chamber*, Amer. J. Roentgenology, **13**, 462, 1925.

94   Gray LH, *The absorption of penetrating radiation*, Proc. Royal Society, London, Series A, **122**, 647, 1929.

95   Mallet L, *Discussion on international units and standards*, Brit. J. Radiology (Röntgen Society Section), **23**, 99, 1927. (See also: Mallet L, *Curietherapie technique physique et posologie, applications aux principeaux cancers*, Ballière: Paris, 1930.)

96   Cade S, *Malignant disease and its treatment by radium*, 2nd edn, John Wright: Bristol, 1948.

97   Hess V, *On the gamma-ray action of extensive flat radium preparations at different distances, with and without absorbing materials*, Physical Review, **19**, 73, 1922.

98   Sievert R, *Eine Methode zur Messung von Roentgen-, Radium-, und Ultra-Strahlung nebst einigen Untersuchungen ueber die Anwendbarweit derselben in der Physik und der Medizin*, Acta Radiologica, Suppl. 14, 1932.

99   Pychlau P, *Ein beitrang zur geschichte der dosimeter mit ionisationskammern*, Medizinische Physik (Hrsg. J. Schütz), 289, 1983.

100   Mayneord WV, *The physics of X-ray therapy*, Churchill: London, 1929.

101   Wilson CW, *Radium therapy: its physical aspects*, Chapman & Hall: London, 1945.

102   Kaye GWC & Binks W, *Dosierung von Gammastrahlen durch Ionisationsmessung*, Strahlentherapie, **56**, 608, 1936.

103   Wood CAP & Boag JW, *Researches on the radiotherapy of oral cancer*, Medical Research Council Special Report Series no. 267, HMSO: London, 1950.

104   Trott NG, Stacey AJ, Ellis RE & Dermetzoglou FM, *The dosimetry of selected procedures using X-rays and radioactive substances*, in: *Medical radionuclides: radiation dose and effects*, Cloutier RJ, Edwards CL & Snyder WS (eds), 157–184, United States Atomic Energy Commission: Oak Ridge, 1970.

105   Bishop PJ, *Evolution of the British Journal of Radiology*, Brit. J. Radiology, **46**, 833–836, 1973.

References: Chapter 23

1   Beck C, *Röntgen ray diagnosis and therapy*, Appleton & Lange: New York, 1904.

2   Kathren RL, *Radiation protection, Medical Physics Handbook 16*, Adam Hilger: Bristol, 1985.

3   Mould RF, *A history of X-rays and radium*, IPC Press: Sutton, 1980. (See also: Mould RF, *Radiation protection in hospitals*, Adam Hilger: Bristol, 1985.)

4   International Commission on Radiological Protection, *Recommendations of the ICRP*, ICRP Report 26, Pergamon: Oxford, 1977.

5   Mayneord WV, *SI units in medical physics*, Contemporary Physics, **17**, 1, 1976. (See also: Spiers FW, *SI units in radiology*, Radiological Protection Bulletin, **18**, 4, 1977.)

6   Flaskamp W & Wintz H, *Über Röntgenschäden und Schäden durch radioaktive Substanzen*, Urban & Schwarzenberg: Berlin, 1930.

7   Pusey WA & Caldwell EW, *Röntgen rays: therapeutics and diagnosis*, WB Saunders: Philadelphia, 1904.

8   Wichmann P, *Radium in der Heilkunde*, Leopold Voss: Hamburg, 1911.

9   Walsh D, *The Röntgen rays in medical work*, Baillière Tindall & Cox: London, 1907.

10 Kassabian MK, *Röntgen rays and electro-therapeutics with chapters on radium and phototherapy*, Lippincott: Philadelphia, 1907.

11 United States Department of Commerce National Bureau of Standards, *Radium protection*, Handbook H23, U.S. Government Printing Office: Washington, DC, 1938.

12 Hopwood FL & Smallman FE, *The use and abuse of radium needles*, Brit. J. Radiology, **3**, 165, 1930.

13 Taft RB, *Radium lost and found*, Furlong: Charleston, 1938.

14 Field CE, *Radium and research: a protest*, Medical Record of New York, **100**, 764–765, 1921.

15 Looney WB, Hasterlik RJ, Brues AM & Skirmont E, *A clinical investigation of the chronic effects of radium salts administered therapeutically 1915–1931*, Amer. J. Roentgenology, **73**, 1006–1037, 1955.

16 Kollinper W, personal communication, 1992.

17 Dautwitz F, *Radiumkuranstalten und Radiumkurote in geologischer, biologischer und klimatischer Beziehung*, in: *Handbuch der Radium-biologie und therapie einschliesslich der anderen radioaktiven elemente*, Lazarus P (ed), 307–320, Bergmann: Wiesbaden, 1918.

18 Pettersson BG, *Improving the management of spent radiation sources*, IAEA Bulletin, **34**, 19–23, 1992.

19 Nori D, personal communication, 1992.

20 Cade S, *Radiation induced cancer in man*, Brit. J. Radiology, **30**, 393, 1957. (See also: Mould RF, *Cancer risks (radiologists, uranium miners, radium dial painters, Hiroshima & Nagasaki)*, in: *Cancer Statistics*, 69, Adam Hilger: Bristol, 1983, and: Mould RF, *Chernobyl: the real story*, (Japanese edition: Nishimura Company), Pergamon: Oxford, 1988.)

21 Morton WJ, *The X-ray or photography of the invisible and its value in surgery*, American Technical Book Co: New York, 1896; Simpkin Marshall Hamilton Kent: London, 1896.

22 Saubermann J, *An address on the progress of radium therapy*, Archives of the Roentgen Ray, **18**, 99, 1913.

23 Mould RF, *Chernobyl: the real story*, Pergamon: Oxford, 1988 (Japanese edition: Nishimura Company).

24 Araki T & Morotani Y (Mayors of the cities of Hiroshima & Nagasaki), *Appeal to the Secretary General of the United Nations*, including: I *Physical destruction due to the atomic bomb*, II *Medical effects of the atomic bomb*, III *Sociological destruction due to the atomic bomb*, IV *Problems for future study*, October 1976.

25 Hersey J, *Hiroshima*, Penguin: London, 1984.

26 Chisholm A, *Faces of Hiroshima*, Cape: London, 1985.

27 Kaplan II, *Irradiation for stimulating or suppressing menstrual function*, New York State Medical Journal, **38**, 626–630, 1937, and: *The treatment of female sterility with X-rays to the ovaries and the pituitary*, Canadian Medical Association Journal, **76**, 43–46, 1951, and: *Third generation follow-up of women treated by X-ray therapy for menstrual dysfunction and sterility 28 years ago, with detailed histories of the grandchildren born to these women*, Amer. J. Obstetrics and Gynaecology, **67**, 484–490, 1954, and: *The treatment of female sterility with X-ray therapy directed to the pituitary and ovaries*, Amer. J. Obstetrics and Gynaecology, **76**, 447–453, 1958, and: *The genetic implications of clinical radiation of women suffering from sterility*, J. National Medical Association, **52**, 13–16, 1958, and: *Genetic effects in children and grandchildren, of women treated for infertility and sterility by Roentgen therapy*, Radiology, **72**, 518–531, 1959.

28 Union Minière du Haut Katanga, *Radium: production, general properties, its applications in therapeutics*, Union Minière du Haut Katanga: Brussels, 1929.

Index

9780367402518